An Introduction to Relativity

General relativity is now an essential part of undergraduate and graduate courses in physics, astrophysics and applied mathematics. This simple, user-friendly introduction to relativity is ideal for a first course in the subject.

The textbook begins with a comprehensive, but simple, review of special relativity, creating a framework from which to launch the ideas of general relativity. After describing the basic theory, it moves on to describe important applications to astrophysics, black-hole physics, and cosmology.

Several worked examples, and numerous figures and images, help students appreciate the underlying concepts. There are also 180 exercises, which test and develop students' understanding of the subject.

The textbook presents all the necessary information and discussion for an elementary approach to relativity. Password-protected solutions to the exercises are available to instructors at www.cambridge.org/9780521735612.

JAYANT V. NARLIKAR is Emeritus Professor at the Inter-University Centre for Astronomy and Astrophysics, Pune, India. He is author of *An Introduction to Cosmology*, now in its third edition (Cambridge University Press, 2002), and has been active in teaching and researching cosmology, theoretical astrophysics, gravitation and relativity for nearly five decades.

An Introduction to Relativity

Jayant V. Narlikar
Inter-University Centre for
Astronomy and Astrophysics
Pune, India

CAMBRIDGE
UNIVERSITY PRESS

University Printing House, Cambridge CB2 8BS, United Kingdom

One Liberty Plaza, 20th Floor, New York, NY 10006, USA

477 Williamstown Road, Port Melbourne, VIC 3207, Australia

314-321, 3rd Floor, Plot 3, Splendor Forum, Jasola District Centre, New Delhi - 110025, India

79 Anson Road, #06-04/06, Singapore 079906

Cambridge University Press is part of the University of Cambridge.

It furthers the University's mission by disseminating knowledge in the pursuit of education, learning and research at the highest international levels of excellence.

www.cambridge.org
Information on this title: www.cambridge.org/9780521735612

First published 2010

A catalogue record for this publication is available from the British Library

Library of Congress Cataloging in Publication data
Narlikar, Jayant Vishnu, 1938–
An introduction to relativity / Jayant V. Narlikar.
    p.   cm.
Includes bibliographical references and index.
ISBN 978-0-521-51497-2 (hardback)
1. General relativity (Physics)   I. Title.
QC173.6.N369   2010
530.11 – dc22   2009035288

ISBN 978-0-521-51497-2 Hardback
ISBN 978-0-521-73561-2 Paperback

Additional resources for this publication at www.cambridge.org/9780521735612

Contents

Contents

Preface

In 1978 I wrote an introductory textbook on general relativity and cosmology, based on my lectures delivered to university audiences. The book was well received and had been in use for about 15–20 years until it went out of print. The present book has been written in response to requests from students as well as teachers of relativity who have missed the earlier text.

An Introduction to Relativity is therefore a fresh rewrite of the 1978 text, updated and perhaps a little enlarged. As I did for the earlier text, I have adopted a simple style, keeping in view a mathematics or physics undergraduate as the prospective reader. The topics covered are what I consider as essential features of the theory of relativity that a beginner ought to know. A more advanced text would be more exhaustive. I have come across texts whose formal and rigorous style or enormous size have been off-putting to a student wishing to know the A, B, C of the subject.

Thus I offer no apology to a critic who may find the book lacking in some of his/her favourite topics. I am sure the readers of this book will be in a position to read and appreciate those topics *after* they have completed this preliminary introduction.

Cambridge University Press published my book *An Introduction to Cosmology*, which was written with a similar view and has been well received. Although the present book contains chapters on cosmology, they are necessarily brief and highlight the role of general relativity. The reader may find it useful to treat the cosmology volume as a companion volume. Indeed, in a few places in this text he/she is directed to this companion volume for further details.

It is a pleasure to acknowledge the encouragement received from Simon Mitton for writing this book. I also thank Vince Higgs, Lindsay Barnes, Laura Clark and their colleagues at Cambridge University Press for their advice and assistance in preparing the manuscript for publication. Help received from my colleagues in Pune, Prem Kumar for figures, Samir Dhurde and Arvind Paranjpye for images and Vyankatesh

Samak for the typescript, has been invaluable. I do hope that teachers and students of relativity will appreciate this rather unpretentious offering!

Jayant V. Narlikar
IUCAA, Pune, India

Chapter 1
The special theory of relativity

1.1 Historical background

1905 is often described as Einstein's *annus mirabilis*: a wonderful year in which he came up with three remarkable ideas. These were the Brownian motion in fluids, the photoelectric effect and the special theory of relativity. Each of these was of a basic nature and also had a wide impact on physics. In this chapter we will be concerned with special relativity, which was arguably the most fundamental of the above three ideas.

It is perhaps a remarkable circumstance that, ever since the initiation of modern science with the works of Galileo, Kepler and Newton, there has emerged a feeling towards the end of each century that the end of physics is near: that is, most in-depth fundamental discoveries have been made and only detailed 'scratching at the surface' remains. This feeling emerged towards the end of the eighteenth century, when Newtonian laws of motion and gravitation, the studies in optics and acoustics, etc. had provided explanations of most observed phenomena. The nineteenth century saw the development of thermodynamics, the growth in understanding of electrodynamics, wave motion, etc., none of which had been expected in the previous century. So the feeling again grew that the end of physics was nigh. As we know, the twentieth century saw the emergence of two theories, fundamental but totally unexpected by the stalwarts of the nineteenth century, viz., relativity and quantum theory. Finally, the success of the attempt to unify electromagnetism with the weak interaction led many twentieth-century physicists to announce that the end of physics was not far off. That hope has not materialized even though the twenty-first century has begun.

While the above feeling of euphoria comes from the successes of the existing paradigm, the real hope of progress lies in those phenomena that seem anomalous, i.e., those that cannot be explained by the current paradigm. We begin our account with the notion of 'ether' or 'aether' (the extra 'a' for distinguishing the substance from the commonly used chemical fluid). Although Newton had (wrongly) resisted the notion that light travels as a wave, during the nineteenth century the concept of light travelling as a wave had become experimentally established through such phenomena as interference, diffraction and polarization. However, this understanding raised the next question: in what medium do these waves travel? For, conditioned by the mechanistic thinking of the Newtonian paradigm, physicists needed a medium whose disturbance would lead to the wave phenomenon. Water waves travel in water, sound waves propagate in a fluid, elastic waves move through an elastic substance ... so light waves also need a medium called aether in which to travel.

The fact that light seemed to propagate through almost a vacuum suggested that the proposed medium must be extremely 'non-intrusive' and so difficult to detect. Indeed, many unsuccessful attempts were made to detect it. The most important such experiment was conducted by Michelson and Morley.

1.2 The Michelson and Morley experiment

The basic idea behind the experiment conducted by A. A. Michelson and E. Morley in 1887 can be understood by invoking the example of a person rowing a boat in a river. Figure 1.1 shows a schematic diagram of a river flowing from left to right with speed v. A boatman who can row his boat at speed c in still water is trying to row along and across the river in different directions. In Figure 1.1(a) he rows in the direction of the current and finds that his net speed in that direction is $c + v$. Likewise (see Figure 1.1(b)), when he rows in the opposite direction his net speed is reduced to $c - v$. What is his speed when he rows across the river in the perpendicular direction as shown in Figure 1.1(c)? Clearly he must row in an oblique direction so that his velocity has a component v in a direction opposite to the current. This will compensate for the flow of the river. The remaining component $\sqrt{c^2 - v^2}$ will take him across the river in a perpendicular direction as shown in Figure 1.1(c).

Suppose now that he does this experiment of rowing down the river a distance d and back the same distance and then rows the same distance perpendicular to the current and back. What is the difference of time τ between the two round trips? The above details lead to the answer that

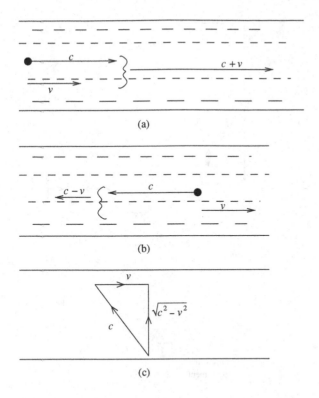

Fig. 1.1. The three cases of a boat being rowed in a river with an intrinsic speed c, the river flowing (from left to right) with speed v: (a) in the direction of the river flow, (b) opposite to that direction and (c) in a direction perpendicular to the flow of the river.

the time for the first trip exceeds that for the second by

$$\tau = \frac{d}{c-v} + \frac{d}{c+v} - \frac{2d}{\sqrt{(c^2-v^2)}} \qquad (1.1)$$

and, for small current speeds ($v \ll c$), we get the answer as

$$\tau \cong \frac{d}{c} \times \frac{v^2}{c^2}. \qquad (1.2)$$

The Michelson–Morley experiment [1] used the Michelson interferometer and is schematically described by Figure 1.2. Light from a source S is made to pass through an inclined glass plate cum mirror P. The plate is inclined at an angle of 45° to the light path. Part of the light from the source passes through the transparent part of the plate and, travelling a distance d_1, falls on a plane mirror A, where it is reflected back. It then passes on to plate P and, getting reflected by the mirror part, it moves towards the viewing telescope. A second ray from the source first gets reflected by the mirror part of the plate P and then, after travelling a distance d_2, gets reflected again at the second mirror B. From there it passes through P and gets into the viewing telescope.

Now consider the apparatus set up so that the first path (length d_1) is in the E–W direction. In a stationary aether the surface of the Earth will

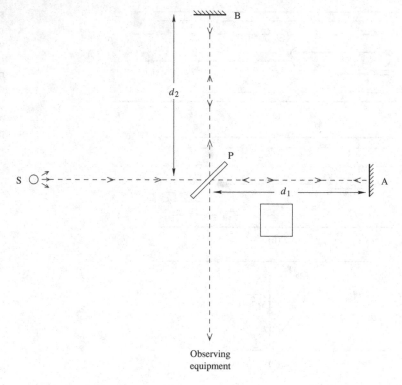

Fig. 1.2. The schematic arrangement of Michelson's interferometer, as described in the text.

have a velocity approximately equal to its orbital velocity of 30 km/s. Thus $(v/c)^2$ is of the order of 10^{-8}. In the actual experiment the apparatus was turned by a right angle so that the E–W and N–S directions of the arms were interchanged. So the calculation for the river-boat crossing can be repeated for both cases and the two times added to give the expected time difference as

$$\tau \cong \frac{d_1 + d_2}{c} \times \frac{v^2}{c^2}. \tag{1.3}$$

Although the effect expected looks very small, the actual sensitivity of the instrument was very good and it was certainly capable of detecting the effect *if indeed it were present*. The experiment was repeated several times. In the case that the Earth was at rest relative to the aether at the time of the experiment, six months later its velocity would be maximum relative to the aether. But an experment performed six months later also gave a null result.

The Michelson–Morley experiment generated a lot of discussion. Did it imply that there was no medium like aether present after all? Physicists not prepared to accept this radical conclusion came up with novel ideas to explain the null result. The most popular of these was the notion of contraction proposed by George Fitzgerald and later worked on by Hendrik Lorentz. Their conclusion as summarized by J. Larmour in a contemporary (pre-relativity) text on electromagnetic theory reads as follows:

> ... if the internal forces of a material system arise wholly from electromagnetic actions between the systems of electrons which constitute the atoms, then the effect of imparting to a steady material system a uniform velocity [v] of translation is to produce a uniform contraction of the system in the direction of motion, of amount $(1 - v^2/c^2)$ ...

It is clear that a factor of this kind would resolve the problem posed by the Michelson–Morley experiment. For, by reducing the length travelled in the E–W direction by the above factor, we arrive at the same time of travel for both directions and hence a null result. Lorentz went further to give an elaborate physical theory to explain why the Fitzgerald contraction takes place.

The Michelson–Morley experiment was explained much more elegantly when Einstein proposed his special theory of relativity. We will return to this point after decribing what ideas led Einstein to propose the theory. As we will see, the Michelson–Morley experiment played no role whatsoever in leading him to relativity.

1.3 The invariance of Maxwell's equations

We now turn to Einstein's own approach to relativity [2], which was motivated by considerations of symmetry of the basic equations of physics, in particular the electromagnetic theory. For he discovered a conflict between Newtonian ideas of space and time and Maxwell's equations, which, since the mid 1860s, had been regarded as the fundamental equations of the electromagnetic theory. An elegant conclusion derived from them was that the electromagnetic fields propagated in space with the speed of light, which we shall henceforth denote by c. It was how this fundamental speed should transform, when seen by two observers in uniform relative motion, that led to the conceptual problems.

The Newtonian dynamics, with all its successes on the Earth and in the Cosmos, relied on what is known as the *Galilean transformation* of space and time as measured by two inertial observers. Let us clarify this notion further. Let O and O$'$ be two inertial observers, i.e., two observers on whom *no force acts*. By Newton's first law of motion both are travelling with uniform velocites in straight lines. Let the speed of O$'$ relative to O be **v**. Without losing the essential physical information we take parallel Cartesian axes centred at O and O$'$ with the X, X' axes parallel to the direction of **v**. We also assume that the respective time coordinates of the two observers were so set that $t = t' = 0$ when O and O' coincided.

Under these conditions the transformation law for spacetime variables for O and O' is given by

$$t' = t, \qquad x' = x - vt, \qquad y' = y, \qquad z' = z. \qquad (1.4)$$

Since v is a constant, the frames of reference move uniformly relative to each other. Laws of physics were expected to be invariant relative to such frames of reference. For example, because of constancy of v, we have equality of the accelerations $\ddot{x}$ and $\ddot{x}'$. Thus Newton's second law of motion is invariant under the Galilean transformation. Indeed, we may state a general expectation that the basic laws of physics should turn out to be invariant under the Galilean transformations. This may be called the *principle of relativity*.

Paving the way to a mechanistic philosophy, Newtonian dynamics nurtured the belief that the basic laws of physics will turn out to be mechanics-based and as such the Galilean transformation would play a key role in them. This belief seemed destined for a setback when applied to Maxwell's equations. Maxwell's equations in Gaussian units and in vacuum (with isolated charges and currents) may be written as follows:

$$\nabla \cdot \mathbf{B} = 0; \qquad \nabla \times \mathbf{E} = -\frac{1}{c}\frac{\partial \mathbf{B}}{\partial t};$$
$$\nabla \cdot \mathbf{D} = 4\pi\rho; \qquad \nabla \times \mathbf{H} = \frac{1}{c}\frac{\partial \mathbf{D}}{\partial t} + \frac{4\pi}{c}\mathbf{j}. \qquad (1.5)$$

Here the fields $\mathbf{B}$, $\mathbf{E}$, $\mathbf{D}$ and $\mathbf{H}$ have their usual meaning and ρ and $\mathbf{j}$ are the charge and current density. We may set $\mathbf{D} = \mathbf{E}$ and $\mathbf{B} = \mathbf{H}$ in this situation. Then we get by a simple manipulation, in the absence of charges and currents,

$$\nabla \times \nabla \times \mathbf{H} \equiv \nabla \nabla \cdot \mathbf{H} - \nabla^2 \mathbf{H}$$
$$= \frac{1}{c}\frac{\partial}{\partial t}\nabla \times \mathbf{E} = -\frac{1}{c^2}\frac{\partial^2 \mathbf{H}}{\partial t^2}. \qquad (1.6)$$

From this we see that $\mathbf{H}$ satisfies the wave equation

$$\Box \mathbf{H} = \mathbf{0}. \qquad (1.7)$$

Similarly $\mathbf{E}$ will also satisfy the wave equation, the operator $\Box$ standing for

$$\Box \equiv \frac{1}{c^2}\frac{\partial^2}{\partial t^2} - \nabla^2.$$

The conclusion drawn from this derivation is this: Maxwell's equations imply that the **E** and **H** fields propagate as waves with the speed c. Unless explicitly stated otherwise, we shall take $c = 1$.[1]

However, this innocent-looking conclusion leads to problems when we compare the experiences of two typical inertial observers, having a uniform relative velocity **v**. Suppose observer O sends out a wave towards observer O′ receding from him at velocity v directed along OO′. Our understanding of Newtonian kinematics will convince us that O′ will see the wave coming towards him with velocity $c - v$. But then we run foul of the *principle of relativity*: that the basic laws of physics are invariant under Galilean transformations. So Maxwell's equations should have the same formal structure for O and O′, with the conclusion that both these observers should see their respective vectors **E** and **H** propagate across space with speed c.

This was the problem Einstein worried about and to exacerbate it he took up the imaginary example of an observer travelling with the speed of the wave. What would such an observer see?

Let us look at the equations from a Galilean standpoint first. The Galilean transformation is given by

$$\mathbf{r}' = \mathbf{r} - \mathbf{v}t, \qquad t' = t. \tag{1.8}$$

Although the general transformation above can be handled, we will take its simplifed version in which O′ is moving away from O along the x-axis and O and O′ coincided when $t' = t = 0$. It is easy to see that the partial derivatives are related as follows:

$$\frac{\partial}{\partial x} = \frac{\partial}{\partial x'}, \qquad \frac{\partial}{\partial y} = \frac{\partial}{\partial y'}, \qquad \frac{\partial}{\partial z} = \frac{\partial}{\partial z'}, \qquad \frac{\partial}{\partial t} = \frac{\partial}{\partial t'} - v\frac{\partial}{\partial x'}.$$

If we apply these transformation formulae to the wave equation (1.7), we find that the form of the equation is changed to

$$\left(\frac{\partial}{\partial t'} - v\frac{\partial}{\partial x'}\right)^2 \mathbf{H} - \nabla'^2\mathbf{H} = \mathbf{0}. \tag{1.9}$$

Clearly Maxwell's equations are not invariant with respect to Galilean transformation. Indeed, if we want the equations to be invariant for all inertial observers, then we need, for example, the speed of light to be invariant for them, as seen from the above example of the wave equation. Can we think of some *other* transformation that will guarantee the above invariances?

In particular, let us ask this question: what is the simplest modification we can make to the Galilean transformation in order to preserve

---

[1] In this book, as a rule, we will choose units such that the speed of light is unity when measured in them.

the form of the wave equation? We consider the answer to this question for the situation of the two inertial observers O and O′ described above. We try linear transformations between their respective space and time coordinates (t, x, y, z) and (t', x', y', z') so as to get the desired answer. So we begin with

$$t' = a_{00}t + a_{01}x, \qquad x' = a_{10}t + a_{11}x, \qquad y' = y, \qquad z' = z. \qquad (1.10)$$

With this transformation, it is not difficult to verify that the wave operator $\Box$ transforms as

$$\Box \equiv \left(a_{00}\frac{\partial}{\partial t'} + a_{10}\frac{\partial}{\partial x'} \right)^2 - \left(a_{01}\frac{\partial}{\partial t'} + a_{11}\frac{\partial}{\partial x'} \right)^2 - \frac{\partial^2}{\partial y'^2} - \frac{\partial^2}{\partial z'^2}. \qquad (1.11)$$

A little algebra tells us that the right-hand side will reduce to the wave operator in the primed coordinates, provided that

$$a_{00}^2 - a_{01}^2 = 1, \qquad a_{10}^2 - a_{11}^2 = -1, \qquad a_{11}a_{01} = a_{10}a_{00}. \qquad (1.12)$$

Now, if we assume that the origin of the frame of reference of O′ is moving with speed v with respect to the frame of O, then setting $x' = 0$ we get $va_{11} = -a_{10}$. Then from (1.12) we get $a_{01} = -va_{00}$. Finally we get the solution to these equations as

$$a_{11} = \gamma, \qquad a_{10} = -v\gamma = a_{01}, \qquad a_{00} = \gamma, \qquad (1.13)$$

where

$$\gamma = (1 - v^2)^{-1/2}. \qquad (1.14)$$

Thus the transformation that preserves the form of the wave equation is made up of the following relations between (t, x, y, z) and (t', x', y', z'), the coordinates of O and O′, respectively:

$$t' = \gamma(t - vx), \qquad x' = \gamma(x - vt), \qquad y' = y, \qquad z' = z. \qquad (1.15)$$

It is easy to invert these relations so as to express the unprimed coordinates in terms of the primed ones. In that case we would find that the relations look formally the same but with $+v$ replacing $-v$:

$$t = \gamma(t' + vx'), \qquad x = \gamma(x' + vt'), \qquad y = y', \qquad z = z'. \qquad (1.16)$$

Physically it means that, if O′ is moving with speed v relative to O, then O is moving with speed $-v$ relative to O′.

A more elaborate algebra will also show that the Maxwell equations are also invariant under the above transformation.

Einstein arrived at this result while considering the hypothetical observer travelling with the light wavefront. He found that such an observer could not exist. (This can be seen in our example below by letting v go to $c = 1$.) In the process he arrived at the above transformation. As we will shortly see, this transformation has echoes of the work Lorentz had done in his attempts to explain the null result of the Michelson–Morley experiment. We will refer to such transformations by the name *Lorentz transformations*, the name given by Henri Poincaré to honour Lorentz for his original ideas in this field.

We also see that the space coordinates and the time coordinate get mixed up in a Lorentz transformation. Thus, for a family of inertial observers moving with different relative velocities, we cannot compartmentalize space and time as separate units. Rather they together form a four-dimensional structure, which we will henceforth call 'spacetime'.

Example 1.3.1 Consider (1.15) with the following definition of θ:

$$v = c \tanh \theta.$$

Then trigonometry leads us to the following transformation laws:

$$t' = t \cosh \theta - x \sinh \theta, \qquad x' = x \cosh \theta - t \sinh \theta, \qquad y' = y,$$
$$z' = z.$$

Compare the first two relations with the rotation of Cartesian axes x, y in two (space) dimensions:

$$x' = x \cos \theta - y \sin \theta, \qquad y' = y \cos \theta + x \sin \theta.$$

We may therefore consider the Lorentz transformation as a rotation through an imaginary angle $i\theta$, if we define an imaginary time coordinate as $T = it$.

1.4 The origin of special relativity

Einstein thus found himself at a crossroads: the Newtonian mechanics was invariant under the Galilean transformation, whereas Maxwell's equations were invariant under the Lorentz transformation. One could try to modify the Maxwell equations and look for invariance of the new equations under the Galilean transformation. Alternatively, one could modify the Newtonian mechanics and make it invariant under the Lorentz transformation. Einstein chose the latter course. We will now highlight his development of the special theory of relativity.

We begin with the introduction of a special class of observers, the *inertial observers* in whose rest frame Newton's first law of motion holds. That is, these observers are under no forces and so move relative to one another with uniform velocities. Notice that there is no explicitly defined frame that could be considered as providing a frame of 'absolute rest'. Thus all inertial observers have equal status and so do their frames, which are the inertial frames. This is in contrast with the Newtonian concept of absolute space, whose rest frame enjoyed a special status. We will comment on it further in Chapter 18 when we discuss Mach's principle.

The *principle of relativity* states that all basic laws of physics are the same for all inertial observers. Notice that this principle has not changed from its Newtonian form; but the inertial observers are now linked by Lorentz rather than Galilean transformations.

When applied to electricity and magnetism this principle tells us that Maxwell's equations are the same for all inertial observers: in particular, the speed of light c, which appears as the wave velocity in these equations, must be the same in all inertial reference frames. We also see that this requirement leads us to the Lorentz transformation. The transformation described by the equations (1.15) is called a 'special Lorentz transformation'. It can be easily generalized to the case in which the observer O' moves with a constant velocity $\mathbf{v}$ in any arbitrary direction. The relevant relations are

$$t' = \gamma[t - (\mathbf{v} \cdot \mathbf{r})], \qquad \mathbf{r}' = \gamma(\mathbf{r}^* - \mathbf{v}t), \tag{1.17}$$

where

$$\mathbf{r}^* = \mathbf{r}/\gamma + (\gamma - 1)\mathbf{v}(\mathbf{v} \cdot \mathbf{r})/\gamma v^2. \tag{1.18}$$

We next look at some of the observable effects of this transformation on some measurements of events in space and time. For it is these effects that tell us what the special theory of relativity is all about.

Example 1.4.1 *Problem.* Show that (1.17) reduces to (1.15) for a special Lorentz transformation.

Solution. In the special Lorentz transformation, $\mathbf{v}$ is in the x-direction. So, if $\mathbf{e}$ is a unit vector in that direction,

$$\mathbf{v} \cdot \mathbf{r} = vx, \qquad \mathbf{r}^* = \mathbf{r}\sqrt{1 - v^2} + \left(\frac{1}{\sqrt{1 - v^2}} - 1\right)v^2 x \frac{1}{\gamma v^2}\mathbf{e},$$

where we have used (1.17) and (1.18). Thus $t' = \gamma(t - vx)$, which is as per (1.15). For the $\mathbf{r}'$ relation, note that the $y-y'$ relation is $y' = y$. Similarly we have $z' = z$. The $x-x'$ relation is

$$x' = x + \gamma(\gamma - 1)v^2 x \cdot \frac{1}{\gamma v^2} - \gamma vt$$

$$= x(1 + \gamma - 1) - \gamma vt = \gamma(x - vt).$$

Thus we recover the special Lorentz transformation.

1.5 The law of addition of velocities

First of all, we notice that the speed of light remains c for all inertial observers. The Michelson–Morley experiment would therefore give zero difference in time gap, not a finite one as was then calculated on the basis of Newtonian kinematics. What does the Lorentz transformation do to the law of addition of velocities?

Let us talk of three inertial observers O, O′ and O″ with frames of reference aligned so that they share the same x-direction while their origins were coinciding at $t = t' = t'' = 0$. We are given that O′ is moving in the x-direction with velocity v_1 relative to O. Likewise, O″ is moving with velocity v_2 relative to O′. So, what is the velocity of O″ relative to O? The Newtonian answer to this question would have been $v_1 + v_2$. Here, however, the result is different.

The Lorentz transformation relating O′ to O is

$$x' = \frac{x - v_1 t}{\sqrt{1 - v_1^2}}, \qquad t' = \frac{t - v_1 x}{\sqrt{1 - v_1^2}}. \tag{1.19}$$

Likewise, the Lorentz transformation linking O″ to O′ is

$$x'' = \frac{x' - v_2 t'}{\sqrt{1 - v_2^2}}, \qquad t' = \frac{t' - v_2 x'}{\sqrt{1 - v_2^2}}. \tag{1.20}$$

Our desired answer is found by combining equations (1.19) and (1.20) so as to express the coordinates (t'', x'') in terms of (t, x). The algebra is simple but a bit tedious and the answer is that O″ moves relative to O as an inertial observer whose velocity v in the x-direction relative to O is given by

$$v = \frac{v_1 + v_2}{1 + v_1 v_2}. \tag{1.21}$$

This is the law of addition of velocities. If the velocities are $\mathbf{v_1}, \mathbf{v_2}$ parallel in any general direction, the formula becomes

$$\mathbf{v} = \frac{\mathbf{v_1} + \mathbf{v_2}}{1 + \mathbf{v_1} \cdot \mathbf{v_2}}. \tag{1.22}$$

From (1.21) it is easy to see that, if one of the velocities is $c = 1$, the resultant is also c. Thus, irrespective of whether a source of light is at rest or moving relative to an observer, the light emitted by it will always have the speed c as measured by the observer.

Example 1.5.1 *Problem.* Relate (1.21) to rotation of axes by imaginary angles.

Solution. From Example 1.3.1, we have $v_1 = \tanh \theta_1$, $v_2 = \tanh \theta_2$, where $i\theta_1$ and $i\theta_2$ are the rotation angles from frame O to O′ and from O′ to O″, respectively. Thus the rotation angle from frame O to O″ is simply $i(\theta_1 + \theta_2)$. The corresponding net velocity is

$$v = \tanh(\theta_1 + \theta_2) = \frac{\tanh\theta_1 + \tanh\theta_2}{1 + \tanh\theta_1 \tanh\theta_2} = \frac{v_1 + v_2}{1 + v_1 v_2}.$$

1.5.1 The Minkowski spacetime

It was Hermann Minkowski [3], one of Einstein's teachers and a distinguished mathematician, who brought elegance into the above picture by pointing out that it was wrong to think of time and space as two separate entities; one should learn to look at the unity of the two. In his own words,

> The views of space and time which I wish to lay before you have sprung from the soil of experimental physics, and therein lies their strength. They are radical. Henceforth space by itself, time by itself, are doomed to fade away into mere shadows and only a kind of union of the two will preserve an independent reality.

Thus one is really dealing with 'spacetime' instead of with 'space' and 'time'. We may use the notation advocated by Minkowski by replacing the time t and the Cartesian space coordinates (x, y, z) by their four-dimensional counterparts:

$$x^0 \equiv ct, \qquad x^1 \equiv x, \qquad x^2 \equiv y, \qquad x^3 \equiv z. \tag{1.23}$$

A Lorentz transformation may therefore be looked upon as a linear spacetime coordinate transformation of the kind below:

$$x'^i = \sum_k A^i_k x^k = A^i_k x^k. \tag{1.24}$$

Here, in the last step, we have dropped the summation symbol in summing over 'k'. Our rule henceforth will be that any expression containing an upper and lower index represented by the same Latin letter is automatically summed for all four values of that letter. The summation in this case is over all values of the index k, viz. 0, 1, 2, 3.

The condition that light has the same velocity in all inertial reference frames may be translated into the invariance of $(c^2t^2 - x^2 - y^2 - z^2)$ in the above transformation, and we may compare this situation with the three-dimensional one where the length square $(x^2 + y^2 + z^2)$ is preserved under coordinate transformation. It is customary to write the above four-dimensional square of the distance from the origin in our new notation, as

$$\eta_{ik}x'^i x'^k = \eta_{ik}A^i{}_m x^m A^k{}_n x^n = \eta_{mn}x^m x^n,$$

which leads us to the general transformation laws of these frames. The transformation coefficients must satisfy the rule given below:

$$A^i{}_m A^k{}_n \eta_{ik} = \eta_{mn}. \qquad (1.25)$$

The 4×4 array η_{ik} is the key factor specifying the measurement of distance in the four-dimensional spacetime and is called the *Minkowski metric*. It has the simple form diag($+1, -1, -1, -1$). The transformations which satisfy the above rule form a group called the *Lorentz group* and denoted by O(1,3).

We define a vector P^i as a four-component entity that transforms under the coordinate transformation (1.24) as follows:

$$P'^i = A^i{}_m \times P^m. \qquad (1.26)$$

Likewise, a second-rank tensor F^{ik} transforms according to the law

$$F'^{ik} = A^i{}_m A^k{}_n \times F^{mn}. \qquad (1.27)$$

We may use the metric η_{ik} to lower an upper index to a lower one, e.g.,

$$A^i \times \eta_{ik} \equiv A_k.$$

Likewise we take the inverse matrix of $||\eta_{ik}||$ to be $||\eta^{ik}||$ so that

$$\eta_{ik}\eta^{kl} = \delta^l_i.$$

Here the delta is the Kronecker delta, which is zero unless the upper and lower index happen to be equal, when its value is unity. It is easy to see that the inverse of the above index-lowering exercise is

$$A_i \times \eta^{ik} \equiv A^k.$$

In addition to vectors and tensors, we have scalars, which retain the same value under all coordinate transformations. Thus $\eta_{ik}P^i P^k$ is a scalar, as can be seen by subjecting it to a Lorentz transformation. A scalar does not have any free index.

If a vector A^k is given we can make a scalar out of it by writing

$$A^2 = \eta_{ik}.A^i.A^k.$$

A is called the magnitude of the vector. If $A^2 > 0$ the vector A^k is said to be 'timelike'. For $A^2 = 0$ we have in A^k a 'null' vector, while for $A^2 < 0$ we have a 'spacelike' vector.

We will use this notation in an extended form (see Chapter 3) when dealing with general relativity. In general we may say that any expression written in terms of vectors and tensors (and, of course, scalars) is guaranteed to be Lorentz invariant. Since the special theory of relativity requires all physics to be the same for all inertial observers, we need all basic physical equations to be Lorentz invariant also. Thus we expect all fundamental physics to be expressed in terms of vectors and tensors.[2]

Example 1.5.2 A tensor is symmetric if any permutation of its indices does not alter its value. Likewise an antisymmetric tensor reverses its value for any odd permutation of its indices.

An example of the latter is ϵ_{ijkl}, defined by

$$\epsilon_{ijkl} = \begin{cases} +1 & \text{if } (i, j, k, l) \text{ is an even permutation of } (1, 2, 3, 4) \\ -1 & \text{if } (i, j, k, l) \text{ is an odd permutation of } (1, 2, 3, 4) \\ 0 & \text{otherwise.} \end{cases}$$

We will encounter this tensor in Chapter 3 in the context of general coordinate transformations.

1.6 Lorentz contraction and time dilatation

The Lorentz transformation generates several paradoxical situations largely because one is normally and intuitively tuned to absolute space measurements and absolute time measurements. We will describe some examples next. For our discussion we will take the Lorentz transformation to be as given in Equations (1.15) and (1.16).

1.6.1 Length contraction

Let us consider the following experiment. Let O' carry a rod of length l as measured by him when the rod is at rest in his reference frame, that is, the frame in which he is at rest. Suppose the rod is laid out along the x'-axis with front end B at $x' = l$ and back end A at $x' = 0$ as shown in Figure 1.3. Suppose that O sets up an experiment of timing when the two ends A and B pass his origin $x = 0$. Evidently the expectation based on Newtonian physics is that the two ends will pass the origin at an interval of l/v. Let us see what the Lorentz transformation (1.15) gives.

When the end B passes the origin of O, as shown in Figure 1.3(a), we have $x' = l$, $x = 0$ and the second of the equations listed in (1.15) gives $l = \gamma(x - vt)$, i.e., for $x = 0$ we get $t = -l/(\gamma v)$. Likewise when the end A passes the origin as shown in Figure 1.3(b), the time in the frame of O is $t = 0$. The time interval between the passages of A and B is therefore $l/(\gamma v)$. Thus, knowing that O' is moving with speed v

---

[2] The inputs provided by quantum mechanics have led to the addition of spinors to this list. We will not have any opportunity to discuss the role of spinors in this text.

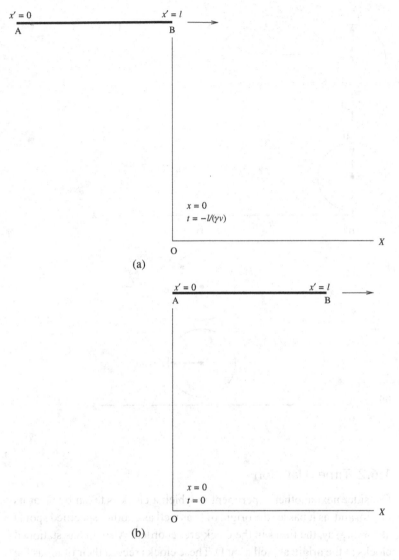

$x' = 0$ $x' = l$
A   B

$x = 0$
$t = -l/(\gamma v)$

O

(a)

$x' = 0$ $x' = l$
A   B

$x = 0$
$t = 0$

O

(b)

Fig. 1.3. Two stages (a) and (b) of the movement of the rod AB as observed by O, as per the arrangement described in the text.

relative to him, O will conclude that the length of the rod is l/γ. Thus the rod appears to him contracted by the factor $\sqrt{1 - v^2}$.

While we may be tempted to identify this with the Fitzgerald–Lorentz idea that an object moving with speed v relative to the aether will contract by just this factor in its direction of motion, such an identification is wrong. The aether contraction was an absolute effect and explained as such on the basis of certain assumptions about the atomic structure of matter. Here the effect is relative. Moreover, to observer O' a similar rod carried by O would appear contracted by the same factor. In short, the effect is symmetric between the two inertial observers.

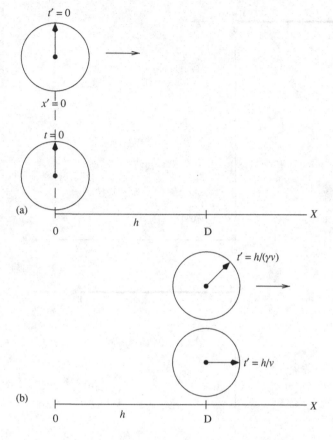

Fig. 1.4. Two stages in the passage of the moving clock of O' are shown above. As described in the text, at stage (a) O' is passing the origin of the stationary observer O whereas in stage (b) this clock is observed by another stationary clock at D.

1.6.2 Time dilatation

Consider next another experiment in which a clock is taken by O' at its origin, and, as it passes the origin of O as well as another specified spot D at $x = h$, say, the times in the clock are recorded. Also, O has stationed clocks at the origin as well as at D. These clocks record their times as the moving clock from O' passes them by. Again our Newtonian expectation is that the time interval recorded by the two clocks of O should equal the time interval between the two readings by the clock of O'. Let us see what the reality is. Figure 1.4 illustrates the experiment.

The clock of O' has coordinate $x' = 0$. When it passes the origin of O, with coordinate $x = 0$, we have from the Lorentz transformation that $t = t' = 0$. This is shown in Figure 1.4(a). Similarly, as shown in Figure 1.4(b), when the clock of O' is at D, we have $x' = 0, x = h$ so that the times when these clocks meet are $t = h/v$ and $t' = (\gamma)^{-1}h/v$. Thus we see that the interval recorded by the moving clock of O' is short by a factor $\sqrt{1 - v^2}$ compared with the times recorded by the two clocks of O. So on the basis of this experiment O will conclude that the clock kept by O' is running slow compared with his clock. This effect is often referred to as *time dilatation*.

However, this experiment could be performed with O' using two clocks and O using one, with the roles of the two observers interchanged. Then O' would find that the clock system used by O is running slow. The experiment as such is not designed such that a symmetric role is played by each inertial frame. Thus there is no paradox involved here. Neither is it the case that one inertial frame is accorded a special status. A look back at the length-contraction experiment described earlier will likewise show that there too no special status is enjoyed by any inertial frame.

1.6.3 Muon decay in cosmic-ray showers

A striking demonstration of time dilatation was given by observations of the muon particles (μ) in cosmic-ray showers. The muon (earlier called the μ meson) is very similar to the electron, except that it is 207 times heavier. The particle is normally unstable and in the laboratory rest frame its decay time is as low as $(2.09 \pm 0.03) \times 10^{-6}$ s. It decays into the electron, a muon neutrino and an electron anti-neutrino.

In cosmic-ray showers the presence of muons would normally be difficult to understand. For they are believed to have been produced at a height of 10 km and, even if they travelled with the speed of light, they could not travel more than a distance of some 600 m before decaying. So how did they manage to survive long enough to be seen so close to Earth's surface?

The puzzle is explained by arguing that, if the muons are travelling with very high speed, their natural clocks as seen by observers on the Earth would run slow. At a speed of $0.995c$ the time-dilatation factor γ is equal to 10. This makes the apparent lifetime of the muon relative to an Earth-based laboratory ten times longer and thus it can travel up to 6 km, instead of 600 m. Therefore in a normal statistical fluctuation we should be able to see some muons coming from even 10 km above sea-level.

1.6.4 Relativity of simultaneity

In Newtonian physics the absolute character of space and time enabled one to give an absolute meaning to the simultaneity of two events at different places. Thus, if event A took place at $x = x_1$ and event B at $x = x_2$, both at time t, then they would be called simultaneous.

In relativistic physics, the above statement needs to be modified. We can describe the above simultaneity as that seen by an inertial observer O. To another inertial observer O′ as defined earlier, the events would not be simultaneous. For event A will occur at

$$t' = \gamma(t - x_1 v),$$

whereas event B will occur at

$$t' = \gamma(t - x_2 v).$$

So the events are *not* simultaneous in the reference frame of O′. As seen by O′, A will occur before or after B depending on whether x_1 exceeds or is less than x_2.

These examples illustrate the tricks time can play with our intuitive perception of physical events. There is another paradox, which we will discuss towards the end of this chapter: the so-called *clock paradox*. Before considering that, we need to discuss how Newtonian mechanics has to be modified in order to accommodate Lorentz invariance, that is, invariance with respect to Lorentz transformation.

1.7 Relativistic mechanics

Let us define the spacetime trajectory of a particle of mass m at rest, denoting its coordinates $x^i \equiv (t, x, y, z)$. Such a trajectory is often called the *world line* of the particle. The three space coordinates are usually denoted by a space vector **r**. In Newtonian physics the three-dimensional velocity of the particle of mass m is denoted by

$$\frac{d\mathbf{r}}{dt} = \mathbf{v}.$$

Here, in relativity, we define a 4-velocity of the particle of mass m by

$$u^i = \frac{dx^i}{ds}. \tag{1.28}$$

Here ds is the 'proper distance' between the two points (t, x, y, z) and $(t + dt, \ x + dx, \ y + dy, \ z + dz)$ on the spacetime trajectory of the particle. Using the fact that

$$ds^2 = dt^2 - (dx^2 + dy^2 + dz^2) = dt^2(1 - v^2) = dt^2/\gamma^2 \tag{1.29}$$

we get the following relationship between the Newtonian 3-velocity and the relativistic 4-velocity:

$$u^i = \gamma \times [1, \mathbf{v}]. \tag{1.30}$$

Notice that the identity (1.29) shows that the four components of u^i are related and so we have only three independent components. The identity may be written as

$$u^i u_i = 1, \qquad \text{i.e.,} \quad u_i \times \frac{\mathrm{d}u^i}{\mathrm{d}s} = 0. \tag{1.31}$$

With this basic definition, we now consider Newton's laws of motion. The first law remains as it is. The second law needs some consideration to make it *Lorentz-invariant*, i.e., invariant under Lorentz transformation. How should force and acceleration be related? In the rest frame of the moving body, suppose its mass is m_0. We assume that the difference introduced by the Lorentz invariance would show up only at large velocities and so the second law should look the same as the Newtonian one in the rest frame of the body. This may be written as

$$\frac{\mathrm{d}}{\mathrm{d}t}\left(m_0 \frac{\mathrm{d}\mathbf{r}}{\mathrm{d}t} \right) = \mathbf{F}, \tag{1.32}$$

where $\mathbf{F}$ is the measure of force in the rest frame of the body. Now, we ask, what is the acceleration formula in a general Lorentz frame, which reduces to the above in the rest frame? To this end we first convert (1.32) into a set of *four* equations instead of three, by adding the fourth component, and then replace $\mathrm{d}t$ by $\mathrm{d}s$. Thus we have the second law as

$$\frac{\mathrm{d}}{\mathrm{d}s}\left(m_0 \frac{\mathrm{d}x^i}{\mathrm{d}s} \right) = G^i, \tag{1.33}$$

where the 4-force G^i is related to the Newtonian force in the following way. Take the spacelike components of the above equation. We get

$$\frac{\mathrm{d}t'}{\mathrm{d}s}\frac{\mathrm{d}}{\mathrm{d}t'}\left(m_0 \frac{\mathrm{d}t'}{\mathrm{d}s}\frac{\mathrm{d}\mathbf{r}'}{\mathrm{d}t'} \right) = \mathbf{G}',$$

say. Here we have assumed that the spacetime coordinates in the general Lorentz frame are $(t', \mathbf{r}')$. This looks like Newton's second law with minimal modification, if we write it as

$$\frac{\mathrm{d}}{\mathrm{d}t'}\left(m \frac{\mathrm{d}\mathbf{r}'}{\mathrm{d}t'} \right) = \mathbf{F}', \tag{1.34}$$

where we define

$$m = m_0 \frac{\mathrm{d}t'}{\mathrm{d}s} \equiv \frac{m_0}{\sqrt{1 - v^2}} = \gamma m_0, \qquad \mathbf{G}' = \gamma \mathbf{F}'. \tag{1.35}$$

This then is the modified version of Newton's second law of motion. We can write it in the four-dimensional language of Minkowski as follows:

$$\frac{\mathrm{d}}{\mathrm{d}s}\left(m_0 \frac{\mathrm{d}x^i}{\mathrm{d}s} \right) = G^i, \tag{1.36}$$

where G^1, G^2, G^3 are the spacelike components which we write as the 3-vector $\mathbf{G}$ and $P = G^0$ is the timelike fourth component. From the above equation we see that this 'extra' component of the relativistic equations of motion implies

$$\frac{\mathrm{d}}{\mathrm{d}s}\left(m_0 \frac{\mathrm{d}t}{\mathrm{d}s}\right) = P. \tag{1.37}$$

We will next study its implication for the mass–energy relation.

1.7.1 The mass–energy relation

Let us re-examine the four-dimensional status given to the force. We have three components in the form of the vector $\mathbf{G}$ defined above as $\mathbf{F}\gamma$. We then have a fourth (time) component, P. Writing the 4-vector for force as

$$F^i = [P, \mathbf{G}],$$

we use the invariance of the expression

$$\eta_{ik} u^i F^k = P\frac{\mathrm{d}t}{\mathrm{d}s} - \mathbf{G} \cdot \frac{\mathrm{d}\mathbf{r}}{\mathrm{d}s}. \tag{1.38}$$

We note two things. First, from (1.31), we get the left-hand side of (1.38) as zero. In the rest frame we therefore have $P = 0$. Secondly, because (1.38) vanishes in one inertial frame, it must do so in all inertial frames. Hence we get

$$P = \mathbf{G} \cdot \mathbf{v}. \tag{1.39}$$

This means that P/γ is $\mathbf{F} \cdot \mathbf{v}$, which is the rate of working of the force $\mathbf{F}$. If we denote by W the energy generated in the particle by these external forces, then

$$\frac{\mathrm{d}W}{\mathrm{d}s} = \frac{\mathrm{d}W}{\mathrm{d}t} \times \frac{\mathrm{d}t}{\mathrm{d}s} = P/\gamma \times \gamma = P. \tag{1.40}$$

However, because of the equation (1.37) satisfied by P, we can write the above as

$$\frac{\mathrm{d}W}{\mathrm{d}s} = \frac{\mathrm{d}}{\mathrm{d}s}(\gamma m_0), \tag{1.41}$$

which, in view of Equation (1.35), integrates to

$$W = m + \text{constant}.$$

The energy generated by external forces being one of motion, we interpret W as kinetic energy and so require it to be zero for the particle at rest. Hence we adjust the constant such that

$$W = (m - m_0)c^2, \tag{1.42}$$

where we have restored c (which had been put equal to unity) so as to remind ourselves of the difference in dimensionality of mass and energy. Arguing that a particle of mass m_0 at rest also possesses energy $m_0 c^2$, Einstein came up with the total energy of a moving particle as

$$E = mc^2. \tag{1.43}$$

This is perhaps the best-known equation of physics, especially if we take lay non-physicists also into consideration! A useful application of this result is found in the Sun. The core of the Sun has hot plasma in which four protons (i.e., nuclei of hydrogen atoms) combine together to make a helium nucleus, through the reaction

$$4\,^1\mathrm{H} \longrightarrow {}^2\mathrm{He} + 2\nu + 2\mathrm{e}^+ + \gamma.$$

That is, the byproducts are neutrinos, positrons and radiation. The mass of the four hydrogen nuclei exceeds the mass of the helium produced. The difference in mass Δm is not lost but is produced in the form of energy $\Delta m\, c^2$. Even after accounting for neutrinos and positrons, the bulk of this energy appears as solar radiation. Thus the Sun shines because of $E = mc^2$!

1.7.2 The linear momentum

In terms of Minkowski's four-dimensional framework, we can write some of the above relations as follows. We start by defining the energy-momentum 4-vector for a particle of rest mass m_0 as

$$P^i = \left(m_0 \frac{\mathrm{d}t}{\mathrm{d}s},\ m_0 \frac{\mathrm{d}\mathbf{r}}{\mathrm{d}s} \right) = (E, \mathbf{p}), \tag{1.44}$$

where $\mathbf{p} = m\mathbf{v}$ is the Newtonian 3-momentum of the particle. Since the Minkowski 4-velocity has unit magnitude, we have

$$E^2 - p^2 = m_0^2, \tag{1.45}$$

where we have put back the condition $c = 1$.

In general interactions between subatomic particles lead to decay, scattering, etc. of particles and there the conservation of 4-momentum generally holds. We will discuss a couple of examples from electrodynamics to illustrate how the law operates.

But first we specify what happens to the *photons*, the particles of light, *vis-à-vis* Equation (1.45). The photons always travel with the speed of light and so one cannot really talk of their 'rest' mass. Nevertheless, a common misnomer is that photons, or all those particles which travel with the speed of light, have a 'zero rest mass'. So if a photon γ of frequency ν is travelling in a direction specified by a unit (spacelike) vector **e** its energy-momentum vector is defined as

$$P_\gamma = [h\nu, h\nu\mathbf{e}].$$ (1.46)

It is easy to see that the magnitude of this momentum is zero, that is $P_\gamma \cdot P_\gamma = 0$. The photon therefore has 'zero' 4-momentum and it is described by a null vector.

1.7.3 The centre-of-mass frame

Consider the situation in electrodynamics in which a particle–antiparticle pair, such as an electron and a positron, is created from radiation. Let us see whether we can create the pair from a single photon.

It is convenient to look at the problem from a special inertial frame: one in which the total 3-momentum of all participating particles is zero. Such a frame is called the *centre-of-mass* frame, because in this frame the centre-of-mass is at rest. How to go from an arbitrary inertial frame to a centre-of-mass frame is shown in the two solved problems at the end of this subsection.

So we take the electron and positron each to have the same rest mass m_0 but equal and opposite 3-velocities $\pm\mathbf{v}$. Thus the 4-momenta for the two particles are

electron e$^-$: $P^- = \gamma m_0 \times [1, \mathbf{v}]$;   positron e$^+$: $P^+ = \gamma m_0 \times [1, -\mathbf{v}]$.

The factor γ has its usual meaning. But, when we add the two momenta, we get the total momentum as

$$P^- + P^+ = 2\gamma m_0 \times [1, 0, 0, 0].$$

If this pair had to have come from just one light photon, the photon must have had the above momentum. But this is impossible, since the photon momentum is a null vector, whereas the above vector is timelike. We can satisfy the above requirement of conservation of 4-momenta if we have *two* photons to work with.

Example 1.7.1 *Problem.* Consider the reaction $\pi^+ + \text{n} \rightarrow \Lambda^0 + \text{K}^+$. The rest masses of the four particles are, respectively, $m_\pi, m_\text{n}, m_\Lambda$ and m_K. The neutron was at rest in the laboratory frame. Show that the total energy in the centre-of-mass frame is $(m_\pi^2 + m_\text{n}^2 + 2m_\text{n}E_\pi)^{1/2}$, where E_π is the pion energy.

Solution. The total 4-momentum is P^i, where

$$P^i = P_\pi^i + P_\text{n}^i = P_\text{K}^i + P_\Lambda^i.$$

In the centre-of-mass frame, the total energy is $E_{\text{total}}^{\text{CM}}$, and the total 3-momentum is $\mathbf{0}$. Since $P_i P^i$ is invariant in all frames, we may equate its values in the lab frame and in the centre-of-mass frame. In the lab frame,

$$P_i P^i = (P_\pi^i + P_n^i)(P_{i\pi} + P_{in})$$
$$= m_\pi^2 + m_n^2 + 2P_\pi^i P_{in} \qquad \text{from (1.45)}.$$

Now $P_\pi^i = (E_\pi, \mathbf{p}_\pi)$, $P_n^i = (E_n, \mathbf{0})$ and so $P_\pi^i P_{in} = E_n E_\pi$. But $E_n = m_n$ since the neutron is at rest. So in the lab frame $P_i P^i = m_\pi^2 + m_n^2 + 2m_n E_\pi$.

In the centre-of-mass frame the total 4-momentum is $(E_{\text{total}}^{\text{CM}}, \mathbf{0})$.

Hence $P_i P^i = (E_{\text{total}}^{\text{CM}})^2 = m_\pi^2 + m_n^2 + 2m_n E_\pi$.

Example 1.7.2 *Problem.* Two particles of rest masses m_1 and m_2 collide. Prior to collision the first particle was at rest while the second was approaching it with speed v_2. What is the energy of the system in the centre-of-mass frame?

Solution. In the laboratory frame, the 3-momentum has only one component $\gamma_2 m_2 v_2$ in the direction of motion. The energy of the system in this frame is $m_1 + \gamma_2 m_2$, where $\gamma_2^{-1} = \sqrt{1 - v_2^2}$.

Let the total energy in the centre-of-mass frame be E_0. The total 3-momentum is zero by definition. Equating $P_i P^i$ in the two frames, we get

$$(m_1 + \gamma_2 m_2)^2 - \gamma_2^2 m_2^2 v_2^2 = E_0^2,$$

i.e., $E_0^2 = m_1^2 + m_2^2 + 2\gamma_2 m_1 m_2$.

1.7.4 Compton scattering

We look at another example from electrodynamics shown in Figure 1.5. We have a photon of frequency v_1 incident on an electron at rest. The electron acquires some momentum from this impact and moves with speed u in a direction making an angle θ with the original direction of the photon. The photon after the impact is scattered in the direction making an angle $-\phi$ with its original direction of motion. We want to relate the frequency v_2 of the scattered photon to the angle of its scattering.

It can be easily verified that the entire scenario is confined to a plane, that determined by the paths of the original and scattered photons. Let m_0 and m denote the rest mass and moving mass of the electron before and after the impact. The 4-momenta of the various particles are as follows.

(1) Before scattering:
   electron momentum $P_e = m_0 \times [1, 0, 0, 0]$.
   photon momentum $P_\gamma = h v_1 \times [1, 1, 0, 0]$

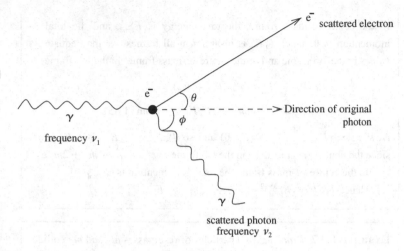

Fig. 1.5. A schematic diagram of Compton scattering. The photon γ coming from the left impacts the stationary electron e^- and scatters it in the direction making angle θ with the original direction of γ. The photon itself is scattered in the direction making angle ϕ with its original direction.

(2) After scattering:

electron momentum $P'_e = m \times [1, \; u \cos \theta, \; u \sin \theta, 0]$

photon momentum $P'_\gamma = h\nu_2 \times [1, \; \cos \phi, \; -\sin \phi, 0]$.

The law of conservation of 4-momenta then reduces to the following three equations:

$$m_0 + h\nu_1 = m + h\nu_2$$

$$h\nu_1 = h\nu_2 \cos \phi + mu \cos \theta \qquad (1.47)$$

$$0 = -h\nu_2 \sin \phi + mu \sin \theta.$$

Upon eliminating θ from these relations, we get

$$m^2 u^2 = h^2 [(\nu_1 - \nu_2 \cos \phi)^2 + \nu_2^2 \sin^2 \phi].$$

Using the relation $m = m_0 / \sqrt{1 - u^2}$ and the first of the three equations in (1.47), we have finally the relation

$$\frac{1}{\nu_2} - \frac{1}{\nu_1} = \frac{h}{m_0}(1 - \cos \phi).$$

That is, in terms of wavelengths,

$$\lambda_2 - \lambda_1 = \frac{h}{m_0}(1 - \cos \phi).$$

This is known as *Compton scattering* because the change of wavelength of the photon on scattering was first measured by A. H. Compton in 1923. The signature of the effect is that the change in wavelength does not depend on the initial wavelength of the photon. It is dependent only on the angle of scattering. The quantity $h/(m_0 c)$ is usually referred to as the *Compton wavelength of the electron*. Its value is 0.0024 nm.

Particle–particle collisions, or decays, are phenomena that have been used extensively to test the validity of the assumptions that lead to special relativity. A few more cases are described in solved problems that follow as well as in the exercises at the end of the chapter.

Example 1.7.3 *Problem.* A particle with rest mass m_0 has 3-momentum **p**. An observer with 3-velocity **u** looks at the particle: what energy would he measure?

Solution. In the rest frame of the observer, $u'^0 = 1$, $u'^\mu = 0$. Suppose he measures the particle's energy to be E'. Suppose also that in his rest frame the 4-momentum of the particle is $P_i' = (E', -\mathbf{p}')$, say. Thus $P_i' u'^i = E'$.

Now evaluate the same invariant in the given laboratory frame. Then the energy of the particle is $P^0 = \sqrt{p^2 + m_0^2}$. The observer has 3-velocity **u**. So his 4-velocity is $\gamma(1, \mathbf{u})$, where $\gamma^{-1} = \sqrt{1 - u^2}$. We therefore have $P_i u^i = \sqrt{p^2 + m_0^2} \times \gamma - \gamma \mathbf{p} \cdot \mathbf{u}$. Therefore the energy measured by the observer is $E' = \gamma(\sqrt{p_0^2 + m_0^2} - \mathbf{p} \cdot \mathbf{u})$.

Example 1.7.4 *Problem.* A particle of rest mass m_0 moving with velocity v collides with a stationery particle of rest mass M and is absorbed by it. Given that energy and momentum are conserved in the collision, find the rest mass and velocity of the composite particle.

Solution. The 4-momentum of the moving particle is $[m_0\gamma, m_0\gamma v, 0, 0]$, where the direction of motion is chosen as the x-axis; $\gamma^{-1} = \sqrt{1 - v^2}$. The 4-momentum of the stationary particle is $[M, 0, 0, 0]$. Thus the 4-momentum of the composite particle is $[M + m_0\gamma, m_0\gamma v, 0, 0]$. The mass of this particle is M_0, say. Then

$$M_0^2 = (M + m_0\gamma)^2 - m_0^2\gamma^2 v^2 = M^2 + 2m_0 M\gamma + m_0^2\gamma^2(1 - v^2)$$
$$= M^2 + 2m_0 M\gamma + m_0^2.$$

The velocity is given by $m_0\gamma v/(M + m_0\gamma)$.

1.8 A uniformly accelerated particle

Let us calculate the motion of a uniformly accelerated particle. That means that the particle has a constant acceleration f_0 in its rest frame. Let us assume that the particle is moving in the x direction and starts with rest at $t = 0$. In the general frame we assume the 4-velocity of the particle to be u^i and so its acceleration will be $du^i/ds \equiv a^i$, say. Now, from the general result

$$\eta_{ik} u^i u^k = 1 \tag{1.48}$$

we get by differentiation with respect to s the result

$$u_i \frac{du^i}{ds} = 0. \tag{1.49}$$

Hence, in the rest frame the acceleration cannot have any component in the time direction, its sole component being f_0 in the x direction. Thus acceleration is a spacelike vector. From the invariance of its magnitude in the general frame we will have components of du^i/ds in the x and t directions such that

$$\eta_{ik} \frac{du^i}{ds} \frac{du^k}{ds} = -f_0^2. \tag{1.50}$$

The negative sign indicates that the vector is spacelike. We have therefore

$$u_i = \gamma \times [1, -v, 0, 0]; \qquad \frac{du^i}{ds} = [a_0, a_1, 0, 0].$$

From Equations (1.49) and (1.50) we therefore have

$$a_0 - v a_1 = 0, \qquad a_0^2 - a_1^2 = -f_0^2.$$

On solving these, we get the result

$$f_0 = \frac{d}{dt} \frac{v}{\sqrt{1 - v^2}}.$$

This equation can be integrated to give

$$v = \frac{f_0 t}{\sqrt{1 + f_0^2 t^2}}, \tag{1.51}$$

where we have used the boundary condition that at $t = 0$, $v = 0$. Now, assuming that at the initial instant the particle was at the origin, we get the integral of this equation as

$$x = \frac{1}{f_0} \left(\sqrt{1 + f_0^2 t^2} - 1 \right). \tag{1.52}$$

For $f_0 t \ll 1$ we have the non-relativistic approximation giving the familiar result from Newtonian dynamics: $v(t) = f_0 t, x = f_0 t^2/2$.

The proper time of the particle is given by

$$\tau = \int_0^t \sqrt{1 - v^2}\, dt = \frac{1}{f_0} \sinh^{-1}(f_0 t). \tag{1.53}$$

As t increases, this grows much more slowly than t. We will use this result to discuss the celebrated 'twin paradox' next.

1.9 The twin paradox

In the early days of special relativity the time-dilatation effect of the kind described earlier in Section 1.6 was considered puzzling and paradoxical largely because one was then accustomed to the Newtonian absolute time. The paradox arose because each of the two inertial observers could apparently argue that the clock of the other was moving more slowly than his. This was clearly logically impossible. However, the resolution of the paradox lay in the fact that in each experiment one observer used *two* clocks in his frame to compare with *one* clock of the other observer moving relative to him. Thus there was no symmetry between the two observers and there was no contradiction if one found the other's clock running slower.

A more sophisticated paradox was invented subsequently to counter this asymmetry. Known as the *twin paradox*, it had two twin brothers, A and B, say, of which B stays at rest in his inertial frame while A takes off in a spacecraft attaining high speed. A goes far and returns after a considerable time as measured by B. But, since A was moving relative to B with high speed, A would be younger than B ... because A's watch would run slower than B's. Also, since A and B (unlike the clocks in the earlier paradox) meet at the same place at the beginning and end of the experiment, the effect must be real. For example, if A accelerates and decelerates but has speed $v = 4c/5$ relative to B most of the time, his watch would run at the rate $\sqrt{1 - v^2}$ compared with B's watch. That is, if 40 years have elapsed as measured by B, only 24 years will have elapsed for A. So A will be younger than B by 16 years!

The paradox is not here. It arises when you argue that, as seen by A, his twin B has travelled away with high speed and come back to the same spot at the same time. Then, by the same argument B should be younger than B by 16 years. So what is the correct situation?

To resolve the paradox, we note that the situation between A and B is not symmetric. B is in an inertial frame whereas A is accelerated and decelerated over stretches of his journey. We can take a simple example that A is uniformly accelerated over a period of 10 years at the end of which he attains a speed of $4c/5$. Then he decelerates uniformly and comes to a halt after 10 years. He reverses his direction of travel and follows the same acceleration/deceleration pattern. Thus he is back with B after 40 years have elapsed by B's watch. How long a time has elapsed according to A's watch?

Using our formulae for uniform acceleration, we get the solution as follows. First, at the end of $t = 10$ years, the speed attained is given by formula (1.51):

$$\frac{v}{c} = \frac{f_0 \times t/c}{\sqrt{1 + f_0^2 t^2/c^2}}.$$ (1.54)

Now setting $v/c = 4/5$, we get the solution as $f_0 t/c = 4/3$. The formula (1.53) gives the elapse of proper time of A as

$$\tau = \frac{c}{f_0} \sinh^{-1}(f_0 t/c) = \frac{3}{4}t \, \sinh^{-1}(4/3).$$

The total proper time of A for the entire journey works out at 32.96 years. Thus A will return to find himself younger than B by about 7 years.

At a deeper level, one may still raise the following question: of A and B, who is entitled to claim the status of an inertial observer? If there is no background to refer to, this may lead to an undecidable proposition. We will come back to this issue towards the end of this book when we discuss Mach's principle.

1.10 Back to electrodynamics

Einstein was led to special relativity through Maxwell's electromagnetic theory. We close this chapter by highlighting a few issues that illustrate the close relationship of electrodynamics, mechanics and special relativity. We begin with Minkowski's four-dimensional view. In terms of the four-dimensional notation, Maxwell's equations become more compact.

Consider a 4-vector A_i as the 4-potential whose timelike component serves as the electrostatic potential ϕ while the spacelike component serves as the electromagnetic potential **A**. The electromagnetic fields are then related to the tensor

$$A_{l;k} - A_{k;l} = F_{kl}.$$ (1.55)

Thus we have

$$[F_{01}, F_{02}, F_{03}]$$

as the electric field and

$$[-F_{23}, F_{13}, -F_{12}]$$

as the magnetic field.

The Maxwell equations then acquire an elegant structure:

$$F_{ik,l} + F_{kl,i} + F_{li,k} \equiv 0, \qquad F^{kl}_{,l} = 4\pi j^k,$$ (1.56)

where the 4-current vector j^k has its zeroth component as the charge density and the remaining three components as the 3-current density. The subscript comma indicates the derivative with respect to the indexed coordinate.

To supplement these equations, we add the four-dimensional version of the Lorentz force equation describing the motion of an electric charge in an electromagnetic field:

$$\frac{\mathrm{d}u^i}{\mathrm{d}s} = \frac{e}{m} F^i{}_k u^k, \tag{1.57}$$

m and e being the mass and charge of the particle.

We will not go into the details of this topic which is covered in detail in graduate-level texts in electrodynamics. We mention one fact, though, which will have relevance to our later work in general relativity. This is the derivation of the above equations from a single action principle, the action being (with $c = 1$)

$$\mathcal{A} = -\frac{1}{16\pi} \int_{\mathcal{V}} F_{ik} F^{ik} \, \mathrm{d}^4 x - \sum \int_{\Gamma} e A_i \, \mathrm{d}x^i - \sum \int_{\Gamma} m \, \mathrm{d}s. \tag{1.58}$$

The action is defined over a spacetime region of volume $\mathcal{V}$ and the particles like m are supposed to move across it along world lines Γ. The variation of the field variables leads to the (Maxwell) field equations while the variation of the particle world lines yields their equations of motion. Later, in Chapter 7, we will generalize this result.

Finally, a plane-wave solution of Maxwell's equation takes the simple form

$$A_m = a_m \times \exp(i k_l \cdot x^l), \tag{1.59}$$

where the null vector k_l has angular frequency ω as the timelike component, the spacelike part being the wavenumber vector $\mathbf{k}$ whose magnitude k is equal to ω/c and whose direction is that of the propagation of the wave. Thus, if the frequency of the wave is ν, then $\omega = 2\pi\nu$ and we may write in the old three-dimensional notation

$$k_l \cdot x^l = 2\pi\nu(t - \mathbf{r} \cdot \mathbf{e}/c),$$

where $\mathbf{e}$ is the unit vector in the direction of propagation of the wave.

1.10.1 The Doppler effect

Using the above notation, we can easily derive the formula for the relativistic Doppler effect, that is, the formula telling us how the frequency (and direction of propagation) of a light source in motion relative to an inertial observer depends on their relative motion. Using our two inertial observers, we wish to know how the frequency and direction of the wave changes between O and O'.

We refer to Equation (1.59) above and note that the quantity in the exponent is the phase of the wave and that this should be invariant between two inertial observers.

Using Equations (1.17) for the Lorentz transformation we get

$$v(t - \mathbf{e} \cdot \mathbf{r}/c) = v'(t' - \mathbf{e}' \cdot \mathbf{r}'/c)$$

$$= v'\gamma[t - \mathbf{v} \cdot \mathbf{r}/c + \mathbf{e}' \cdot (\mathbf{r}^* - \mathbf{v}t)/c]. \qquad (1.60)$$

By equating coefficients of t and $\mathbf{r}$ on both sides we get

$$v = v'\gamma(1 - \mathbf{e}' \cdot \mathbf{v}/c),$$

$$\mathbf{e} = \frac{\mathbf{e}'^* - \mathbf{v}/c}{1 - \mathbf{e}' \cdot \mathbf{v}/c}. \qquad (1.61)$$

Here $\mathbf{e}'^*$ is given in terms of $\mathbf{e}'$ just as r^* is given in terms of $\mathbf{r}$ by (1.18).

If we take without loss of generality the space components of these vectors as $\mathbf{v} = (v, 0, 0), \mathbf{e} = (\cos\theta, \sin\theta, 0), \mathbf{e}' = (\cos\theta', \sin\theta', 0)$, then we can reduce Equations (1.61) to

$$v = \gamma v' \times (1 - v\cos\theta'/c) \qquad (1.62)$$

and

$$\cos\theta = \frac{\cos\theta' - v/c}{1 - v\cos\theta'/c},$$

$$\sin\theta = \frac{\sin\theta'}{\gamma(1 - v\cos\theta'/c)}. \qquad (1.63)$$

In the case of radial motion away from the observer, we get the answer as

$$v = v'\sqrt{\frac{c - v}{c + v}}. \qquad (1.64)$$

That is, the source of light has a reduced apparent frequency as seen by the observer. This phenomenon is known as *redshift*, since in the visible spectrum the red colour lies at the lowest-frequency end.

The formula (1.63) is used to explain the phenomenon of aberration as seen in the example that follows.

Example 1.10.1 The formula (1.63) is useful in the measurement of the change in the apparent direction of a light source when viewed from a moving frame, with different velocities. The angle θ in one frame changes to θ' as seen in the text. The change in the direction is called *aberration*.

In the case of the Earth, the direction of its motion changes to the opposite after six months (half the orbit). So in (1.63) v changes to $-v$, leading to change in the direction (θ) of the source. The net shift in direction, $\sim v \sin \theta / c$, is of the order of 10^{-4}, but can be measured. The aberration of the star γ *Draconis* was first measured in 1725 by James Bradley and this was the first proof of Galileo's firm belief that the Earth moves.

1.11 Conclusion

This brings us to the end of a 'crash course' on special relativity. The reader may wish to study the subject at greater depth than dealt with here and to this end the References [4, 5, 6] may be worth a look. We have, however, built a framework from which to launch the study of a more elaborate theory that deals with arbitrarily accelerated observers and the modification of the Newtonian framework of gravitation. This is the *general theory of relativity* which determines the main interest of this book.

Exercises

1. Find whether the following vectors are timelike, spacelike or null.

(i) $A^0 = 4$, $A^1 = 3$, $A^2 = 2$, $A^3 = 1$.
(ii)   The tangent to the circle $x^2 + y^2 = 1$, $z = 0$, $t = 0$.
(iii)   The normal to the hyperboloid $x^2 + y^2 + z^2 - c^2 t^2 = 1$.
(iv)   The tangent to the curve parametrized by λ, where $x^1 = \int r \sin \theta \, d\lambda$, $x^2 = \int r \cos \theta \, d\lambda$, $x^3 = \int z \, d\lambda$ and $x^0 = \int \sqrt{r^2 + z^2} \, d\lambda$. The r, θ, z are arbitrary functions of λ.

2. From the application of a special Lorentz transformation work out how the electric and magnetic fields transform in vacuum.

3. A rod moves with velocity $3c/5$ in a straight line relative to an inertial frame S. In its rest frame the rod makes an angle of $60°$ with the forward direction of its motion. Show that in the frame S the rod appears to make an angle $\cot^{-1}(4/(5\sqrt{3}))$ with the direction of motion.

4. A mirror is moving with speed v in the x direction with its plane surface normal to it. In this frame a photon travelling in the x–y plane is incident on the mirror surface at angle θ to the normal. Show that, as seen from this frame, the reflected photon makes an angle $\bar{\theta}$ with the normal to the mirror, where

$$\cos \bar{\theta} = \frac{\cos \theta + \cos \alpha}{1 + \cos \theta \cos \alpha}$$

and $\cos \alpha = 2v/(1 + v^2)$.

5. In a Compton-scattering experiment, a photon was scattered in a direction making an angle of 60° with its original direction. Show that the wavelength of the photon will have increased by $h/(2m_0c)$.

6. Show that, for the Maxwell equations, the quantities $B^2 - E^2$ and $\mathbf{B} \cdot \mathbf{E}$ are invarient under the Lorentz transformation.

7. Show that if $\mathbf{E} \cdot \mathbf{B} = 0$ there is a Lorentz transformation that makes either $\mathbf{E}$ or $\mathbf{B}$ equal to zero.

8. An electric charge q of mass m at rest moves in a circular orbit around a magnetic field B perpendicular to the orbit. The charge takes time $2\pi/\omega$ and the radius of the orbit is R. Show that

$$B = \frac{m\omega}{q\sqrt{1-v^2}} = \frac{m\omega}{q\sqrt{1-\omega^2 R^2}}.$$

9. A source of light is moving towards an observer with speed v such that its direction of motion makes an angle θ with the line of sight to the source. If there is zero Doppler shift, find θ.

10. From the observation formula derived in the text show that a source viewed from the Earth today and six months later will show a shift in direction equal to $2v/c \times \sin\theta$, where θ is the angle the direction to the source makes with the Earth's motion. Estimate the order of magnitude of the effect.

11. Can an electron by itself absorb or emit a single photon? If the answer is 'Yes', show one example. If the answer is 'No', prove it.

12. A particle of rest mass M_0 disintegrates into three particles of rest masses M_1, M_2 and M_3, respectively. If, in the rest frame of the original particle, the particles of rest masses M_2 and M_3 move at right angles and have equal energies, calculate the energy of each particle and show that the following inequalities must be satisfied:

$$M_1^2 < \frac{1}{2}(M_0 - 2M_2)^2 + \frac{1}{2}(M_0 - 2M_3)^2,$$

$$M_2^2 + M_3^2 < \frac{1}{2}(M_0 - M_1)^2.$$

13. Two electrons are approaching each other, each with energy γmc^2 in the laboratory frame. What is the energy of one electron as seen in the rest frame of the other?

14. A traveller through interstellar space took off from the Earth at $t = 0$ with constant acceleration f and, after $t = t_1$, continued moving at the acquired speed until $t = t_2$. Then he decelerated with constant deceleration f, until he came to rest at $t = t_2 + t_1$. He then reversed the trajectory to return to Earth at $t = 2(t_1 + t_2)$. What duration for this journey was registered on his own clock?

15. A speeding spaceship went through a red light at a traffic junction in interstellar space. When stopped by cops, the driver said 'But I saw only green lights'.

Could he be telling the truth? If so, how will you estimate his approximate velocity?

16. A train 100 m long is approaching a tunnel 75 m long, at a speed of $0.8c$. The tunnel keeper has instructions to close both (entrance and exit) ends of the tunnel *simultaneously* when the rear end of the train enters the tunnel. How long after the tunnel is sealed does the engine strike the exit door of the tunnel? For the engine driver, the tunnel appears shrunk to what length? Can the shrunken tunnel accommodate the train? How do you resolve the contradiction between the tunnel keeper's and the engine driver's version? (This example usually generates considerable discussion.)

17. An electron at rest is hit by an approaching photon with energy equal to the rest energy of the electron. Show that in the centre-of-mass frame of the system the electron is moving with speed half that of light.

Chapter 2
From the special to the general theory of relativity

2.1 Space, time and gravitation

The special theory of relativity reviewed in the last chapter marked a major advance in physics. The basic assumption that the fundamental laws of physics are invariant for all inertial observers looks at first sight a reasonable premise. However, as we saw in Chapter 1, its application to Maxwell's equations of electromagnetic theory led to a drastic revision of how such observers make and relate their measurements of space and time. One consequence was that the Newtonian notions of absolute space and absolute time had to be abandoned and replaced by a unified entity of *spacetime*. The Galilean transformation relating the space and time measurements of two inertial observers had to be replaced by the Lorentz transformation. Strange and non-intuitive though the consequences of this transformation were, as we saw in Chapter 1, several experiments confirmed them.

In spite of these successes, Einstein felt that the special theory addressed limited issues. For example, what was the nature of physical laws when viewed *not* in the inertial frames of reference, but in an *accelerated* one? Was there some more general principle that, when applied to these laws, preserved their form? Intuitively Einstein felt that some such situation must prevail. But that required a formalism more general than that provided by the Lorentz transformation.

On another matter, of the two classical theories of physics known in the first decade of the twentieth century, the electromagnetic theory had played a major role in the genesis of special relativity. The invariance of Maxwell's equations under the Lorentz transformation is linked with the invariance of the speed of light for all inertial observers. The light speed thus has a special status in spacetime and causality is preserved by demanding that *all* physical interactions travel with speeds not exceeding this speed. Nevertheless, this fiat was broken by the Newtonian law of gravitation, the other known basic interaction. The gravitational attraction seemed to be instantaneous across space: that is, it seemed to propagate with *infinite* speed. Just as Newtonian dynamics had to be adapted to suit the new rules of spacetime measurements, the Newtonian law of gravitation also required a suitable adaptation.

Hermann Bondi had highlighted the conflict between the law of gravitation and special relativity by the following example. Imagine that by 'some magic' the Sun is removed from its place. How and when will we on the Earth come to know of the event? Because sunlight takes about 500 seconds to reach us, we would come to know of the Sun's absence after that period. However, the disappearance of the Sun's gravity will be 'instantaneous' and the Earth would move off its usual trajectory 'immediately'. Thus gravitational interaction would tell us of the event 500 seconds before light would do so. This thought experiment, though physically impossible, illustrates the point.

One may think that it is relatively simple to adapt Newtonian gravitation to suit special relativity. But the reality is different! For example, consider the Laplace equation

$$\nabla^2 \phi = 4\pi G \rho, \tag{2.1}$$

where ϕ is the gravitational potential and ρ is the mass density. If one wishes to make the interaction travel with the speed of light, one may change the '∇^2' operator on the left-hand side to the wave operator '$\Box$'. This can be easily done. But look at the right-hand side. Since special relativity teaches us that $E = Mc^2$, we need to include in ρ all the energy densities also. Now, Newtonian theory tells us that gravitation itself has energy, and so we need to include it on the right-hand side. The energy density of gravitation has a form something like

$$\rho_\phi = -[(\nabla \phi)^2 - \dot{\phi}^2]/(8\pi G). \tag{2.2}$$

In other words the modified equation (2.1) has a $(\nabla \phi)^2$-dependent term on the right-hand side. Thus the problem has become non-linear. One may further ask whether the self-action from ϕ will not change its value further. Indeed it will! In fact Equation (2.1) becomes even more complicated. We will not get further into this issue here since our purpose, to demonstrate that the original Newtonian law becomes non-linear if we try to make it consistent with special relativity, has already been served.

While this example illustrates that the Newtonian law needs to be modified in the presence of special relativity, the reverse is also true.

We need to modify special relativity in the presence of gravity. For we have looked at how special relativity functions in the framework of inertial observers. As defined in Chapter 1, these observers are moving under no force. Can one find such observers in reality? The answer is 'Not in the presence of gravitation!'. For it is not possible to isolate a finite-sized region inside which there is no gravitational force for a finite time. What about astronauts in spaceships apparently floating in a gravity-free region? Even there gravity is not altogether absent: for, wherever there is matter, there is a gravitational effect, howsoever small it may be. So we need a theory that allows for the ever-present gravity. We will return to this issue shortly. The theory that Einstein came up with to address this issue, the *general theory of relativity*, represents perhaps the most remarkable flight of imagination in science.

Every major scientific theory carries its own mark of distinction. The distinctive feature of Newtonian gravitation is the radial inverse-square law. To those uninitiated in the laws of dynamics, the fact that a planet goes *around* the Sun under a force of attraction *towards* the Sun comes as a surprise. Yet this is a natural consequence of the inverse-square law. The major achievement of Maxwell's electromagnetic theory was the unification of electricity and magnetism and the demonstration that light itself is an electromagnetic wave. The unique place held by the speed of light characterizes Einstein's special theory of relativity, while quantum mechanics can point to the uncertainty principle as the crucial feature that sets it apart from classical mechanics.

To what distinctive feature can general relativity lay its own special claim? A clue to the answer to this question is provided in the title of this section.

Let us compare gravitation with electricity. We know that two unlike electric charges attract each other through the Coulomb inverse-square law, just as any two masses attract each other gravitationally by the Newtonian inverse-square law. To this extent, electricity and gravitation are similar. However, we can go no further! We also know that two like electric charges repel each other and that this property seems to have no parallel in gravitation. Every bit of matter attracts every other bit and, as yet, we do not have any instance of gravitational repulsion.

We can express this difference between electricity and gravitation in another, more practical way. The existence of repulsion as well as attraction with positive and negative charges enables us to construct a closed chamber whose interior is completely sealed from any outside electrical or magnetic influence. Not so with gravitation! We cannot point to any region of space as being totally free of external gravitational influences. Gravitation is permanent: it cannot be switched off at will.

This ever-present nature of gravitation plays a key role in Einstein's general theory of relativity. Einstein argued that, because of its permanence, gravitation must be related to some intrinsic feature of space and time (which are also permanent!). With a master stroke of genius, he identified this feature as the *geometry* of space and time. He suggested that any effects we ascribe to gravitation actually arise because the *geometry* of space and time is 'unusual'. Let us now try to understand what is meant by the word 'unusual' and how this property of space and time leads to gravitational effects – for therein lies the distinctive characteristic that sets general relativity apart from other physical theories.

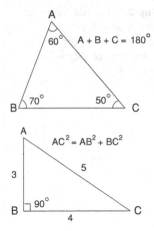

Fig. 2.1. Some familiar results of Euclid's geometry: above, the three angles of a triangle add up to 180°; below, Pythagoras' theorem that the square of the hypotenuse of a right-angled triangle equals the sum of the squares of its other two sides.

2.2 Non-Euclidean geometries

The 'usual' geometry of space, the geometry that we learn at school and apply in so many ways, is the geometry whose foundations were laid by the Greek mathematician Euclid *c.* 300 BC. Euclidean geometry is a logical structure wherein theorems about triangles, parallelograms, circles and so on are proved on the basis of postulates that are taken as self-evident. Thus the results shown in Figure 2.1 follow as theorems in Euclid's geometry, being based on the original postulates of Euclid, and the validity of these results appears to be borne out by measurements of lengths and angles in physical space.

Postulates are assumptions that are regarded as self-evident and are not expected to be 'proved'. One such postulate is illustrated in Figure 2.2. Here we have a straight line *l* with a point P outside it. How many straight lines can we draw through P parallel to *l*? Our experience suggests that the answer is 'only one'. But can this expectation be further proved? In Euclid's geometry this is taken as a postulate (sometimes known as the *parallel postulate*) and many of its theorems are based on it. One such theorem is that the three angles of a triangle add up to two right angles.

It was only in the last century that mathematicians realized that there is nothing sacrosanct about Euclid's postulates. Provided that they are not mutually contradictory, any new set of postulates can lead to a new and consistent type of geometry. Indeed, as the work of such mathematicians as Gauss (1777–1855), Bolyai (1802–1860), Lobatchevsky (1793–1856) and Riemann (1826–1866) showed, a host of such new geometries can be constructed. These are collectively called *non-Euclidean geometries*. For instance, the parallel postulate can be changed: one may assume that there is *no* straight line through P parallel to *l*, or one may assume that *several* lines can be drawn through P parallel to *l*. Geometries using these revised postulates will be non-Euclidean.

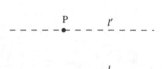

Fig. 2.2. The parallel postulate of Euclid, described in the text.

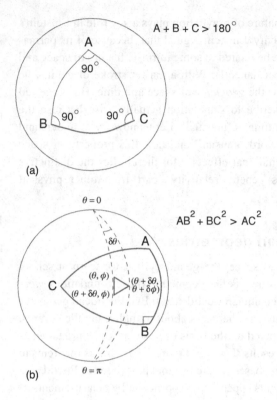

In this sense, the geometry on the surface of a sphere is non-Euclidean. If we define a straight line on the surface of a sphere as the line of shortest distance between two points, it is easy to see that these lines are arcs of great circles. Because any two great circles inter-sect, there are no parallel lines in this geometry: *no* line can be drawn through P parallel to *l*. Figure 2.3 demonstrates how the theorem about the sum of the three angles of a triangle breaks down in this case.

We may also mention in passing that Euclidean straight lines such as latitude lines drawn on a flat map of the Earth *do not* represent paths of shortest distance. Rather such paths drawn on flat maps look curved. Air-craft pilots choosing to fly along the shortest routes follow these paths.

We will now explore the possibility that Einstein advocated, namely that the effect of gravity can be described not through the conventional Newtonian interpretation of a force but by ascribing a non-Euclidean character to the geometry of spacetime.

2.3 Gravity, geometry and dynamics

The Einsteinian concept of non-Euclidean geometry of space extended to space and time or *spacetime* can be illustrated by an example in dynamics. Figure 2.4 shows the spacetime trajectory followed by a stone thrown vertically upwards with an initial velocity v. Counting time t from the instant of throw, the height y of the stone at any instant t is given by

$$y = vt - \frac{1}{2}gt^2, \tag{2.3}$$

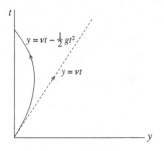

Fig. 2.4. As explained in the text, in the Newtonian dynamics the dotted trajectory describes straight-line motion without acceleration. The curved continuous line shows the trajectory wherein acceleration is induced by gravity. According to Einstein only this trajectory has real status and it describes uniform motion in a straight line in the spacetime 'curved' by gravity.

where g is the acceleration due to gravity. This is a Newtonian equation and may be interpreted as follows. *If there were no gravity*, the stone would have continued to move in a straight line with uniform speed v directed upwards, as indicated by the first term only on the right-hand side of the above equation. In Figure 2.4 this trajectory is shown by a dotted straight line. By contrast the actual trajectory is shown by the parabolic continuous line. In the Newtonian framework, the dotted trajectory illustrates the first law of motion, whereas the continuous one arises from the second law of motion because of the application of the force of gravity.

At this juncture a sceptic may argue that the continuous line of Figure 2.4 clearly demonstrates a curved trajectory with a varying velocity. How can we call it a straight-line motion with uniform speed? The answer is that, if the rules of geometry are changed, so are the definitions of straight lines and measurements of velocities. In Figure 2.3, for example, the apparently curved arcs on the sphere are in fact straight lines in the spherical geometry. Similarly, in this case, the geometry of spacetime is changed in such a way that the trajectory that is curved in Euclidean geometry becomes straight! Recall the example of the flat map of the Earth in the previous section. Similarly, the constancy of velocity is to be determined not by the rules of Euclidean measurements but by the prevailing non-Euclidean geometry.

According to Einstein, the interpretation of this phenomenon would be as follows. First we have to recognize that the gravity of the Earth has a permanent existence. It cannot be 'turned on' or 'turned off'. So we should not talk of the dotted trajectory, dealing as it does with a possibility that cannot happen. The continuous trajectory is the only trajectory that we have to interpret. Only we now assume that the geometry of spacetime in which the stone is moving is *non-Euclidean*, being rendered so by the presence of Earth's gravity. So the apparently parabolic-looking trajectory is actually describing 'straight line motion with a uniform velocity' but in a non-Euclidean spacetime.

To summarize, therefore, Einstein's way of describing gravity is to do away with the notion that it is a force. Rather the trick is to find a suitably non-Euclidean spacetime geometry in which matter under no force moves in 'straight-line trajectories with uniform speed' as measured in terms of the rules of the new geometry.

We need a mathematical apparatus to describe these ideas precisely. Also, if spacetime is 'curved' because of its non-Euclidean geometry we need suitable machinery to describe physics in it. For, as we just saw, even the simple concept of velocity requires re-definition. So we will begin our discussion of general relativity by setting up the framework needed for these purposes. Chapter 3 accordingly starts the process.

Exercises

1. An optical photon has wavelength 500 nm. Estimate its gravitational mass.

2. Two airports exist on the same latitude θ but on longitudes differing by 180°. Show that an aircraft connecting them flying along the latitude θ will exceed the shortest path by $2R[\theta - \pi \sin^2\theta/2]$. R is the radius of the Earth.

3. A triangle on the spherical Earth has its vertices at specified latitude (l) and longitude (L) as follows: vertex A at $l = 45°$, $L = 180°$; vertex B at $l = 0°$, $L = 120°$ and vertex C at $l = 0°$, $L = 240°$. By what amount does the sum of the three angles of the triangle ABC exceed 180°?

4. Assuming that the Newtonian potential at an external point at distance R from the centre of a spherical mass distribution is GM/R, estimate a correction to it to first order using formula (2.2) for gravitational energy.

5. Consider a sphere of uniform density ρ and radius R. Calculate the Newtonian gravitational potential ϕ for this sphere. Compare the gravitational energy source $(\nabla\phi)^2/(8\pi G)$ with ρ.

6. A projectile is ejected with vertical speed v from the surface of a planet of mass M and radius R. Show that it comes to rest at a distance r from the centre of the planet, where

$$r = \frac{R}{1 - \dfrac{Rv^2}{2GM}},$$

provided that the denominator is positive. Put $v = c$ in the above example and deduce the condition between M and R such that the planet acts like a Newtonian black hole.

Chapter 3
Vectors and tensors

3.1 The spacetime metric

The classical definition of 'geometry' is that it is a science of measurements of distances and angles. Let us first recall a familiar result from special relativity in the following form. Let (x, y, z) denote a Cartesian coordinate system and t the time measured by an observer O at rest in an inertial frame, that is by an observer who is acted on by no force. Let two neighbouring events in space and time be labelled by the coordinates (x, y, z, t) and $(x + dx, y + dy, z + dz, t + dt)$. The resulting analogue of the Pythagorean theorem is as follows. The square of the 'distance' between the two events is given by

$$ds^2 = c^2\,dt^2 - dx^2 - dy^2 - dz^2. \tag{3.1}$$

The distance ds is invariant under Lorentz transformation in the sense that another inertial observer O' using a different coordinate system (x', y', z', t') to measure this distance will find the same answer.

However, when we make a transition from special to general relativity and quantify Einstein's idea that the geometry of space and time is unusual in the presence of gravitation, we abandon the simple form of (3.1) in favour of a more complicated form. The more complicated form is still quadratic, and we may state it formally as follows:

$$ds^2 = \sum_{i,k=0}^{3} g_{ik}\,dx^i\,dx^k. \tag{3.2}$$

Here we have modified the notation as follows. The coordinates are now called x^i, with $i = 1, 2, 3$ representing the three space coordinates and $i = 0$ the time coordinate. The coefficients g_{ik} are functions of x^i with the property that the matrix $\|g_{ik}\|$ has the signature -2.[1]

The expressions for ds^2 are often referred to as the 'line element' or the 'metric'. Thus we have the general expectation that the spacetime metric has signature -2. We shall denote the determinant of the matrix $\|g_{ik}\|$ by g and its inverse matrix by $\|g^{ik}\|$. It is easy to verify that g is negative.

Clearly, the geometry of spacetime in which the basic invariant distance is given by (3.2) instead of by (3.1) is going to be more complicated to describe. Its properties will depend on the functions g_{ik}. But do these complications arise simply because of a choice of coordinates, or do they indicate a spacetime with a geometry genuinely different from that used in special relativity?

Example 3.1.1 Consider for example the form

$$ds^2 = c^2 \, dt^2 - [dr^2 + r^2(d\theta^2 + \sin^2\theta \, d\phi^2)],$$

which looks more complicated than (3.1). Does it describe some new geometry? A little investigation will show that it is obtainable from (3.1) by the coordinate transformation

$$x = r \cos\theta, \qquad y = r \sin\theta \cos\phi, \qquad z = r \sin\theta \sin\phi.$$

We do not expect that a fundamental change in the properties of spacetime, such as its geometry, should be brought about by such a change of coordinates. However, consider another example.

Let us take the geometry on the surface of a sphere Σ of radius a. If we consider the sphere as embedded in a three-dimensional space with the Cartesian coordinates x, y, z, we may write the equation of the surface of the sphere as

$$x^2 + y^2 + z^2 = a^2.$$

However, can we study the geometry on this surface without recourse to the embedding space? For describing the geometry on the surface of the sphere it is more convenient to use coordinates *intrinsic* to the surface of the sphere. Such coordinates are available and are like the latitude and longitude used by geographers to locate a point on the Earth. More specifically,

$$x = a \cos\theta, \qquad y = a \sin\theta \cos\phi, \qquad z = a \sin\theta \sin\phi,$$

---

[1] This means that, if at any spacetime point the quadratic form (3.2) is diagonalized, it has one square term with a positive coefficient and three square terms with negative coefficients. The signature equals the number of positive terms minus the number of negative terms.

so that for any (θ, ϕ) with $0 \leq \theta \leq \pi$ and $0 \leq \phi < 2\pi$ we can locate a point (x, y, z) on the surface of the sphere. Spherical trigonometry tells us how to measure and relate the angles, sides and so on of triangles drawn on this surface. The rules of Euclid's geometry do not apply to these measurements.

In our above example, the square of the distance between two neighbouring points (θ, ϕ) and $(\theta + d\theta, \phi + d\phi)$ is given by

$$d\sigma^2 = [dx^2 + dy^2 + dz^2]_\Sigma = a^2(d\theta^2 + \sin^2\theta \, d\phi^2). \qquad (3.3)$$

Thus we have here another example of g_{ik} that are not all constants. In this as well as the previous example, this property is shared. However, in the earlier case the geometry was Euclidean, whereas here it is not. Simply having a coordinate-dependent g_{ik} does not therefore convey the physical reality. So the mathematical formalism that we build up should be such that it can distinguish between real effects and coordinate effects. In a qualitative way we can see that the essential information must survive even when we change from one coordinate system to another. In order to extract such information, we must devise machinery that tells us *what things remain unchanged under coordinate transformations*. Such machinery is provided by the invariants, the vectors and the tensors, which we shall now study.

3.2 Scalars and vectors

Let us first introduce the summation convention which was already used in a limited way in Section 1.5.1. We will frequently encounter sums like

$$\sum_{i=0}^{3} A_i B^i, \qquad \sum_{k=0}^{3} A_{ik} B^k, \qquad \sum_{i,k=0}^{3} P_{ik}\xi^i \xi^k, \ldots. \qquad (3.4)$$

It is convenient in such cases to drop the summation symbol $\sum$ and write these quantities as

$$A_i B^i, \qquad A_{ik} B^k, \qquad P_{ik}\xi^i \xi^k, \ldots, \qquad (3.5)$$

the rule being that, *whenever an index appears once as a subscript and once as a superscript in the same expression, it is automatically summed over all the values* (from 0 to 3). Thus we can rewrite (3.2) in the more compact form

$$ds^2 = g_{ik} \, dx^i \, dx^k. \qquad (3.6)$$

A note of caution is needed here: the summation convention does not apply under any other circumstances. Thus it does not apply to quantities like

$$A_i B_i, \ A_{ik} B_i C_i, \ldots, \tag{3.7}$$

wherein repeated indices do not follow the rule of appearing only twice, once up and once down. However, such expressions fortunately do not arise in most relativistic calculations. Indeed, the appearance of such 'monster' expressions is a warning that we have made a mistake in our index manipulation. At this stage the appearance of subscripts and superscripts may seem somewhat arbitrary. We ask the reader to be patient: they will be properly introduced into the formalism very shortly.

We will assume that the Latin indices $i, j, k, \ldots$ will run over all four values 0, 1, 2, 3. On some (rather infrequent) occasions we may want to refer to index values 1, 2, 3 only, which are usually reserved for space components, and we will use Greek indices $\mu, \nu, \ldots$ to represent these. Thus $A_\mu B^\mu$ will equal $A_1 B^1 + A_2 B^2 + A_3 B^3$.

It is worth pointing out here that many other textbooks use the convention of denoting the spacetime coordinates by Greek indices λ, μ, ν, etc. and the space coordinates by Latin indices i, j, k, etc. Also many authors prefer to write (3.1) with the opposite sign for the right-hand side. In that case the signature of the metric is $+2$. Likewise, in some texts, time is treated as coordinate number 4 instead of 0, as it is here. These differences are of a 'cosmetic' nature and do not affect the 'physics' being described. We caution the reader to check these differences before comparing expressions from different sources.

We now consider a simple example in two-dimensional Euclidean space, i.e., in a space where Euclid's geometry holds. Let, as in Figure 3.1,

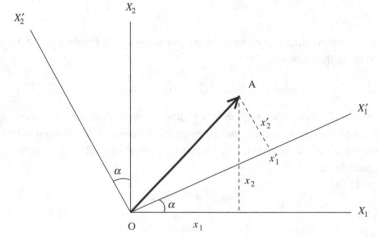

Fig. 3.1. A change of Cartesian coordinates changes the components of a vector, although the vector remains unchanged. Here we see the effect of rotation of axes around the origin.

OX_1 and OX_2 denote two Cartesian coordinate axes corresponding to coordinates x_1 and x_2, respectively. Suppose we have a vector **A** with two components A_1 and A_2 in these directions. We now change coordinates by rotating the axes by an angle α. The new coordinates x_1' and x_2' are given in terms of the old ones by the formulae

$$x_1' = x_1 \cos \alpha + x_2 \sin \alpha, \qquad x_2' = x_2 \cos \alpha - x_1 \sin \alpha. \qquad (3.8)$$

Notice that under this transformation the components of **A** also transform in a similar fashion:

$$A_1' = A_1 \cos \alpha + A_2 \sin \alpha, \qquad A_2' = A_2 \cos \alpha - A_1 \sin \alpha. \qquad (3.9)$$

Now in the usual definition we associate a vector with a magnitude and a direction. The above equations keep track of the direction of the vector, ensuring that two different observers, one using the unprimed coordinates and the other using the primed ones, are talking about the same entity even though they are measuring different components relative to their axes. They also agree on the *magnitude* of the vector, since it is easy to verify that

$$A_1^2 + A_2^2 = A_1'^2 + A_2'^2. \qquad (3.10)$$

In short, these transformation laws preserve the physical essentials of a vector, namely its magnitude and direction. We will be guided by this simple example in generalizing the concept of a vector under *any* coordinate transformation.

3.2.1 Scalars

We now introduce spacetime as a manifold $\mathcal{M}$ of $3 + 1$ dimensions in which a typical point P is specified by four coordinates x^i. We shall in general talk about geometrical entities of $\mathcal{M}$ or of physical quantities defined in $\mathcal{M}$, which are continuous and differentiable (at least twice) with respect to x^i. It is within this manifold that we now proceed to describe our geometry–physics relationship. We begin with the simplest physical notion.

A *scalar* or an *invariant* does not change under any change of coordinates. Thus if $\phi(x^i)$ is a function of coordinates, then it is invariant provided that it retains its value under a transformation from x^i to new coordinates x'^i:

$$\phi(x^i) = \phi[x^i(x'^k)] = \phi'(x'^k). \qquad (3.11)$$

Note that the form of the function may change, but its value does not. Note further that the infinitesimal square of distance (3.6) is a scalar quantity. In our example of vectors in two dimensions, we had

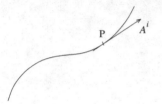

Fig. 3.2. The tangent to the curve at P acts like a contravariant vector.

encountered the property that the *magnitude* of a vector does not change under the coordinate transformation representing rotation of axes. It is therefore a scalar.

3.2.2 Contravariant vectors

Suppose we are given a curve Γ in space and time, which is parametrized by λ. (See Figure 3.2.) Thus, the points along the curve have coordinates

$$x^i \equiv x^i(\lambda), \tag{3.12}$$

where x^i are given functions of λ. The direction of the tangent to Γ at any point P on it is given by a vector with four components,

$$A^i \equiv \frac{dx^i}{d\lambda}. \tag{3.13}$$

Notice that the direction of a tangent to the curve is an invariant concept: a change of coordinates should not alter this concept, although its four components in the new coordinates will be different. Suppose the new coordinates are x'^i and the new components are A'^i. Then

$$A'^i \equiv \frac{dx'^i}{d\lambda}. \tag{3.14}$$

As stated earlier, we will assume that the transformation functions

$$x^i = x^i(x'^k), \qquad x'^k = x'^k(x^i) \tag{3.15}$$

are continuous and possess at least second derivatives. It is then easy to see that A'^i and A^i are related by the linear transformation

$$A'^k = \frac{\partial x'^k}{\partial x^i} A^i. \tag{3.16}$$

We use (3.16) as the transformation law for *any* vector A^i. Quantities in general that transform according to the above linear law are called *contravariant vectors*. The four components of a contravariant vector are specified by a superscript.

Example 3.2.1 Consider the curve parametrized by

$$x^0 = \text{constant}, \qquad x^1 = \text{constant}, \qquad x^2 = \lambda, \qquad x^3 = \lambda^2.$$

The tangent to this curve is specified by the contravariant vector A^i with components

$$A^0 = 0, \qquad A^1 = 0, \qquad A^2 = 1, \qquad A^3 = 2\lambda.$$

A comparison with the two-dimensional Cartesian example will show that the transformation law (3.16) used above is a generalization of the law (3.9) used there. In that simple example the coordinate transformation was linear and so the coefficients $\partial x'^i/\partial x^k$ and $\partial x^k/\partial x'^i$ were constants, $\cos\alpha$, $\sin\alpha$, etc.

3.2.3 Covariant vectors

Consider next a scalar function $\phi(x^k)$. The equation

$$\phi(x^k) = \text{constant} \qquad (3.17)$$

describes a hypersurface (that is, a surface of three dimensions) Σ, whose normal at a typical point Q has the direction given by the four quantities

$$B_i = \frac{\partial\phi}{\partial x^i}. \qquad (3.18)$$

(See Figure 3.3.) Again, the concept of a normal to a hypersurface should be independent of the coordinates used. Under the coordinate transformation (3.15), the new components are

$$B_i' = \frac{\partial\phi}{\partial x'^i}.$$

It is easy to see that $B_i' \leftrightarrow B_i$ is a linear transformation:

$$B_k' = \frac{\partial x^i}{\partial x'^k} B_i. \qquad (3.19)$$

Again, we generalize (3.19) as a transformation law of any vector B_i. Quantities that transform according to this rule are called *covariant vectors*.

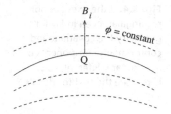

Fig. 3.3. The normal to the hypersurface at Q acts like a covariant vector.

Example 3.2.2 The normal to the unit sphere given by

$$\phi \equiv (x^1)^2 + (x^2)^2 + (x^3)^2 = 1$$

has the covariant components

$$B_0 = 0, \qquad B_1 = 2x^1, \qquad B_2 = 2x^2, \qquad B_3 = 2x^3.$$

Consider oblique coordinate axes OX_1 and OX_2 inclined at an acute angle β in a two-dimensional Euclidean plane. Figure 3.4 illustrates the situation. Let a vector $\mathbf{A}$ be shown by arrow OP. How do we specify the two components of this vector? There are two obvious ways. One is to draw straight lines through P parallel to the axes intersecting them at R_1 and R_2, respectively. The lengths OR_1 and OR_2 then specify

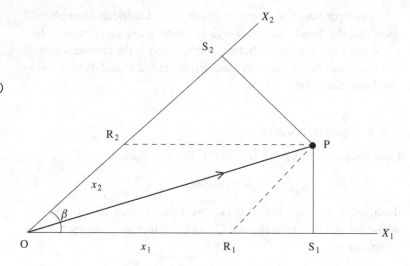

Fig. 3.4. If the axes are not rectangular, even in Euclid's geometry the covariant and contravariant components of a vector are different, as seen here. (See the text for details.)

the *contravariant* components of the vector. For these components are in the directions *tangential* to the coordinate lines $x_2 =$ constant and $x_1 =$ constant. The second way is to drop perpendiculars PS_1 and PS_2 from P on to the two coordinate axes. The lengths OS_1 and OS_2 then represent the *covariant* components of the vector, since they are given by the intercepts of normals to the coordinate axes. As will be appreciated, the two ways of describing the vector coincide if we choose rectangular Cartesian coordinates.

We will return to this example later.

3.3 Tensors

The concept of a vector can be generalized to that of a tensor. Imagine a product of two contravariant vectors A^i and B^k. The 4×4 quantities $A^i B^k$ describe a tensor. Since we know from (3.16) how A^i and B^k transform, we can work out how their product transforms, and apply the rule to a general tensor with two contravariant indices. Thus a contravariant tensor of rank 2 is characterized by the following transformation law:

$$T'^{ik} = \frac{\partial x'^i}{\partial x^m} \frac{\partial x'^k}{\partial x^n} T^{mn}. \tag{3.20}$$

A covariant tensor of rank 2 is likewise characterized by the transformation law

$$T'_{ik} = \frac{\partial x^m}{\partial x'^i} \frac{\partial x^n}{\partial x'^k} T_{mn}. \tag{3.21}$$

It is also possible to have *mixed* tensors. Thus T^i_k is a mixed tensor of rank 2, with one contravariant index and one covariant index. It transforms as

$$T'^i_k = \frac{\partial x'^i}{\partial x^m} \frac{\partial x^n}{\partial x'^k} T^m_n. \tag{3.22}$$

Again, these concepts are easily generalized to tensors of rank higher than 2. The rule is to introduce a transformation factor $\partial x'^i / \partial x^m$ for each contravariant index i and a factor $\partial x^n / \partial x'^k$ for each covariant index k. In general, a mixed tensor of rank $r = p + q$ may have p contravariant indices and q covariant indices.

Trivially, we may consider a scalar as a tensor of rank 0 and a vector as a tensor of rank 1.

Example 3.3.1 The quantities g_{ik} transform as a covariant tensor. This result follows from the assumption that ds^2 as given by (3.6) is invariant. For

$$\begin{aligned}
ds^2 &= g_{ik} \, dx^i \, dx^k \\
&= g_{ik} \left(\frac{\partial x^i}{\partial x'^m} \, dx'^m \right) \left(\frac{\partial x^k}{\partial x'^n} \, dx'^n \right) \\
&= \left(g_{ik} \frac{\partial x^i}{\partial x'^m} \frac{\partial x^k}{\partial x'^n} \right) dx'^m \, dx'^n \\
&= g'_{mn} \, dx'^m \, dx'^n;
\end{aligned}$$

that is,

$$g'_{mn} = \frac{\partial x^i}{\partial x'^m} \frac{\partial x^k}{\partial x'^n} g_{ik}.$$

Example. The *Kronecker delta* defined by

$$\delta^i_k = 1 \text{ if } i = k, \qquad \text{otherwise } \delta^i_k = 0$$

is a mixed tensor of rank 2. This can be easily proved using (3.22) and the identity

$$\frac{\partial x^i}{\partial x^l} = \delta^i_l.$$

Example. Define $\|g^{ik}\|$ to be the inverse matrix of $\|g_{ik}\|$, assuming that g is the determinant of $\|g_{ik}\| \neq 0$. Thus we have

$$g_{ik} g^{kl} = \delta^l_i.$$

We now show that g^{ik} transforms as a contravariant tensor of rank 2. We use the result just derived, namely that g_{ik} transforms as a covariant tensor so that

$$g'_{ik} = \frac{\partial x^m}{\partial x'^i} \frac{\partial x^n}{\partial x'^k} \cdot g_{mn}.$$

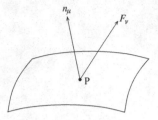

Fig. 3.5. An example of a second-rank tensor. In the example illustrated, the stress tensor at P relates F_ν, the stress force at P, to the direction n_μ of the local normal to the surface at P. In general the two directions are not parallel.

Now define

$$B'^{kl} = g^{pq} \frac{\partial x'^k}{\partial x^p} \frac{\partial x'^l}{\partial x^q}$$

and consider the product

$$g'_{ik} B'^{kl} = \frac{\partial x^m}{\partial x'^i} \frac{\partial x^n}{\partial x'^k} \frac{\partial x'^k}{\partial x^p} \frac{\partial x'^l}{\partial x^q} g^{pq} g_{mn}$$

$$= g_{mn} g^{pq} \delta^n_p \frac{\partial x^m}{\partial x'^i} \frac{\partial x'^l}{\partial x^q}$$

$$= g_{mp} g^{pq} \frac{\partial x^m}{\partial x'^i} \frac{\partial x'^l}{\partial x^q}$$

$$= \delta^l_i.$$

In other words B'^{kl} is the inverse of the matrix g'_{ik}. This proves the result.

Example. A physical example for tensors is found in the three-dimensional space, when discussing deformation of substances. Figure 3.5 illustrates the surface Σ of such a substance, which has normal n_μ at a typical point P. If the surface is subjected to stress, the resulting force on an element of surface around P will be F_ν, different in direction from the normal n_μ, but related to it by the linear tensor relation

$$F_\nu = T_{\mu\nu} n_\mu,$$

where $T_{\mu\nu}$ is the stress tensor. If the stress is isotropic, then

$$T_{\mu\nu} = p \delta_{\mu\nu},$$

where p is the pressure which produces a force *normal* to the surface Σ. (Notice that we have not used the upper/lower indices since we are discussing Cartesian tensors in three dimensions.)

Example. In dynamics we encounter the moment-of-inertia tensor $I_{\mu\nu}$ of a massive extended body, which is a second-rank tensor in three-dimensional space. If ω_μ is the angular velocity of the body then its angular momentum is given by the vector $I_{\mu\nu} \omega_\mu$.

3.3.1 Contraction

The operation of *contraction* consists of identifying a lower index with an upper index in a mixed tensor. This procedure reduces the rank of the tensor by 2, since the repeated index implies a sum over all of its four values.

Thus $A^i B_k$ is a tensor of rank 2 if A^i and B_k are vectors. The identification $i = k$ gives a *scalar*:

$$A^i B_i = A^0 B_0 + A^1 B_1 + A^2 B_2 + A^3 B_3.$$

As in special relativity, we define a vector A^i to be *spacelike, timelike* or *null* according to

$$g_{ik} A^i A^k < 0, \qquad g_{ik} A^i A^k > 0, \qquad \text{or} \qquad g_{ik} A^i A^k = 0.$$

It is convenient to define associated tensors by the relations

$$A_i = g_{ik} A^k, \qquad A^k = g^{ik} A_i. \tag{3.23}$$

Thus $g_{ik} A^i A^k = A_k A^k$. The operations embodied in (3.23) are called *lowering* and *raising* the indices. We may frequently refer to A^i and A_i as the same object.

Example 3.3.2 Let us go back to the example of oblique axes on a plane given on page 48. We now use the coordinates as x^1 and x^2 to keep within our general convention. We have the line element given by

$$ds^2 = (dx^1)^2 + (dx^2)^2 + 2 \cos \beta \, dx^1 \, dx^2.$$

(Although, being spacelike distances, all these terms should be negative, we have omitted the negative signs everywhere to simplify the discussion, which in any case is not affected by this change of sign.) From the above we have the following components of the metric tensor and its inverse:

$$g_{11} = 1, \qquad g_{12} = \cos \beta, \qquad g_{22} = 1;$$
$$g^{11} = \operatorname{cosec}^2 \beta = g^{22}, \qquad g^{12} = -\operatorname{cosec}^2 \beta \cos \beta.$$

From this it is easy to see that, if the contravariant components of a vector are A^1 and A^2, respectively, then the covariant components are

$$A_1 = A^1 + A^2 \cos \beta, \qquad A_2 = A^2 + A^1 \cos \beta;$$

These are the same components as those we had derived from geometrical considerations.

Example 3.3.3 *Problem.* A curve is specified by the following coordinates in terms of parameter t:

$$x^0 = \sqrt{3}ct, \qquad x^1 = ct, \qquad x^2 = ct_0 \cos\left(\frac{t}{t_0}\right), \qquad x^3 = ct_0 \sin\left(\frac{t}{t_0}\right).$$

Determine whether the tangent vector at a typical point is spacelike, timelike or null. The spacetime is Minkowskian.

Solution. The tangent vector has components

$$A^0 = \sqrt{3}c, \qquad A^1 = c, \qquad A^2 = -c \sin\left(\frac{t}{t_0}\right), \qquad A^3 = c \cos\left(\frac{t}{t_0}\right).$$

Therefore

$$A^i A_i = (A^0)^2 - (A^1)^2 - (A^2)^2 - (A^3)^2$$

$$= 3c^2 - c^2 - c^2 \sin^2\left(\frac{t}{t_0}\right) - c^2 \cos^2\left(\frac{t}{t_0}\right)$$

$$= c^2 > 0.$$

So the tangent vector is timelike.

3.3.2 The quotient law

From the above manipulations of tensors it is clear (and can easily be proved) that the product of two tensors is a tensor. A reverse result is sometimes useful in deducing that a certain quantity is a tensor. This result is known as the *quotient law*. It states that, if a relation such as

$$PQ = R \tag{3.24}$$

holds in all coordinate frames, where P is an *arbitrary* tensor of rank m and R a tensor of rank $m + n$, then Q is a tensor of rank n. The reader may try to formulate a proof of this statement.

3.3.3 Symmetric and antisymmetric tensors

If tensors S_{ik} and A_{ik} satisfy the relations

$$S_{ik} = S_{ki}, \qquad A_{ik} = -A_{ki}, \tag{3.25}$$

then they are respectively *symmetric* and *antisymmetric* tensors of rank 2. These ideas can be generalized to higher-rank tensors, and we will encounter specific tensors having the properties of symmetry and anti-symmetry with respect to some or all indices.

Example 3.3.4 g_{ik} and g^{ik} are symmetric tensors.

Example. Consider the symbol ϵ_{ijkl} with the following properties:

$$\epsilon_{ijkl} = +1 \text{ if } (ijkl) \text{ is an even permutation of } (0123),$$

$$\epsilon_{ijkl} = -1 \text{ if } (ijkl) \text{ is an odd permutation of } (0123),$$

$$\epsilon_{ijkl} = 0 \text{ otherwise.}$$

We will now show that

$$e_{ijkl} = \sqrt{-g}\epsilon_{ijkl}$$

transforms as a tensor.

First take the determinant of the transformation law of g_{mn} given in Example 3.3.1. Let J denote the Jacobian $|\partial x^i/\partial x'^m|$. Then, using the rule that the determinant of a product of matrices is equal to the product of their determinants, we get

$$g' = J^2 g.$$

However, we have from the algebraic definition of a determinant

$$\epsilon_{mnpq}J = \epsilon_{ijkl}\frac{\partial x^i}{\partial x'^m}\frac{\partial x^j}{\partial x'^n}\frac{\partial x^k}{\partial x'^p}\frac{\partial x^l}{\partial x'^q}.$$

Just write out the full expansion of J as a sum of products of its elements and this result will be clear! Using the above two relations, the result follows: e_{ijkl} is a tensor that is totally antisymmetric. Strictly speaking, however, e_{ijkl} is a *pseudotensor*, since it changes sign under transformations involving reflection, such as $x'^0 = -x^0$, $x'^1 = x^1$, $x'^2 = x^2$ and $x'^3 = x^3$.

3.3.4 Totally symmetric and antisymmetric tensors

Consider a tensor $T_{i_1 i_2 \ldots i_n}$ of rank n. From this we construct a tensor

$$S_{i_1 \ldots i_n} = \frac{1}{n!}\sum_P T_{Pi_1 Pi_2 \ldots Pi_n}, \tag{3.26}$$

where the sum is over all permutations P of $(1, 2, \ldots, n)$. Evidently, if we permute the indices of S in any way, its value does not change. Such a tensor is called a totally symmetric tensor.

Likewise, if we write $(-1)^P = +1$ for an even permutation and $(-1)^P = -1$ for an odd permutation, then the sum

$$A_{i_1 i_2 \ldots i_n} = \frac{1}{n!}\sum_P (-1)^P T_{Pi_1 \ldots Pi_n} \tag{3.27}$$

gives us a totally antisymmetric tensor. An odd permutation of $(i_1, \ldots, i_n)$ changes the sign of $A_{i_1 \ldots i_n}$, whereas an even permutation does not.

If A_{ik} is any tensor we symmetrize it by writing

$$A_{(ik)} = \frac{1}{2}(A_{ik} + A_{ki}). \tag{3.28}$$

Similarly, we antisymmetrize it by writing

$$A_{[ik]} = \frac{1}{2}(A_{ik} - A_{ki}). \tag{3.29}$$

We can easily extend these concepts to tensors of higher rank as indicated in (3.27) and (3.26). Note the convention of writing (ik) for symmetrizing with respect to indices (i, k) and $[ik]$ for antisymmetrizing.

Example 3.3.5 *Problem.* Using g_{ik} twice, construct totally symmetric and totally antisymmetric tensors of rank 4.

Solution. Write $T_{iklm} = g_{ik}g_{lm}$.
Even permutations of *iklm* are
iklm, ilmk, imkl, lkmi, likm, lmik, kmli, klim, kiml, mkil, mlki, milk.
Odd permutations of *iklm* are likewise
kilm, limk, mikl, klmi, ilkm, mlik, mkli, lkim, ikml, kmil, lmki, imlk.
Thus the totally symmetric tensor is

$$S_{iklm} = \frac{1}{24}[g_{ik}g_{lm} + g_{il}g_{mk} + g_{im}g_{kl} + g_{lk}g_{mi} + g_{li}g_{km} + g_{lm}g_{ik}$$

$$+ g_{km}g_{li} + g_{kl}g_{im} + g_{ki}g_{ml} + g_{mk}g_{il} + g_{ml}g_{ki} + g_{mi}g_{lk}]$$

$$+ \frac{1}{24}[g_{ki}g_{lm} + g_{li}g_{mk} + g_{mi}g_{kl} + g_{kl}g_{mi} + g_{il}g_{km} + g_{ml}g_{ik}$$

$$+ g_{mk}g_{li} + g_{lk}g_{im} + g_{ik}g_{ml} + g_{km}g_{il} + g_{lm}g_{ki} + g_{im}g_{lk}].$$

Using the symmetry $g_{ik} = g_{ki}$, we can simplify S_{iklm} to

$$S_{iklm} = \frac{1}{3}[g_{ik}g_{lm} + g_{il}g_{mk} + g_{im}g_{lk}].$$

Insofar as the totally antisymmetric tensor is concerned, it will be the difference of the two expressions in square brackets given above for S_{iklm}. This difference is zero. So there is no totally antisymmetric tensor that can be constructed this way!

Problem. If F_{ik} is an antisymmetric tensor then show that

$$Z_{ikl} \equiv \frac{\partial F_{ik}}{\partial x^l} + \frac{\partial F_{kl}}{\partial x^i} + \frac{\partial F_{li}}{\partial x^k}$$

is a third-rank tensor.

Solution. Consider the tensor transformation law for F_{ik} below:

$$F'_{mn} = \frac{\partial x^i}{\partial x'^m} \frac{\partial x^k}{\partial x'^n} F_{ik}.$$

Differentiate with respect to x^l, and use the relation

$$\frac{\partial}{\partial x^l} \equiv \frac{\partial x'^p}{\partial x^l} \frac{\partial}{\partial x'^p}.$$

Thus we get

$$\frac{\partial x'^p}{\partial x^l} \frac{\partial F'_{mn}}{\partial x'^p} = \frac{\partial x^i}{\partial x'^m} \frac{\partial x^k}{\partial x'^n} \frac{\partial F_{ik}}{\partial x^l}$$

$$+ \frac{\partial x'^p}{\partial x^l} \frac{\partial^2 x^i}{\partial x'^p \partial x'^m} \frac{\partial x^k}{\partial x'^n} F_{ik}$$

$$+ \frac{\partial x^i}{\partial x'^m} \frac{\partial^2 x^k}{\partial x'^p \partial x'^n} \frac{\partial x'^p}{\partial x^l} F_{ik}.$$

Multiply both sides by $\partial x^l / \partial x'^q$ and use the result $\partial x'^p / \partial x'^q = \delta^p_q$. Then we get

$$\frac{\partial F'_{mn}}{\partial x'^q} = \frac{\partial x^l}{\partial x'^q} \frac{\partial x^i}{\partial x'^m} \frac{\partial x^k}{\partial x'^n} \frac{\partial F_{ik}}{\partial x^l} + \frac{\partial^2 x^i}{\partial x'^q \partial x'^m} \frac{\partial x^k}{\partial x'^n} F_{ik}$$

$$+ \frac{\partial^2 x^k}{\partial x'^q \partial x'^n} \frac{\partial x^i}{\partial x'^m} F_{ik}.$$

We will have similar expressions from the second and third expressions on the right-hand side of the relation for Z_{ikl}:

$$\frac{\partial F'_{nq}}{\partial x'^m} = \frac{\partial x^i}{\partial x'^m} \frac{\partial x^k}{\partial x'^n} \frac{\partial x^l}{\partial x'^q} \frac{\partial F_{kl}}{\partial x^i} + \frac{\partial^2 x^k}{\partial x'^m \partial x'^n} \frac{\partial x^l}{\partial x'^q} F_{kl}$$

$$+ \frac{\partial^2 x^l}{\partial x'^m \partial x'^q} \frac{\partial x^k}{\partial x'^n} F_{kl}.$$

$$\frac{\partial F'_{qm}}{\partial x'^n} = \frac{\partial x^k}{\partial x'^n} \frac{\partial x^l}{\partial x'^q} \frac{\partial x^i}{\partial x'^m} \frac{\partial F_{li}}{\partial x^k} + \frac{\partial^2 x^l}{\partial x'^n \partial x'^q} \frac{\partial x^i}{\partial x'^m} F_{li}$$

$$+ \frac{\partial^2 x^i}{\partial x'^n \partial x'^m} \frac{\partial x^l}{\partial x'^q} F_{li}.$$

When all three equations are added the antisymmetry of F_{ik} ensures that the terms involving second derivatives cancel out and the result follows.

Problem. Prove that

$$\epsilon^{ijkl} \epsilon_{ijrs} = 2(\delta^k_{\ r} \delta^l_{\ s} - \delta^k_{\ s} \delta^l_{\ r}).$$

Solution. To prove this result we may resort to the properties of determinants. It easy to verify that

$$\epsilon^{ijkl} \epsilon_{pqrs} = \begin{vmatrix} \delta^i_p & \delta^i_q & \delta^i_r & \delta^i_s \\ \delta^j_p & \delta^j_q & \delta^j_r & \delta^j_s \\ \delta^k_p & \delta^k_q & \delta^k_r & \delta^k_s \\ \delta^l_p & \delta^l_q & \delta^l_r & \delta^l_s \end{vmatrix}.$$

Putting $i = p$, $j = q$ gives

$$\epsilon^{ijkl} \epsilon_{ijrs} = \begin{vmatrix} 4 & \delta^i_j & \delta^i_r & \delta^i_s \\ \delta^j_i & 4 & \delta^j_r & \delta^j_s \\ \delta^k_i & \delta^k_j & \delta^k_r & \delta^k_s \\ \delta^l_i & \delta^l_j & \delta^l_r & \delta^l_s \end{vmatrix}.$$

The result follows on evaluating the determinant. (It is simpler to use the expansion formula in terms of products of 2×2 determinants in the top two rows and the bottom two rows.)

3.4 Concluding remarks

We end this chapter with a note that the vectors and tensors described here are definable in *any* coordinate frame. Thus we are not restricted to inertial frames or to linear transformations between such frames. Clearly this machinery will be useful to us in general relativity, where we aim to describe physics and dynamics in any general reference frame.

However, we need to proceed further along this track before addressing those issues. We have to progress from the *tensor algebra* described above to *tensor calculus*, since equations of physics and dynamics require us to differentiate quantities with respect to space and time coordinates. Having done that, we also have to describe the essential features of a non-Euclidean spacetime. We will therefore first take up the question of how to describe a physically meaningful derivative of a tensor.

Exercises

1. Which of the following expressions are invalid with respect to the summation convention: (a) $A_{ij} B^{jk} A_{jl}$, (b) $g_{ik} g^{ik}$, (c) $R_{ik} g_{ik}$, (d) $e_{iklm} e^{iklm}$ and (e) $T^{ik} g^k_l$? Simplify those expressions that are valid.

2. A_{ik} is a tensor such that $\| A_{ik} \|$ is non-singular. Show that the components of the inverse matrix transform as a tensor. (An example of this result is the tensor g^{ik}.)

3. If A_i is a vector, show that $F_{lm} \equiv A_{m,l} - A_{l,m}$ is a second-rank tensor.

4. Show that $e^{ijkl} e_{ijkr} = 6 \delta^l_r$.

5. If ξ^i is a vector field, deduce from first principles that

$$\phi_{mn} = \xi^i \frac{\partial g_{mn}}{\partial x^i} + g_{mi} \frac{\partial \xi^i}{\partial x^n} + g_{ni} \frac{\partial \xi^i}{\partial x^m}$$

is a tensor field.

6. Find a coordinate transformation of the form

$$R = R(r, t), \qquad T = T(r, t),$$

which will transform the line element

$$ds^2 = dt^2 - S^2(t)\left[\frac{dr^2}{1 - kr^2} + r^2(d\theta^2 + \sin^2\theta\, d\phi^2)\right],$$

where $S(t)$ is a function of t and k is a constant, to the form

$$ds^2 = e^\nu\, dT^2 - e^\lambda\, dR^2 - R^2(d\theta^2 + \sin^2\theta\, d\phi^2).$$

Deduce that

$$e^{-\lambda} = 1 - r^2\left[k + \left(\frac{dS}{dt}\right)^2\right].$$

7. A surface of revolution is generated by rotating the parabola

$$y^2 = 4ax$$

about the x-axis. By writing $x = at^2$, $y = 2at$ or otherwise, show that the line element on this surface is given by

$$ds^2 = 4a^2[t^2\, d\phi^2 + (1 + t^2)dt^2].$$

8. In the Kerr spacetime, the line element is given by

$$ds^2 = \left(\frac{R^2 - 2MGR + h^2}{R^2 + h^2 \cos^2\theta}\right)(dT - h \sin\theta\, d\phi)^2$$

$$- \frac{\sin^2\theta}{h^2 \cos^2\theta + R^2}[(R^2 + h^2)d\phi - h\, dT]^2$$

$$- \left(\frac{R^2 + h^2 \cos^2\theta}{R^2 - 2GMR + h^2}\right)dR^2 - (r^2 + h^2 \cos^2\theta)d\theta^2.$$

An observer in this spacetime has constant R, θ, ϕ coordinates. Show that the world line of this observer has a timelike tangent, provided that

$$R > GM + (G^2 M^2 - h^2 \cos^2\theta)^{1/2}.$$

9. Given that a_{ik} and b_{ik} are two symmetric tensors satisfying the relation

$$a_{ij}b_{kl} - a_{il}b_{jk} + a_{jk}b_{il} - a_{kl}b_{ij} = 0,$$

show that $a_{ij} = \rho b_{ij}$, where ρ is a scalar.

10. From an antisymmetric tensor F_{ik} is constructed its dual

$$F^{*lm} = \frac{1}{2}e^{iklm} F_{ik}.$$

Show that the dual of F^{*lm} is F_{ik}. Is this result valid for all types of tensors?

11. Show that the tensor $\theta_{ik} = g_{ik} - U_i U_k$ when multiplied by any vector V^k projects it into a 3-surface orthogonal to the unit timelike vector U_i.

12. Define $V^*_{jkl} = e_{ijkl} V^i$. Then show that $V_i V^i = \frac{1}{6} V^*_{jkl} V^{*jkl}$.

13. Show that the proper volume element $dV = \sqrt{-g}\, dx^0\, dx^1\, dx^2\, dx^3$ transforms as a scalar under general coordinate transformations.

14. In Minkowski spacetime two focal points A and B are identified with Cartesian coordinates $x = a$, $y = 0$, $z = 0$ and $x = -a$, $y = 0$, $z = 0$, t being the time coordinate. Let P be any point in space at $t =$ constant and let r_1 and r_2 be the distances of P from A and B. Let $\xi = \frac{1}{2}(r_1 + r_2)$ and $\eta = \frac{1}{2}(r_1 - r_2)$. Suppose the azimuthal coordinate for the plane PAB around the axis AB is ϕ. Then the Minkowski line element in these coordinates is

$$ds^2 = c^2\, dt^2 - (\xi^2 - \eta^2)\left(\frac{d\xi^2}{\xi^2 - a^2} + \frac{d\eta^2}{a^2 - \eta^2} \right) - \frac{(\xi^2 - a^2)(a^2 - \eta^2)}{a^2}\, d\phi^2.$$

What surfaces do $\xi =$ constant and $\eta =$ constant represent?

15. Show by a coordinate transformation that the spacetime metric

$$ds^2 = c^2\, dt^2 - c^2 t^2 \left[\frac{dr^2}{1 + r^2} + r^2(d\theta^2 + \sin^2\theta\, d\phi^2) \right]$$

represents the Minkowski spacetime.

16. Show that $\epsilon^{iklm} \epsilon_{lmpq} \epsilon^{pq}_{ik} \equiv 0$, but $\epsilon^{iklm} \epsilon_{lmpq} \epsilon^{pqrs} \epsilon_{rsik} = 96$. Can you formulate a general rule for the closed cycle of n such symbols multiplied together, where n can be odd or even?

Chapter 4
Covariant differentiation

4.1 The concept of general covariance

We begin this chapter by introducing the idea of a *field* in physics (to be distinguished from the 'field' in algebra that mathematicians talk about). The idea was popularized by Michael Faraday in the context of the electric and magnetic fields. Figure 4.1 shows what happens when iron filings are sprinkled in the vicinity of a bar magnet. The filings get distributed in a pattern somewhat like that in this figure. Faraday called these curves *lines of force*. If we imagine a magnetic pole placed anywhere on one of these lines, it will move along that line, being guided by the magnetic force on it. The lines of force therefore represent the 'magnetic field' **B** both in strength and direction at any point in the vicinity of the magnet. In short the magnet generates a 'field' of **B**-vectors all around it, representing the force exerted by it on another magnetic pole.[1]

We generalize the concept of a vector field by defining a vector function of spacetime variables, so that at each point a vector is defined. This idea may further be generalized by having tensor fields as functions of spacetime coordinates. Thus we could argue that equations of physics involve fields related by partial differential equations. If we additionally require that these equations do not change their *form* under changes of coordinates (thereby being the same for all observers), then they should be represented by *tensor fields*. Thus general relativity assumes as a

---

[1] In reality there is no magnetic pole existing in isolation. But the concept of a line of force can nevertheless be explained this way.

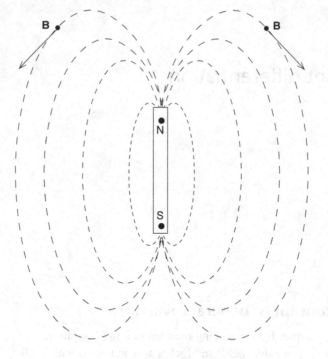

Fig. 4.1. Lines of force of a magnet obtained by sprinkling iron filings. The tangent to a line of force at any point indicates the direction of the magnetic field **B**. The field strength is high where such lines appear crowded, as, for example, near the poles of the magnet.

basic postulate that fundamental physics is described by such fields. This premise is stated in the following form: *the laws of physics are generally covariant*. This postulate greatly restricts the form of a physics equation.

Example 4.1.1 Consider the differential equation

$$\frac{\partial^2 \phi}{\partial t^2} = \frac{\partial^2 \phi}{\partial x^2} + 2\frac{\partial^2 \phi}{\partial y^2} + \frac{\partial \phi}{\partial x}\frac{\partial \phi}{\partial z}.$$

This equation is *not* generally covariant; that is, it *does not* preserve the above form under a change of coordinates. On the other hand, the wave equation $\Box \phi = 0$ preserves its form, which is why it is found in various branches of physics.

The mathematical implication of general covariance for spacetime geometry can be likewise understood. If we are to deal with non-Euclidean geometry, we will at some stage encounter the concept of *curvature of spacetime*. This is an example of the intrinsic properties of spacetime that require description independently of the choice of coordinates. Such concepts are best described in terms of vectors and tensors.

Thus, having established the need for tensors in formulating space-time structure as well as the physics in it, we have to ensure that any differential equations involving them should also be generally covariant, i.e., not depending on any specific choice of coordinates. In short, they also should be expressible as vectors and tensors. We will soon find that this is a non-trivial requirement, for the partial derivative of a tensor need not be a tensor.

4.2 Parallel transport

We begin the discussion of vector derivatives with the example of a vector field. Let $B_i(x^k)$ be a covariant vector field whose four components transform according to the rule in (3.19) at each point (x^k) where it is defined. Suppose B_i is a differentiable function of (x^k). Do the partial derivatives $\partial B_i/\partial x^k$ transform as a tensor?

We have already seen that the derivatives $\partial\phi/\partial x^k$ of a scalar transform as a vector. So at first sight the answer to the above question might be 'yes'. Indeed, in special relativity we do encounter such results. For example, if A_i is the 4-potential of the electromagnetic field (described in the four-dimensional language of special relativity), then $\partial A_i/\partial x^k$, for Cartesian coordinates (x, y, z) and the time t of (3.1), do transform as a tensor. In our more general spacetime with an arbitrary coordinate system, however, the answer to the above question is in the negative.

This result is easily verified by differentiating (3.19) with respect to x'^m. We get (by writing $\partial/\partial x'^m$ as $\partial x^n/\partial x'^m \cdot \partial/\partial x^n$)

$$\frac{\partial B'_k}{\partial x'^m} = \frac{\partial x^i}{\partial x'^k}\frac{\partial x^n}{\partial x'^m}\frac{\partial B_i}{\partial x^n} + \frac{\partial^2 x^i}{\partial x'^m\,\partial x'^k}B_i. \tag{4.1}$$

Thus, whereas the first term on the right-hand side does appear in the right form to make $\partial B_i/\partial x^n$ a tensor, the second term spoils the effect. It also gives a clue as to why this happens. The second derivative

$$\frac{\partial^2 x^i}{\partial x'^m\,\partial x'^k}$$

is expected to be non-zero because, in general, the transformation coefficients in Equation (3.19) vary with position in spacetime. When we seek to construct the derivative $\partial B_i/\partial x^n$, we have to define it as a limit:

$$\frac{\partial B_i}{\partial x^n} = \lim_{\delta x^n\to 0}\left[\frac{B^i(x^k+\delta x^k) - B^i(x^k)}{\delta x^n}\right].$$

However, the two terms in the numerator transform as vectors at two different points, and because of the variation of the transformation coefficients with position their difference is not expected to be a vector. (The

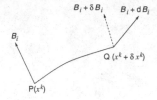

Fig. 4.2. The parallel transport of the vector B_i at P to Q shows, in general, different components $B_i + \delta B_i$ at Q (see the dotted vector). In the case of a vector field, the components specified at Q, namely $B_i + dB_i$, would be different from those (at Q) obtained by parallel transport.

difference of two vectors is a vector provided that both are so defined at the *same point*.)

This situation is illustrated in Figure 4.2. P and Q are the two neighbouring points (x^k) and $(x^k + \delta x^k)$, with the vectors B_i shown there with continuous arrows. In order to describe the change in the vector from P to Q, we must somehow measure the difference at *the same point*. How can this be achieved?

This is achieved by a device known as *parallel transport*. Assume that the vector B_i at P is moved from P to Q, parallel to itself, that is, as if its magnitude and direction did not change. In Figure 4.2 this is shown by a dotted vector at Q. The difference between the vector $B_i(x^k + \delta x^k)$ and this dotted vector is a vector at Q and this tells us the real physical difference in the vector from P to Q. So we may after all be able to define a process of differentiation of vectors, provided that we know what happens to B_i during a parallel transport from P to Q.

First we have to note that the dotted vector at Q need not have the same components as the undotted vector at P. It is only with Cartesian coordinates that the components are the same.

Example 4.2.1 Consider the Euclidean plane with a polar coordinate system. A vector A at a point P with coordinates (r, θ) has components A_r and A_θ in the radial and transverse directions. If we now move the vector parallel to itself from P to a neighbouring point Q with polar coordinates $(r + \delta r, \theta + \delta\theta)$, as shown in Figure 4.3, the radial and transverse directions at Q will not be parallel to those at P. Hence after parallel transport of A from P to Q its radial and transverse components at Q will be different from A_r and A_θ.

A simple calculation using the geometry of infinitesimal rotation shows that the components of A at Q are $A_r + \delta\theta \, A_\theta$ and $A_\theta - \delta\theta \, A_r$.

Taking a cue from this example for our general case, we see that the changes in the components of B_i through parallel transport will be proportional to the original components B_i, and also to the displacement δx^k in position from P to Q. We may express the change as a linear function of both these quantities and the most general form that we can have for it is

$$\delta B_i = \Gamma^l_{ik} B_l \, \delta x^k, \tag{4.2}$$

where the coefficients Γ^l_{ik} are, in general, functions of space and time. These quantities are called the *three-index symbols* or the *Christoffel symbols*.

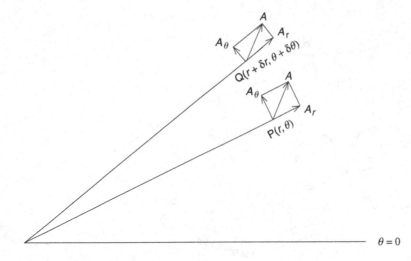

Fig. 4.3. The vector A at P is parallely transported to Q; but its components in the polar coordinates (r, θ) are different at P and Q. This is because (unlike the Cartesian coordinates) the local directions of r = constant and θ = constant change on going from P to Q.

Notice that the introduction of (4.2) involves something new in addition to the introduction of the metric. The metric tells us how to measure distance between neighbouring points, whereas Γ^i_{kl} in (4.2) tells us how to define parallel vectors at neighbouring points. This property of connecting neighbouring vectors through the concept of local parallelism is often called the *affine connection* of spacetime.

The reader may be worried at this stage as to how parallelism can be assumed when, as we saw in Chapter 2, the concept of parallel lines in non-Euclidean geometries is non-trivial. So we clarify that we are talking here of *local* parallelism, i.e., of parallelism over infinitesimal distances. Indeed, as we will elaborate later in Chapter 5, the 'local' region in the neighbourhood of any observer can be approximated by 'flat' space or spacetime, where the concepts of Euclid's geometry hold.

There is a practical way of describing parallel propagation in the following fashion.

We take the example of a sphere. Suppose Γ is a curve drawn on the spherical surface connecting points P_1 and P_2. The arrow shown in Figure 4.4 represents in magnitude as well as direction a vector $\mathbf{A}_1$ at P_1. How do we transport it parallely to P_2 along Γ? Imagine a plane touching the sphere at P_1, with the vector $\mathbf{A}_1$ mapped at the corresponding point Q_1 on the plane. (By mapping, we mean that the magnitude and direction of the original vector on the sphere and the mapped vector on the tangent plane should match.) Let the vector in the plane be called $\tilde{\mathbf{A}}_1$. Now carefully roll the sphere on the plane so that it keeps touching it along the successive points of Γ. When you reach P_2, stop there. Let the corresponding point on the plane be Q_2. Draw a vector $\tilde{\mathbf{A}}_2$ at Q_2 parallel

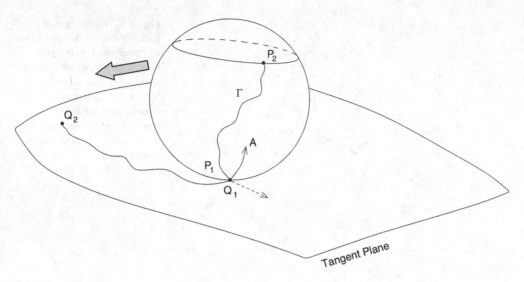

Fig. 4.4. A practical way of finding the directions of a vector **A** parallely transported along a specified curve (Γ) from P_1 to P_2 on a sphere. (See the text for details.)

to the starting vector $\tilde{\mathbf{A}}_1$ *on the plane* at Q_1. Next map this vector onto vector $\mathbf{A}_2$ at P_2 on the sphere. This will be the required parallely transported vector at P_2. This method can, in principle, be used for other surfaces also.

4.3 The covariant derivative

Returning to (4.2), we see that the difference between the continuous and the dotted vectors at Q is given by

$$B_i(x^k + \delta x^k) - [B_i(x^k) + \delta B_i] = \left(\frac{\partial B_i}{\partial x^k} - \Gamma_{ik}^l B_l \right) \delta x^k. \qquad (4.3)$$

We may accordingly redefine the physically meaningful derivative of a vector by

$$B_{i;k} \equiv \frac{\partial B_i}{\partial x^k} - \Gamma_{ik}^l B_l \equiv B_{i,k} - \Gamma_{ik}^l B_l. \qquad (4.4)$$

This derivative, by definition, must transform as a tensor. It is called the *covariant derivative* and will be denoted by a semicolon, as against the ordinary derivative, which will be denoted by a comma.

If $B_{i;k}$ must transform as a tensor, the coefficients Γ_{kl}^i have to transform according to the following law:

$$\Gamma_{kl}'^i = \frac{\partial x'^i}{\partial x^m} \frac{\partial x^n}{\partial x'^k} \frac{\partial x^p}{\partial x'^l} \Gamma_{np}^m + \frac{\partial^2 x^p}{\partial x'^k \partial x'^l} \frac{\partial x'^i}{\partial x^p}. \qquad (4.5)$$

This result can be verified after some straightforward but tedious calculation.

Example 4.3.1 *Problem.* Γ^i_{kl} and $\tilde{\Gamma}^i_{kl}$ are two affine connections defined in a spacetime with the same metric tensor. Show that $\tilde{\Gamma}^i_{kl} - \Gamma^i_{kl}$ transforms as a tensor.

Solution. Under a general transformation from x^i to x'^i coordinates we have, from Equation (4.5),

$$\Gamma'^i_{kl} = \frac{\partial x'^i}{\partial x^m} \frac{\partial x^n}{\partial x'^k} \frac{\partial x^p}{\partial x'^l} \Gamma^m_{np} + \frac{\partial^2 x^p}{\partial x'^k \partial x'^l} \cdot \frac{\partial x'^i}{\partial x^p}$$

and also

$$\tilde{\Gamma}'^i_{kl} = \frac{\partial x'^i}{\partial x^m} \frac{\partial x^n}{\partial x'^k} \frac{\partial x^p}{\partial x'^l} \tilde{\Gamma}^m_{np} + \frac{\partial^2 x^p}{\partial x'^k \partial x'^l} \cdot \frac{\partial x'^i}{\partial x^p}.$$

On taking the difference of these two relations, we get

$$(\tilde{\Gamma}'^i_{kl} - \Gamma'^i_{kl}) = \frac{\partial x'^i}{\partial x^m} \frac{\partial x^n}{\partial x'^k} \frac{\partial x^p}{\partial x'^l}(\tilde{\Gamma}^m_{np} - \Gamma^m_{np}).$$

This is the transformation law of a third-rank mixed tensor. Hence the result follows.

A scalar, of course, does not change under parallel transport, which is why $\partial \phi / \partial x^k$ transform as a vector. If we use this result we see that, for any arbitrary vector fields A^i and B_i, $(A^i B_i)_{,k}$ is a vector. This property enables us to construct the covariant derivative of a *contravariant* vector A^i as follows.

We have $(A^i B_i)_{;k} \equiv A^i_{;k} B_i + A^i B_{i;k} = (A^i B_i)_{,k} \equiv A^i_{,k} B_i + A^i B_{i,k}$ and using (4.4) we get $A^i_{;k} B_i = A^i_{,k} B_i + A^i \Gamma^l_{ik} B_l$. Since B_i is an arbitrary vector, we must have

$$A^i_{;k} \equiv \frac{\partial A^i}{\partial x^k} + \Gamma^i_{lk} A^l \equiv A^i_{,k} + \Gamma^i_{lk} A^l. \tag{4.6}$$

The rule of covariant differentiation of a tensor of arbitrary rank is easily obtained: we introduce a $(+\Gamma)$ term for each contravariant index as in (4.6) and a $(-\Gamma)$ term for each covariant index as in (4.4). Thus, for the metric tensor we have

$$g_{ik;l} = \frac{\partial g_{ik}}{\partial x^l} - \Gamma^p_{il} g_{pk} - \Gamma^p_{kl} g_{ip}. \tag{4.7}$$

4.4 Riemannian geometry

Einstein used the non-Euclidean geometry developed by Riemann to describe his theory of gravitation. The Riemannian geometry introduces the additional specification that

$$\Gamma^i_{kl} = \Gamma^i_{lk}; \qquad g_{ik;l} \equiv 0. \tag{4.8}$$

Note that, as defined in the previous section, the affine connection need not satisfy these conditions. Indeed, geometries for which the above relations are not satisfied also exist. For the theory of relativity and Riemannian geometry, however, these conditions are *additionally assumed*.

Going back to (4.7) we see that $g_{ik;l} = 0$ gives us 40 linear equations for the 40 unknowns Γ^i_{kl}. These equations have a unique solution. For, from (4.7) and (4.8), we get

$$\Gamma_{k|il} + \Gamma_{i|kl} = g_{ik,l}, \tag{4.9}$$

where

$$\Gamma_{k|il} = g_{pk}\Gamma^p_{il}. \tag{4.10}$$

Rotate the indices cyclically to obtain two more relations:

$$\Gamma_{l|ki} + \Gamma_{k|li} = g_{kl,i}, \qquad \Gamma_{i|lk} + \Gamma_{l|ik} = g_{li,k}.$$

Next use the symmetry condition in (4.8) to eliminate $\Gamma_{l|ki} = \Gamma_{l|ik}$ and $\Gamma_{k|il} = \Gamma_{k|li}$ from the above three relations to get

$$2\Gamma_{i|kl} = g_{ik,l} + g_{li,k} - g_{kl,i}.$$

On raising the index i we get the required solution:

$$\Gamma^i_{kl} = \frac{1}{2}g^{im}\left(\frac{\partial g_{mk}}{\partial x^l} + \frac{\partial g_{lm}}{\partial x^k} - \frac{\partial g_{kl}}{\partial x^m}\right). \tag{4.11}$$

In other words, once the metric tensor is known, the Christoffel symbols are fully determined.

Example 4.4.1 *Problem.* Show that, if the metric g_{ik} is diagonal, then

$$\Gamma^a_{aa} = \frac{\partial}{\partial x^a} \ln \sqrt{|g_{aa}|},$$

where the summation convention is suspended.

Solution. We have $g = g_{00} \cdot g_{11} \cdot g_{22} \cdot g_{33}$, and $g^{00} = 1/g_{00}, g^{11} = 1/g_{11}$, etc.,

$$\Gamma^a_{aa} = g^{aa}\Gamma_{a|aa} = \frac{1}{g_{aa}} \cdot \frac{1}{2}\frac{\partial g_{aa}}{\partial x^a} = \frac{1}{2}\frac{1}{g_{aa}}\frac{\partial g_{aa}}{\partial x^a}.$$

Taking $g_{00} > 0$, this gives

$$\Gamma^0_{00} = \frac{1}{2}\frac{1}{g_{00}}\frac{\partial g_{00}}{\partial x^0} = \frac{\partial}{\partial x^0} \ln \sqrt{g_{00}}.$$

For the spacelike case we have $g_{\mu\mu} < 0$ and we get similarly

$$\Gamma^1_{11} = \frac{1}{2}\frac{1}{g_{11}}\frac{\partial g_{11}}{\partial x^1} = \frac{\partial}{\partial x^1} \ln \sqrt{-g_{11}}.$$

Hence the result follows.

4.5 Some useful identities

We next consider some particular identities relating to the Christoffel symbols, that are useful in various manipulations. If we differentiate the determinant of the metric tensor we get

$$dg = gg^{ik}\, dg_{ik}. \tag{4.12}$$

This relation is useful in expressing some combinations of Γ^i_{kl} and covariant derivatives in relatively simple forms. Thus, using (4.9) and (4.10), it is possible to prove the following relations:

$$\Gamma^l_{il} = \frac{1}{\sqrt{-g}} \frac{\partial}{\partial x^i}(\sqrt{-g}),$$

$$\Gamma^l_{ik}g^{ik} = -\frac{1}{\sqrt{-g}} \frac{\partial}{\partial x^m}(\sqrt{-g}g^{ml}),$$

$$A^i_{;i} = \frac{1}{\sqrt{-g}} \frac{\partial}{\partial x^i}(\sqrt{-g}A^i), \tag{4.13}$$

$$F^{ik}_{;k} = \frac{1}{\sqrt{-g}} \frac{\partial}{\partial x^k}(\sqrt{-g}F^{ik}) \qquad \text{for } F^{ik} = -F^{ki}.$$

(Here A^i and F^{ik} are respectively vector and tensor fields.) For example, to prove the first relation note that (4.11) gives, with $k \equiv i$,

$$\Gamma^i_{il} = \frac{1}{2}g^{im}(g_{mi,l} + g_{lm,i} - g_{il,m}).$$

Since $(g_{lm,i} - g_{il,m})$ is antisymmetric in (i, m), its product with the symmetric g^{im} vanishes. The result then follows when we recall (4.12).

Example 4.5.1 *Problem.* If a vector ξ^i satisfies the relation $\xi^l_{,m}g_{ln} + \xi^l_{,n}g_{lm} + g_{mn,l}\xi^l = 0$, show that $\xi_{i;m} + \xi_{m;i} = 0$.

Solution. Consider $\xi^l_{;m} = \xi^l_{,m} - \Gamma^l_{mp}\xi^p$. We then have

$$g_{ln}(\xi^l_{;m} - \Gamma^l_{mp}\xi^p) + (\xi^l_{;n} - \Gamma^l_{np}\xi^p)g_{lm} + g_{mn,l}\xi^l = 0,$$

i.e.,

$$\xi_{n;m} + \xi_{m;n} + g_{mn,l}\xi^l - \Gamma_{n|mp}\xi^p - \Gamma_{m|np}\xi^p = 0.$$

Since $\Gamma_{n|mp} + \Gamma_{m|np} \equiv g_{mn,p}$, the result follows.

Problem. Show that $g^{im}_{,n} = -g^{mk}g^{il}g_{kl,n}$ and hence deduce that $g^{im}_{;n} \equiv 0$.

Solution. Differentiate with respect to x^n the identity $g^{im}g_{ml} = \delta^i_l$, to get

$$g^{im}_{,n}g_{ml} + g^{im}g_{ml,n} = 0.$$

Multiply by g^{lk} and use the identity $g_{ml}g^{lk} = \delta^k_m$ to get

$$g^{ik}_{,n} + g^{im}g^{lk}g_{ml,n} = 0.$$

Change the index k to m, l to k and m to l to get the required answer.

4.6 Locally inertial coordinate systems

The symmetry condition in (4.8) enables us to choose special coordinates in which the Christoffel symbols all vanish at any given point. Suppose we start with $\Gamma^m_{np} \neq 0$ in the coordinate system (x^i) at point P. Let the coordinates of P be given x^i_P. Now define new coordinates in the neighbourhood of P by

$$x'^k = -\frac{1}{2}\Gamma^k_{nm}(x^n - x^n_P)(x^m - x^m_P). \tag{4.14}$$

Then we have at P

$$x'^i_P = 0, \qquad \frac{\partial x'^i}{\partial x^m} = 0, \qquad \frac{\partial^2 x'^i}{\partial x^n\, \partial x^m} = -\Gamma^i_{nm},$$

with the result that from (4.5)

$$\Gamma'^i_{mn}|_P = 0.$$

Further, by a linear transformation we can arrange to have a coordinate system with

$$g_{ik} = \eta_{ik} = \text{diag}(+1, -1, -1, -1), \qquad \Gamma^i_{kl} = 0 \tag{4.15}$$

at our chosen point P. Such a coordinate system is called a *locally inertial* coordinate system, for reasons that will become clear later. Apart from its physical implications in general relativity, the locally inertial coordinate system is often useful as a mathematical device for simplifying calculations. We also warn the reader that the operative word is 'local': the simplifications implied in (4.15) *cannot be achieved globally*. What prevents us from achieving a globally inertial coordinate system? In seeking an answer to this question we encounter the most crucial aspect in which a non-Euclidean geometry differs from its Euclidean counterpart.

Exercises

1. Show that if $\Gamma^i_{kl} \neq \Gamma^i_{lk}$ the condition $g_{ik;l} \equiv 0$ implies that

$$\Gamma^i_{(kl)} = \frac{1}{2}g^{im}(g_{mk,l} + g_{ml,k} - g_{kl,m}) + g^{im}g_{kn}\Gamma^n_{[lm]}$$
$$+ g^{im}g_{ln}\Gamma^n_{[km]}.$$

2. Under a *conformal transformation* g_{ik} changes to $e^{2\sigma}g_{ik}$, where σ is a real twice-differentiable function of spacetime coordinates. Show that the new Christoffel symbols are given by

$$\tilde{\Gamma}^i_{kl} = \Gamma^i_{kl} + \delta^i_k\sigma_{,l} + \delta^i_l\sigma_{,k} - g_{kl}g^{im}\sigma_{,m}.$$

3. For a symmetric tensor field A_{ik} show that

$$A^k_{i;k} = \frac{1}{\sqrt{-g}} \frac{\partial}{\partial x^k} (A_i{}^k \sqrt{-g}) - \frac{1}{2} A^{lk} \frac{\partial g_{lk}}{\partial x^i}$$

$$= \frac{1}{\sqrt{-g}} \frac{\partial}{\partial x^k} (A_i{}^k \sqrt{-g}) + \frac{1}{2} A_{lk} \frac{\partial g^{lk}}{\partial x^i}.$$

4. Show that, for a scalar field ϕ, the wave operator takes the form

$$\Box \phi = g^{ik} \phi_{;ik} = \frac{1}{\sqrt{-g}} \frac{\partial}{\partial x^k} \left(\sqrt{-g} g^{ik} \frac{\partial \phi}{\partial x^i} \right).$$

5. Show that, to arrive at a locally inertial system, it is necessary to have $\Gamma^i_{kl} = \Gamma^i_{lk}$.

6. Show that

$$\frac{\partial g^{ik}}{\partial x^l} = -\Gamma^i_{ml} g^{mk} - \Gamma^k_{ml} g^{mi}.$$

7. In polar coordinates (r, θ), the radial and transverse accelerations are $(\ddot{r} - r\dot{\theta}^2)$ and $r\ddot{\theta} + 2\dot{r}\dot{\theta}$. Try to relate these expressions to the notion of covariant differentiation described in this chapter.

8. Set up equations of parallel propagation of a vector along a latitude line $\theta = 60°$ on a unit sphere. By what angle has a vector initially directed along the longitude line at zero longitude (ϕ) turned by the time it has gone half-way round the latitude circle?

9. Show that Maxwell's equations are invariant under the conformal transformation $g_{ik} \rightarrow g_{ik} e^{2\sigma}$.

10. A vector is moved parallel to itself along the line $t =$ constant on the paraboloid of revolution whose metric is given by (see Exercise 7 of Chapter 3)

$$ds^2 = 4a^2[(1 + t^2)dt^2 + t^2 d\phi^2].$$

Initially the vector was lying tangentially to the line $t =$ constant. Show that on moving round once from $\phi = 0$ to $\phi = 2\pi$ the vector makes an angle $2\pi \sqrt{t^2 + 1}$ with its initial direction.

Chapter 5
Curvature of spacetime

5.1 Parallel propagation around finite curves

Figure 5.1 repeats the previous example of non-Euclidean geometry on the surface of a sphere which we discussed in Section 2.2 of Chapter 2. We have the triangle ABC of Figure 2.3 whose three angles are each 90°. Consider what happens to a vector (shown by a dotted arrow) as it is parallely transported along the three sides of this triangle. As shown in Figure 5.1, this vector is originally perpendicular to AB when it starts its journey at A. When it reaches B it lies along CB; it keeps pointing along this line as it moves from B to C. At C it is again perpendicular to AC. So, as it moves along CA from C to A, it maintains this perpendicularity, with the result that when it arrives at A it is pointing along AB. In other words, one circuit around this triangle has resulted in a change of direction of the vector by 90°, although at each stage it was being moved parallel to itself!

A similar experiment with a triangle drawn on a flat piece of paper will tell us that there is no resulting change in the direction of the vector when it moves parallel to itself around the triangle. So our spherical triangle behaves differently from the flat Euclidean triangle.

The phenomenon illustrated in Figure 5.1 can also be described as follows. If we had moved our vector from A to C along two different routes – (i) along AC and (ii) along AB followed by BC – we would have found it pointing in two different directions in the two cases. In fact, if we had taken any arbitrary curves from A to C we would have found that the outcome of parallel transport of a vector from A to C varies from curve to curve; that is, the outcome depends on the path of transport from A to C.

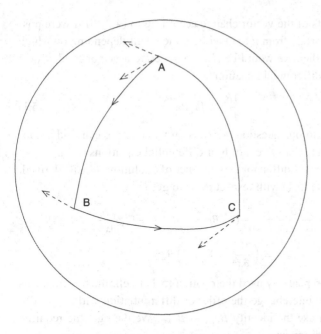

Fig. 5.1. A figure illustrating the parallel-transport problems on a spherical surface described in the text.

Recall that we had introduced the concept of parallel transport with the proviso that we would apply it to transport along an infinitesimal path. We are now witnessing the consequence of breaking that fiat and taking the concept to finite lengths. The above example shows that the result of parallel transport is *path-dependent*.

This is one of the properties that distinguishes a curved space from a flat space. Let us consider it in more general terms for our four-dimensional spacetime. Let a vector B_i at P be transported parallely to Q and let us ask for the condition that the answer should be *independent* of the curve joining P to Q. (See Figure 5.2.) We have seen that, under parallel transport from a point $\{x^i\}$ to a neighbouring point $\{x^i + \delta x^i\}$,

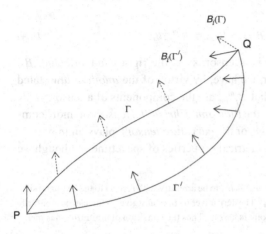

Fig. 5.2. A vector **B** at P, transported along two curves Γ and Γ' to Q, ends up having different directions at Q, as shown here. The dotted vector was obtained by moving along Γ and the continuous one by moving along Γ'.

the components of the vector change according to (4.2). If it were possible to transport B_i from P to Q without the result depending on which path is taken, then we would be able to generate a vector field $B^i(x^k)$ satisfying the differential equation

$$\frac{\partial B_i}{\partial x^k} = \Gamma^l_{ik} B_l. \tag{5.1}$$

So the answer to our question depends on whether we can find a nontrivial solution to the system of four differential equations (5.1).

A necessary condition for the existence of a solution is easily derived. We differentiate (5.1) with respect to x^n to get

$$\frac{\partial^2 B_i}{\partial x^n \, \partial x^k} = \frac{\partial}{\partial x^n}(\Gamma^l_{ik} B_l) = \frac{\partial \Gamma^l_{ik}}{\partial x^n} B_l + \Gamma^l_{ik} \frac{\partial B_l}{\partial x^n}$$

$$= \left(\frac{\partial \Gamma^m_{ik}}{\partial x^n} + \Gamma^l_{ik} \Gamma^m_{ln} \right) B_m.$$

Here we have repeatedly used the relation (5.1) to eliminate derivatives of B_i. We now interchange the order of differentiation with respect to x^n and x^k and use the identity $B_{i,nk} \equiv B_{i,kn}$. We then get the required necessary condition as

$$R_i{}^m{}_{kn} B_m = 0. \tag{5.2}$$

Here the four-indexed symbol R, as defined in (5.3) below, is *independent* of the vector B_m and so we conclude that the spacetime must satisfy the condition

$$R_i{}^m{}_{kn} \equiv \frac{\partial \Gamma^m_{ik}}{\partial x^n} - \frac{\partial \Gamma^m_{in}}{\partial x^k} + \Gamma^l_{ik} \Gamma^m_{ln} - \Gamma^l_{in} \Gamma^m_{lk} = 0. \tag{5.3}$$

It is not obvious simply from the above expression that $R_i{}^m{}_{kn}$ should be a tensor. Yet our result, in order to be significant, must clearly hold whatever coordinates we employ to derive it. So we do expect $R_i{}^m{}_{kn}$ to be a tensor. A simple calculation shows that, for any twice differentiable vector field B_i,

$$B_{i;nk} - B_{i;kn} \equiv R_i{}^m{}_{kn} B_m \tag{5.4}$$

Since the left-hand side is a tensor, so is the right-hand side and, B_m being an arbitrary vector, we have, by virtue of the *quotient law* stated in Chapter 2, the result that $R_i{}^m{}_{kn}$ are the components of a tensor.[1]

This tensor, known as the *Riemann–Christoffel tensor* (or, more commonly, the *Riemann tensor*, or the *curvature tensor*), plays an important role in specifying the geometrical properties of spacetime. Although we

---

[1] The quotient law requires the vector B_m to be arbitrary with respect to the left-hand side. Isn't $B_{i;nk}$ connected with B_m? The derivatives of B_i at any given point can be arbitrarily specified, even if B_i at that point is known. Thus the condition of arbitrariness is met.

have derived (5.4) as a necessary condition, a slightly more sophisticated technique shows that (5.4) is also the sufficient condition that a vector field $B_i(x^k)$ can be defined over the spacetime by parallel transport. We will not, however, go into the detailed mathematical proof here. The interested reader may look up Reference [7] listed at the end.

Spacetime is said to be *flat* if its Riemann tensor vanishes everywhere. Otherwise, it is said to be *curved*. In the curved spacetime the identity (5.4) can be generalized to a tensor of any rank by including a term containing the Riemann tensor for each free index of the given tensor.

Example 5.1.1 *Problem*. A contravariant vector on the surface of a unit 2-sphere with polar coordinates θ, ϕ ($\theta = 0$ being the north pole and $\theta = \pi/2$ the equator) is parallely transported along the equator from $\phi = 0$ to $\phi = \pi/2$, then similarly transported along the meridian ($\phi = $ constant) from $\theta = \pi/2$ to $\theta = \pi/3$ and then along the latitude ($\theta = $ constant) from $\phi = \pi/2$ to $\phi = 0$. Finally, it is transported similarly along the meridian back to the starting point $\theta = \pi/2$, $\phi = 0$. Show that, if the affine connection is Riemannian, the vector now makes an angle $\pi/4$ with its initial direction. (See Figure 5.3.)

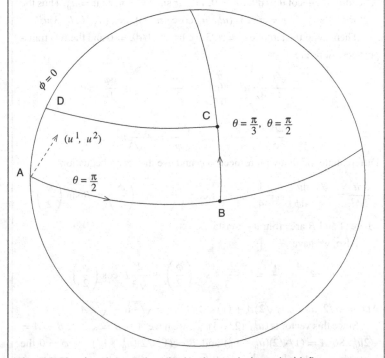

Fig. 5.3. The closed circuit described in the text is shown in this figure.

Solution. On the unit sphere we have

$$ds^2 = d\theta^2 + \sin^2\theta \, d\phi^2.$$

Thus with $x^1 = \theta$, $x^2 = \phi$ we have $g_{11} = 1$, $g_{22} = \sin^2\theta$. The non-zero Christoffel symbols are $\Gamma_{1|22} = -\sin\theta\cos\theta$, $\Gamma_{2|12} = \sin\theta\cos\theta$, i.e., $\Gamma_{22}^1 = -\sin\theta\cos\theta$ and $\Gamma_{12}^2 = \cot\theta$.

Let a unit vector (u^1, u^2) be parallely transported along a curve $\theta = \theta(\lambda)$, $\phi = \phi(\lambda)$, where λ is a parameter. Then the transport equations are

$$\frac{du^1}{d\lambda} + \Gamma_{22}^1 u^2 \frac{d\phi}{d\lambda} \equiv \frac{du^1}{d\lambda} - \sin\theta\cos\theta \frac{d\phi}{d\lambda} u^2 = 0$$

and

$$\frac{du^2}{d\lambda} + \Gamma_{12}^2 \left(u^1 \frac{d\phi}{d\lambda} + u^2 \frac{d\theta}{d\lambda} \right) = 0,$$

i.e.,

$$\frac{du^2}{d\lambda} + \cot\theta \left(u^1 \frac{d\phi}{d\lambda} + u^2 \frac{d\theta}{d\lambda} \right) = 0.$$

The equator is given by $\theta = \pi/2$. So along the equator $du^1/d\lambda = 0$, $du^2/d\lambda = 0$, i.e., $u^1 = \text{constant} = u_0^1$ and $u^2 = u_0^2$, where (u_0^1, u_0^2) is the starting value of the vector. Let us assume that it is a unit vector to start with.

Along the meridian $\phi = \pi/2$, $d\phi/d\lambda = 0$ so that $u^1 = \text{constant} = u_0^1$. Also, $du^2/d\lambda + \cot\theta \, u^2 \, d\theta/d\lambda = 0$, i.e., $u^2 \sin\theta = \text{constant} = u_0^2$. Thus the vector at $\theta = \pi/3$, $\phi = \pi/2$ is $(u_0^1, u_0^2 \csc(\pi/3))$, i.e., $(u_0^1, (2/\sqrt{3})u_0^2)$.

Then along the latitude $\theta = \pi/3$ we have $d\theta/d\lambda = 0$ and the two transport equations are

$$\frac{du^1}{d\lambda} - \frac{\sqrt{3}}{4} \frac{d\phi}{d\lambda} u^2 = 0, \qquad \frac{du^2}{d\lambda} + \frac{1}{\sqrt{3}} u^1 \frac{d\phi}{d\lambda} = 0,$$

i.e.,

$$\frac{du^1}{d\phi} = \frac{\sqrt{3}}{4} u^2, \qquad \frac{du^2}{d\phi} = -\frac{1}{\sqrt{3}} u^1.$$

Differentiate the first with respect to ϕ and use the second equation:

$$\frac{d^2 u^1}{d\phi^2} = \frac{\sqrt{3}}{4} \frac{du^2}{d\phi} = -\frac{1}{4} u^1, \qquad \text{i.e., } u^1 = A \cos\left(\frac{\phi}{2}\right) + B \sin\left(\frac{\phi}{2}\right),$$

where A and B are arbitrary constants.

Also, we have

$$u^2 = \frac{4}{\sqrt{3}} \frac{du^1}{d\phi} = -\frac{2}{\sqrt{3}} A \sin\left(\frac{\phi}{2}\right) + \frac{2}{\sqrt{3}} B \cos\left(\frac{\phi}{2}\right).$$

At $\phi = \pi/2$, $u^1 = (1/\sqrt{2})A + (1/\sqrt{2})B$, $u^2 = \sqrt{\frac{2}{3}}B - \sqrt{\frac{2}{3}}A$.

Since this vector is $(u_0^1, (2/\sqrt{3})u_0^2)$, we have $A + B = \sqrt{2}u_0^1$, $B - A = \sqrt{2}u_0^2$. So $A = (1/\sqrt{2})(u_0^1 - u_0^2)$ and $B = (1/\sqrt{2})(u_0^1 + u_0^2)$. At $\phi = 0$ the vector is given by $u^1 = (u_0^1 - u_0^2)/\sqrt{2}$, $u^2 = \sqrt{\frac{2}{3}}(u_0^1 + u_0^2)$.

For transport along $\phi = 0$ the equations are $u^1 =$ constant, $u^2 \sin \theta =$ constant. So we have $u^1 = (u_0^1 - u_0^2)/\sqrt{2}$, $u^2 \sin \theta = \sqrt{\frac{2}{3}}(u_0^1 + u_0^2)(\sqrt{3}/2) = (u_0^1 + u_0^2)/\sqrt{2}$. Thus at $\theta = \pi/2$ we have $u^1 = (u_0^1 - u_0^2)/\sqrt{2}$, $u^2 = (u_0^1 + u_0^2)/\sqrt{2}$. It can easily be verified that (u^1, u^2) is a unit vector. The angle made by this vector with the initial vector (u_0^1, u_0^2) is given by ψ, where

$$\cos \psi = g_{11} u_0^1 u^1 + g_{22} u_0^2 u^2 = u_0^1 u^1 + u_0^2 u^2$$

$$= \frac{1}{\sqrt{2}} \left\{ u_0^1 (u_0^1 - u_0^2) + u_0^2 (u_0^1 + u_0^2) \right\}$$

$$= \frac{1}{\sqrt{2}},$$

for (u_0^1, u_0^2) is a unit vector. Hence $\psi = \pi/4$.

5.1.1 Symmetries of R_{iklm}

It is more convenient to lower the second index of the Riemann tensor to study its symmetry properties. Since the symmetry or antisymmetry of a tensor does not depend on what coordinates are used, it is more convenient to write (5.3) in the locally inertial coordinates (4.15). We then get

$$R_{iklm} = \frac{1}{2}(g_{kl,im} + g_{im,kl} - g_{km,il} - g_{il,km}). \tag{5.5}$$

From this expression the following symmetries are immediately obvious:

$$R_{iklm} = -R_{kilm} = -R_{ikml} = R_{lmik}. \tag{5.6}$$

We also get relations of the following type:

$$R_{iklm} + R_{imkl} + R_{ilmk} \equiv 0. \tag{5.7}$$

If we take all these symmetries into account, we find that of the $4^4 = 256$ components of the Riemann tensor, only 20 at most are independent! For, consider the first pair (i, k) in R_{iklm}. It has altogether six independent combinations, in view of the antisymmetry of R_{iklm} with respect to (i, k). Similarly the last pair (l, m) has six independent values. Since, from (5.6), $R_{iklm} = R_{lmik}$ we have altogether $6 \times 7/2 = 21$ independent components of this tensor. However, Equation (5.7) generates one more constraint, reducing the above number to 20. Moreover, we will soon see that there are identities linking their derivatives too.

5.1.2 The Ricci and Einstein tensors

By the process of contraction we can construct lower-rank tensors from R_{iklm}. The tensor

$$R_{in} \equiv R_i{}^m{}_{mn} \tag{5.8}$$

is called the *Ricci tensor*. If we use the locally inertial coordinate system, we immediately see that $R_{in} = R_{ni}$. In a general frame we get

$$R_{in} = \frac{\partial^2 \ln\sqrt{-g}}{\partial x^i \, \partial x^n} - \frac{\partial \Gamma^m_{in}}{\partial x^m} + \Gamma^l_{im}\Gamma^m_{ln} - \Gamma^l_{in}\frac{\partial}{\partial x^l}\ln\sqrt{-g}. \tag{5.9}$$

Owing to the symmetries of (5.6) there are no other independent second-rank tensors that can be constructed out of R_{iklm}.

By further contraction we get a scalar:

$$R = g^{ik}R_{ik} \equiv R^k{}_k. \tag{5.10}$$

R is called the *scalar curvature*. The tensor

$$G_{ik} \equiv R_{ik} - \frac{1}{2}g_{ik}R \tag{5.11}$$

will turn out to have a special role to play in Einstein's general relativity. This tensor is called the *Einstein tensor*.

Example 5.1.2 *Problem.* If F_{ik} is an antisymmetric tensor, then $F^{ik}{}_{;ik} \equiv 0$.

We have from antisymmetry

$$F^{ik}{}_{;ik} = -F^{ki}{}_{;ik} = -F^{ik}{}_{;ki}.$$

However,

$$F^{ik}{}_{;ik} - F^{ik}{}_{;ki} = R^i{}_{mki}F^{mk} - R^k{}_{mki}F^{im}$$

$$= R_{mk}F^{mk} + R_{mi}F^{im} = 0.$$

Thus $F^{ik}{}_{;ik} = F^{ik}{}_{;ki}$. Hence the result follows.

Example 5.1.3 *Problem.* A two-dimensional space has a metric given by $ds^2 = g_{11}(dx^1)^2 + g_{22}(dx^2)^2$. Show that $R_{11}g_{22} = R_{22}g_{11}$ and $R_{12} = 0$, and that the Einstein tensor is identically zero.

Solution. We have $g^{11} = 1/g_{11}$, $g^{22} = 1/g_{22}$ and $g = g_{11}g_{22}$. The non-zero Christoffel symbols are

$$\Gamma^1_{11} = \frac{g_{11,1}}{2g_{11}}, \qquad \Gamma^1_{21} = \frac{g_{11,2}}{2g_{11}}, \qquad \Gamma^2_{11} = -\frac{g_{11,2}}{2g_{22}}, \qquad \Gamma^2_{12} = \frac{g_{22,1}}{2g_{22}},$$

$$\Gamma^2_{22} = \frac{g_{22,2}}{2g_{22}}, \qquad \Gamma^1_{22} = -\frac{g_{22,1}}{2g_{11}}.$$

A simple but tedious calculation then gives $R_{12} = 0$ and

$$R_{11} = \frac{g_{22,11} + g_{11,22}}{2g_{22}} - \frac{(g_{22,1})^2}{4g_{22}^2} - \frac{(g_{11,2})^2}{4g_{11}g_{22}}$$

$$- \frac{g_{11,2}g_{22,2}}{4g_{22}^2} - \frac{g_{11,1}g_{22,1}}{4g_{11}g_{22}}.$$

On writing R_{11}/g_{11} one can easily check by interchanging 1 and 2 that the outcome equals R_{22}/g_{22}. Thus the required result follows.

5.1.3 Bianchi identities

The expression (5.5) suggests another symmetry for the components of R_{iklm}. This symmetry is not algebraic but involves calculus. In covariant language we may express it as follows:

$$R_{iklm;n} + R_{iknl;m} + R_{ikmn;l} \equiv 0. \tag{5.12}$$

These relations are known as the *Bianchi identities*. Their proof is most easily given in the locally inertial system as in (5.5). Simply write the expressions for the three Rs in (5.12) in terms of the third derivatives of the metric tensor.

But multiplying (5.12) by $g^{im}g^{kn}$ and using (5.8)–(5.10), we can deduce from these identities another that is of importance to relativity:

$$\left(R^{ik} - \frac{1}{2}g^{ik}R \right)_{;k} \equiv 0. \tag{5.13}$$

In other words, *the Einstein tensor G^{ik} has zero divergence identically*.

Example 5.1.4 *Problem.* Show that, if $R_{iklm} = K(g_{il}g_{km} - g_{im}g_{kl})$, then $K = $ constant.

Solution. We have $R_{kl} = g^{im}K(g_{il}g_{km} - g_{im}g_{kl}) = -3Kg_{kl}$.
Therefore $R = -12K$ and the Einstein tensor is

$$G_{ik} = -3Kg_{ik} + 6Kg_{ik} = 3Kg_{ik}$$

Since $G^{ik}_{;k} \equiv 0$, we have

$$(3Kg^{ik})_{;k} = 0,$$

i.e.,

$$K_{,i} = 0,$$

since $g^{ik}_{;k} \equiv 0$. Thus $K = $ constant.

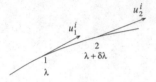

Fig. 5.4. The tangent vector to a geodesic does not change its direction. In the figure the tangent vectors at points 1 and 2 are technically parallel, once we take the non-Euclidean geometry into account.

5.2 Geodesics

So far we have talked about non-Euclidean geometries without mentioning whether, in general, they have the equivalents of straight lines in Euclidean geometry. We now show how equivalent concepts do exist in the Riemannian geometry under consideration here.

There are two properties of a straight line that can be generalized: the property of 'straightness' and the property of 'shortest distance'. Straightness means that, as we move along the line, its direction does not change. Let us see how we can generalize this concept first.

Let $x^i(\lambda)$ be the parametric representation of a curve in spacetime. Its tangent vector is given by

$$u^i = \frac{dx^i}{d\lambda}. \tag{5.14}$$

Our straightness criterion demands that u^i should not change as it moves along the curve. (See Figure 5.4.) In going from λ to $\lambda + \delta\lambda$, the change in u^i is given by

$$\Delta u^i = \frac{du^i}{d\lambda}\,\delta\lambda + \Gamma^i_{kl}u^k\,\delta x^l.$$

The second expression on the right-hand side arises from the change produced by parallel transport through a coordinate displacement δx^l. However, since the displacement arises from λ changing to $\lambda + \delta\lambda$, we have $\delta x^l = u^l\,\delta\lambda$. Therefore the condition of no change of direction u^i implies $\Delta u^i = 0$; that is,

$$\frac{du^i}{d\lambda} + \Gamma^i_{kl}u^k u^l = 0. \tag{5.15}$$

This is the condition that our curve must satisfy in order to be straight.

The second property of a straight line in Euclidean geometry is that it is the curve of *shortest distance* between two points. Let us generalize this property in the following way. Let the curve, parametrized by λ, connect two points P_1 and P_2 of spacetime, with parameters λ_1 and λ_2, respectively. Then the 'distance' of P_2 from P_1 is defined as

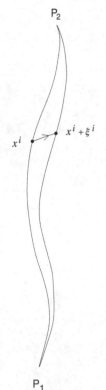

Fig. 5.5. For a geodesic connecting P_1 and P_2, all lines joining these points will have the same length for small displacement ξ^i.

$$s(P_2, P_1) = \int_{\lambda_1}^{\lambda_2} \left(g_{ik}\frac{dx^i}{d\lambda}\frac{dx^k}{d\lambda} \right)^{1/2} d\lambda \equiv \int_{\lambda_1}^{\lambda_2} L\,d\lambda, \tag{5.16}$$

say. We now demand that $s(P_2, P_1)$ be 'stationary' for small displacements of the curve connecting P_1 and P_2, with these displacements vanishing at P_1 and P_2. (See Figure 5.5.)

This is a standard problem in the calculus of variations, and its solution leads to the familiar Euler–Lagrange equations

$$\frac{d}{d\lambda}\left(\frac{\partial L}{\partial \dot{x}^i}\right) - \frac{\partial L}{\partial x^i} = 0, \tag{5.17}$$

where $\dot{x}^i \equiv dx^i/d\lambda$ and $L \equiv [g_{ik}(dx^i/d\lambda)(dx^k/d\lambda)]^{1/2}$ is a function of x^i and $\dot{x}^i$. It is easy to see that (5.17) leads to

$$\frac{d}{d\lambda}\left(g_{ik}\frac{1}{L}\frac{dx^k}{d\lambda}\right) - \frac{1}{2}g_{mn,i}\frac{1}{L}\frac{dx^m}{d\lambda}\frac{dx^n}{d\lambda} = 0.$$

If we substitute

$$ds = L\,d\lambda \tag{5.18}$$

and use (4.9), we get the above equation in the form

$$\frac{d^2x^i}{ds^2} + \Gamma^i_{kl}\frac{dx^k}{ds}\frac{dx^l}{ds} = 0. \tag{5.19}$$

There are a few loose ends to be sorted out in the above derivation. First, L would be real only for timelike curves. Thus, if we want to use a real parameter along the curve, then for spacelike curves we must replace ds by

$$d\sigma = i\,ds, \qquad i = \sqrt{-1}. \tag{5.20}$$

For null curves, $L = 0$. The above treatment therefore breaks down. It is then more convenient to replace the integral (5.16) by another, namely,

$$I = \int_{\lambda_1}^{\lambda_2} L^2\,d\lambda, \tag{5.21}$$

and consider $\delta I = 0$. We can always choose a new parameter $\lambda' = \lambda'(\lambda)$ such that the equation of the curve takes the same form as (5.19), with λ' replacing s.

It is easy to see that (5.19) is the same as (5.15). Although s in (5.19) has the special meaning 'length along the curve' while λ in (5.15) appears to be general, it is not difficult to see that, if (5.15) is satisfied, λ must be a constant multiple of s. This is because (5.15) has the first integral

$$g_{ik}\frac{dx^i}{d\lambda}\frac{dx^k}{d\lambda} = C, \qquad C = \text{constant.} \tag{5.22}$$

These curves of 'stationary distance' are called *geodesics*. For timelike curves $C > 0$ and for spacelike curves $C < 0$, while for null curves $C = 0$. λ is called an *affine parameter*.

Example 5.2.1 Let us calculate the null geodesic from $t = 0, r = 0$ to the point $t = T, r = R, \theta = \theta_1, \phi = \phi_1$ in the de Sitter spacetime

$$ds^2 = c^2\,dt^2 - e^{2Ht}[dr^2 + r^2(d\theta^2 + \sin^2\theta\,d\phi^2)],$$

where $H =$ constant. It is not difficult to verify that the θ and ϕ equations of (5.19) are satisfied by $\theta = \theta_1$, $\phi = \phi_1$, both θ_1 and ϕ_1 being constants. That is, our straight line moves in the fixed (θ, ϕ) direction. The t equation simplifies to

$$\frac{d^2t}{d\lambda^2} + \frac{H}{c^2}e^{2Ht}\left(\frac{dr}{d\lambda}\right)^2 = 0.$$

The first integral (5.22) gives, on the other hand, for $ds = 0$

$$c^2\left(\frac{dt}{d\lambda}\right)^2 = e^{2Ht}\left(\frac{dr}{d\lambda}\right)^2.$$

The two equations can be easily solved to give

$$t = \frac{1}{H}\ln\left(1 + \frac{\lambda}{\lambda_0}\right), \qquad r = \frac{c}{H}\frac{\lambda}{\lambda + \lambda_0},$$

where λ_0 is determined from the boundary condition that when $r = R$, $t = T$. Note that a solution is possible only if R and T are related by the condition

$$R = \frac{c}{H}(1 - e^{-HT}).$$

5.3 Geodesic deviation

We end this chapter by describing another geometrical feature that distinguishes a flat spacetime from a curved one. Again we take the spherical surface as illustrative of a curved space.

Imagine, as in Figure 5.6, longitude lines drawn between the poles on a spherical Earth. We know and can verify by using Equation (5.19) that the longitude lines are geodesics. Now consider two points P and Q on two neighbouring lines of this set, located at the same distance from the nearest pole. As both P and Q move away from the pole the distance between them at first increases. However, the rate of increase is not uniform; it is rapid at first but slows down until it is maximum when P and Q are on the equator. Thereafter the distance PQ decreases to zero as the other pole is reached.

This behaviour is different from that for geodesics drawn on the flat space of a Euclidean plane. Figure 5.7 shows straight lines drawn from a point O with points P and Q on two neighbouring geodesics, i.e., straight lines. In this case it is easy to verify that the distance PQ increases at a uniform rate with respect to the distance of the pair from O.

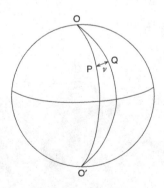

Fig. 5.6. On a spherical surface the separation PQ between two neighbouring longitudes increases on going from pole O to the equator but decreases as we move from the equator to the other pole O'.

So one may say that the rate at which two neighbouring geodesics deviate from each other gives us information on the curvature of space and time. The general result valid for Riemannian geometries is derived below.

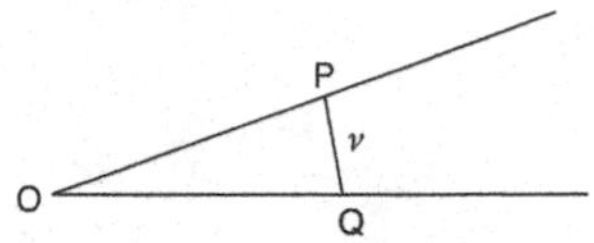

Let a bundle of geodesics in a general Riemannian spacetime be specified by a parameter μ, so that a typical point on the μ-geodesic has the coordinates $x^k(\lambda, \mu)$, λ being the affine parameter, as shown in Figure 5.8. The vector $v^k = \partial x^k/\partial \mu$ denotes the rate of deviation from one geodesic to another across the bundle. We first show that

$$v^k_{;l}u^l = u^k_{;l}v^l, \qquad \text{where } u^k = \partial x^k/\partial \lambda. \tag{5.23}$$

The proof is simple from first principles. We have

$$v^k_{;l}u^l = \frac{\partial v^k}{\partial x^l}u^l + \Gamma^k_{lm}v^m u^l = \frac{\partial^2 x^k}{\partial \lambda \, \partial \mu} + \Gamma^k_{lm}v^m u^l.$$

Similarly, we also have

$$u^k_{;l}v^l = \frac{\partial u^k}{\partial x^l}v^l + \Gamma^k_{lm}u^m v^l = \frac{\partial^2 x^k}{\partial \mu \, \partial \lambda} + \Gamma^k_{lm}u^m v^l.$$

Since the order of partial differentiation with respect to λ and μ can be interchanged and also because $\Gamma^k_{lm} = \Gamma^k_{ml}$, the result follows.

We next show that

$$\mathrm{d}^2 v^k/\mathrm{d}\lambda^2 + R^k_{lmn}u^l v^m u^n = 0. \tag{5.24}$$

In view of the equality proved above, the first term on the left-hand side may be written as

$$(v^k_{;l}u^l)_{;m}u^m = (u^k_{;l}v^l)_{;m}u^m = u^k_{;lm}v^l u^m + u^k_{;l}v^l_{;m}u^m.$$

Fig. 5.7. On a flat surface the separation PQ between two geodesics (straight lines as drawn here) increases uniformly as P and Q move further from the origin O of the geodesics.

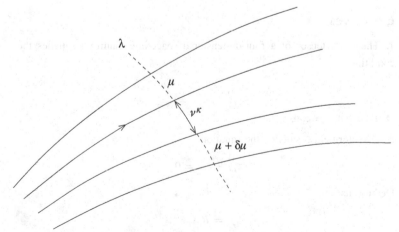

Fig. 5.8. In the bundle of geodesics the λ-parameter increases as one moves to the right on any geodesic. The parameter μ increases in the orthogonal direction. The separation v between two neighbouring geodesics tells us whether they are coming closer or moving apart. See the text for details.

Using the identity (5.4), the first term on the right-hand side may be replaced by

$$u^k{}_{;ml} v^l u^m - R^k{}_{ilm} u^i v^l u^m.$$

Therefore we have

$$\mathrm{d}^2 v^k / \mathrm{d}\lambda^2 + R^k{}_{ilm} u^i v^l u^m = u^k{}_{;ml} v^l u^m + u^k{}_{;l} v^l{}_{;m} u^m. \tag{5.25}$$

Consider the first term on the right-hand side of the above equation. By interchanging the dummy indices l, m it can be rewritten as

$$u^k{}_{lm} v^m u^l = [u^k{}_{;l} u^l]_{;m} v^m - u^k{}_{;l} u^l{}_{;m} v^m.$$

Now the first term on the right-hand side of the above equation vanishes since u^k is the tangent vector to a geodesic. In the second term use the identity (5.23) to replace $u^l{}_{;m} v^m$ by $v^l{}_{;m} u^m$. This term then cancels out the second term on the right-hand side of Equation (5.25). Thus we get zero on the right-hand side. This is the equation of *geodesic deviation*.

The appearance of the Riemann tensor is an indication that we are looking at the effect of curvature. We will have occasion to return to this equation in the context of gravitational effects on motion.

5.4 Concluding remarks

This is our introduction to non-Euclidean geometries insofar as they relate to general relativity. In the following chapter we will discuss spacetime symmetries. It may be somewhat mathematical. Those impatient for applications to physics may wish to skip it and go to Chapter 7. The results derived in Chapter 6 do, however, have important applications to specific problems.

Exercises

1. The Ricci tensor of a four-dimensional spacetime manifold satisfies the condition

$$R_{ik} = f g_{ik}.$$

Deduce that $f = \text{constant}$.

2. A vector field ξ_i satisfies the equations

$$\xi_{i;k} + \xi_{k;i} = 0.$$

Deduce that

$$\xi_{l;ik} = R_{likm} \xi^m.$$

3. In the spacetime whose metric is given by

$$ds^2 = e^{\phi}(dx^4) - e^{-\phi}(x^2)^2(dx^3)^2 - e^{\lambda}\{(dx^1)^2 + (dx^2)^2\},$$

where ϕ and λ are functions of x^1 and x^2 only, show that, provided $R_{ik} = 0$, for $i = k = 0$,

$$\frac{\partial^2 \phi}{(\partial x^1)^2} + \frac{\partial^2 \phi}{(\partial x^2)^2} + \frac{1}{x^2}\frac{\partial \phi}{\partial x^2} = 0.$$

4. Write down the equations of a null geodesic in the spacetime given by the line element

$$ds^2 = dt^2 - 2e^{x^1}\,dt\,dx^2 - (dx^1)^2 + \frac{1}{2}e^{2x^1}(dx^2)^2 - (dx^3)^2$$

and show that the following is a first integral of them:

$$2x^2\frac{dx^1}{d\lambda} - \left[3 + \frac{1}{2}(x^2)^2 e^{2x^1}\right]\frac{dx^2}{d\lambda} - [(x^2)^2 e^{x^1} + 2e^{-x^1}]\frac{dt}{d\lambda} = \text{constant},$$

where λ is an affine parameter.

5. Two metrics $g^{(1)}_{ik}$ and $g^{(2)}_{ik}$ on a given spacetime give the same geodesic curves. Show that their respective Christoffel symbols $\Gamma^{(1)i}_{kl}$ and $\Gamma^{(2)i}_{kl}$ satisfy a relation of the form

$$\Gamma^{(1)i}_{kl} - \Gamma^{(2)i}_{kl} = \delta^i_k V_l + \delta^i_l V_k,$$

where V_k are the components of a vector.

6. A vector is parallelly propagated round a spherical triangle ABC. Show that, at the end of the round, the vector makes an angle $(A + B + C - \pi)$ with its original direction.

7. Consider the conformal transformation

$$g^*_{ik} = g_{ik}e^{2\sigma}.$$

Show that under such a tranformation the wave equation

$$\Box\phi + \frac{1}{6}R\phi = 0$$

remains invariant. (Here R is the scalar curvature.)

8. Show that, under a conformal transformation, the *Weyl tensor*

$$C_{iklm} = R_{iklm} - \frac{1}{2}(g_{ik}R_{lm} - g_{im}R_{kl} - g_{kl}R_{im} + g_{lm}R_{ik})$$

$$+ \frac{1}{6}(g_{il}g_{km} - g_{im}g_{kl})R$$

is invariant. Deduce that, if the metric is conformal to the flat spacetime metric, the Weyl tensor vanishes.

9. For the metric

$$ds^2 = c^2 \, dt^2 - dr^2 - r^2(d\theta^2 + \sin^2\theta \, d\varphi^2)$$

verify that $R_{iklm} = 0$.

10. Show that, if a geodesic is timelike over a finite part, then it is timelike throughout.

11. Show that, if $\bar{g}_{ik} = g_{ik} \exp \zeta$, then

$$\bar{R}_{ik} = R_{ik} + 2(\zeta_{;ik} - \zeta_{;i}\zeta_{ik}) + g_{ik}(\Box\zeta + 2\zeta_{;l}\zeta^{;l}).$$

Chapter 6
Spacetime symmetries

6.1 Introduction

In Euclidean geometry or in the pseudo-Euclidean spacetime of special relativity, the geometrical properties are invariant under translations and rotations. The same is not necessarily true of the non-Euclidean spacetimes of general relativity. As we shall see in Chapter 8, the spacetime geometry is intimately related to the distribution of gravitating matter (and energy). A completely general spacetime arising from an arbitrary distribution of gravitating objects will not have any symmetries at all. Such cases are difficult to solve as solutions of Einstein's gravitational equations. It is, however, easier to solve problems where mass distributions have certain symmetries. For example, a point mass in an otherwise empty space is expected to generate a solution that has spherical symmetry about that point. Cases like these may be looked upon as approximations to reality. A similar approach is adopted in Newtonian gravitation. For example, as a first approximation the gravitating masses in the Solar System (the Sun and the planets) are treated as spherical distributions. In this chapter we will look at certain symmetric spacetimes that will be of use in solving specific problems in general relativity. The main question that we shall begin with is that of how to identify a symmetry in a given spacetime. How do we discover an intrinsic property like symmetry, when given the spacetime metric?

We will have occasion to use symmetric and antisymmetric tensors. To facilitate their writing as well as recognition of the nature of their symmetry we will write indices (ik) for a symmetric tensor and $[ik]$ for an antisymmetric one. Thus, if T_{ik} is *any* tensor,

$$S_{ik} = \frac{1}{2}(T_{ik} + T_{ki}) = T_{(ik)}; \qquad A_{ik} = \frac{1}{2}(T_{ik} - T_{ki}) = T_{[ik]}.$$

This notation was introduced in Chapter 3.

6.2 Displacement of spacetime

It is worth recalling here the stress put on circles as special curves by the Greek philosopher Aristotle (384–322 BC). Aristotle argued that the displacement symmetry displayed by circles was unique: no other curve had it. By displacement symmetry, we mean the following. Take any finite arc of the circle and move it so as to place it anywhere else on the circle. It will lie congruently on the corresponding part of the circle. Because of this symmetry Aristotle felt that circles had a special role to play in the behaviour of natural phenomena.

We will adopt a similar criterion in our specification of the symmetry of the a spacetime manifold. Suppose x^i are the coordinates and g_{ik} are the components of the metric tensor specifying a spacetime manifold $\mathcal{M}$. Let P be a typical point with coordinates x_P^k. We may make a 'copy' of $\mathcal{M}$, called $\mathcal{M}'$, and imagine that $\mathcal{M}$ is placed congruently on $\mathcal{M}'$.

Imagine now an infinitesimal displacement of $\mathcal{M}$ so that each point moves over to a new place. Such a displacement may be described by the relation

$$x^i \rightarrow x^i + \xi^i, \tag{6.1}$$

where ξ^i is an infinitesimal vector field. Equation (6.1) implies that the point P with coordinate x_P^i now moves over to a position that is occupied by a point P' in the manifold $\mathcal{M}'$ with coordinates $x_P^i + \xi^i(x_P^i)$. Figure 6.1 illustrates this move.

Fig. 6.1. In the shift described in the text the point P of $\mathcal{M}$ falls on P' of $\mathcal{M}'$.

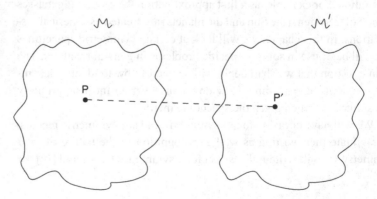

A simple example of such a displacement is an infinitesimal translation or a rotation. In the three-dimensional Euclidean space we can consider the rotation of a spherical surface about its centre. In Figure 6.2, the point P after rotation moves over to P'. However, under such a displacement the new surface is indistinguishable from the old one. We now ask the following question: what should be the condition on ξ^i for this to happen in the displacement given by Equation (6.1)?

To find this condition let us consider the two spacetimes in the above problem. The point P' of $\mathcal{M}'$ coincides with the point P of $\mathcal{M}$ (see Figure 6.1). Since the coordinate system was carried along when P was displaced to its new position, P continues to have coordinates x^i_p in $\mathcal{M}$. P', on the other hand, has coordinates $x^i_P + \xi^i(x^i_P)$ in $\mathcal{M}'$. Suppose in $\mathcal{M}'$ we now introduce a new coordinate system given by

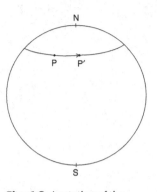

Fig. 6.2. A rotation of the sphere about the NS polar axis takes point P to P'.

$$x'^i = x^i - \xi^i. \tag{6.2}$$

Under this transformation P' in $\mathcal{M}'$ will have coordinates $x'^i = x^i_P$, the same as the coordinates of P in $\mathcal{M}$; and this must be true for all the coinciding points of $\mathcal{M}$ and $\mathcal{M}'$. But what about the spacetime metric at the corresponding points?

The metric tensor at P in $\mathcal{M}$ is $g_{ik}(x^l_P)$. The metric tensor at P' in the old coordinate system was $g_{ik}(x^l_P + \xi^l_P)$, where $\xi^l_P = \xi^l(x^i_P)$. In the new coordinate system this is transformed to

$$g'_{mn} = \left[\frac{\partial x^i}{\partial x'^m} \frac{\partial x^k}{\partial x'^n} \right]_{P'} g_{ik}(x^l_P + \xi^l_P). \tag{6.3}$$

Since ξ^i is infinitesimal we can use the following approximations which ignore errors of second and higher order in ξ^i and its derivatives:

$$\left. \frac{\partial x^i}{\partial x'^m} \right|_{P'} \cong \delta^i_m + \xi^i_{P,m},$$

$$g_{ik}(x^l + \xi^l_P) \cong g_{ik}(x^l) + \xi^l_P g_{ik,l}(x^l_p).$$

Then it is easy to see that, to first order in ξ^i,

$$g'_{mn}(P') = g_{mn}(P) + [\xi^l g_{mn,l} + \xi^l_{,m} g_{ln} + \xi^l_{,n} g_{lm}]_P. \tag{6.4}$$

From Equation (6.4) we see that $\mathcal{M}$ and $\mathcal{M}'$ become geometrically indistinguishable at the coinciding points P and P' if the expression in square brackets vanishes. Since P is any typical point of $\mathcal{M}$ this relation must hold everywhere. Thus ξ^i must satisfy the set of equations

$$\xi^l g_{mn,l} + \xi^l_{,m} g_{ln} + \xi^l_{,n} g_{lm} = 0. \tag{6.5}$$

Using the Riemannian affine connection, these equations can be rewritten as

$$\xi_{m;n} + \xi_{n;m} = 0. \tag{6.6}$$

Equations (6.5) or (6.6) are known as *Killing's equations* and the vector field ξ^i is known as a *Killing vector field*. In general, for a heterogeneous spacetime a non-trivial solution of (6.6) will not exist.

If the spacetime does admit a Killing vector field we can consider its displacement under (6.1). Then, as shown above, the displaced spacetime is indistinguishable from its original state (*vide* the example of the sphere under rotation). The existence of such a displacement is an indicator of symmetry. A displacement of this type is often referred to as an *isometry*. Aristotle had precisely this concept in mind in his choice of circles.

Example 6.2.1 In terms of the spherical polar coordinates θ, ϕ the line element on the surface of a unit sphere is given by

$$ds^2 = d\theta^2 + \sin^2\theta \, d\phi^2.$$

The Killing vector $\xi^i = (\xi^\theta, \xi^\phi)$ satisfies Equations (6.5) above, which are explicitly as follows:

(i) $\dfrac{\partial \xi^\theta}{\partial \theta} = 0,$ (ii) $\dfrac{\partial \xi^\theta}{\partial \phi} + \sin^2\theta \dfrac{\partial \xi^\phi}{\partial \theta} = 0,$ (iii) $\dfrac{\partial \xi^\phi}{\partial \phi} + \cot\theta \, \xi^\theta = 0.$

From (i) we get $\xi^\theta = f(\phi)$, where f is an arbitrary function of ϕ. Then (ii) gives

$$\sin^2\theta \, \frac{\partial \xi^\phi}{\partial \theta} = -f'(\phi), \qquad \text{i.e., } \xi^\phi = f'(\phi)\cot\theta + g(\phi),$$

where $f'(\phi) \equiv df/d\phi$ and $g(\phi)$ is an arbitrary function of ϕ. On substituting for ξ^θ and ξ^ϕ in (iii) we get

$$g'(\phi) + [f''(\phi)\cot\theta + f(\phi)\cot\theta] = 0.$$

Since this must hold for all θ and ϕ, we have

$$g'(\phi) = 0, \qquad f''(\phi) + f(\phi) = 0.$$

Thus the most general solution of the Killing equations in this case is

$$f(\phi) = A\sin\phi + B\cos\phi, \qquad g(\phi) = C;$$

$$\xi^\theta = A\sin\phi + B\cos\phi, \qquad \xi^\phi = (A\cos\phi - B\sin\phi)\cot\theta + C,$$

where A, B and C are arbitrary constants. Thus there are three linearly independent Killing vectors.

6.3 Some properties of Killing vectors

We now discuss some general properties of the Killing equation and its solutions.

Integrability. Using the formulae (5.4) and (5.6), we at once deduce a simple consequence of Equation (6.6):

$$2\xi_{m;np} \equiv (\xi_{m;np} - \xi_{m;pn}) + (\xi_{p;nm} - \xi_{p;mn}) + (\xi_{n;pm} - \xi_{n;mp}),$$

i.e.,

$$\xi_{m;np} = -R^l{}_{pmn}\xi_l. \tag{6.7}$$

From Equation (6.7) we see that, if ξ_l and its derivatives $\xi_{l;m}$ are known at a typical point P, we can determine all higher derivatives of ξ_l at P and hence the entire function ξ_l in a neighbourhood of P, by Taylor expansion. Thus, provided that Equations (6.6) and (6.7) have a solution, we can formally write it in the form

$$\xi_m(x^l) = A^n{}_m(X, P)\xi_n(P) + B_m{}^{pq}(X, P)\xi_{p;q}(P), \tag{6.8}$$

where X is a general point and the quantities $A^n{}_m$ and $B_m{}^{pq}$ depend on the global properties of spacetime, i.e., on g_{mn}, and on the points P and X. By virtue of Equation (6.6), $B_m{}^{pq} = -B_m{}^{qp}$. In a spacetime of n dimensions there are up to n independent quantities $\xi_n(P)$ and up to $\frac{1}{2}n(n-1)$ independent quantities $\xi_{p;q}(P)$ because of the antisymmetry implied by the Killing equations. Thus there are in general up to $n + \frac{1}{2}n(n-1) = \frac{1}{2}n(n+1)$ linearly independent Killing vectors in a spacetime of n dimensions.

What are the conditions for Equation (6.7) to be integrable? From Equation (6.7) we get

$$-\xi_{m;npq} = R^l{}_{pmn;q}\xi_l + R^l{}_{pmn}\xi_{l;q},$$

$$-\xi_{m;nqp} = R^l{}_{qmn;p}\xi_l + R^l{}_{qmn}\xi_{l;p}.$$

Taking the difference of these and using Equation (5.4), we get

$$\xi_{m;npq} - \xi_{m;nqp} = R^l{}_{mpq}\xi_{l;n} + R^l{}_{npq}\xi_{m;l}.$$

From this follows the result

$$\xi_l(R^l{}_{qmn;p} - R^l{}_{pmn;q}) + R^l{}_{qmn}\xi_{l;p} - R^l{}_{pmn}\xi_{l;q}$$
$$- R^l{}_{mpq}\xi_{l;n} - R^l{}_{npq}\xi_{m;l} = 0. \tag{6.9}$$

These are the conditions for integrability, which by relating ξ_l and $\xi_{l;m}$ impose restrictions on how many Killing vectors can exist at a given point of spacetime.

Example 6.3.1 *Problem.* ξ^i is a timelike Killing vector and $u^i = \phi \xi^i$ with ϕ so chosen that u^i is a unit vector. Show that

$$u^i u_{k;i} = \frac{\partial}{\partial x^k}(\ln \phi).$$

Solution. On writing $\xi_i = u_i \phi^{-1}$, we get $\xi_{i;k} = u_{i;k}\phi^{-1} - u_i \phi_{,k}\phi^{-2}$. From the Killing equations we get for the above relation

$$\phi(u_{i;k} + u_{k;i}) - (u_i \phi_{,k} + u_k \phi_{,i}) = 0.$$

Multiply by u^i and use $u^i u_i = 1$ and $u^i u_{i;k} = 0$. Then

$$\phi u^i u_{k;i} = \phi_{,k} + u^i u_k \phi_{,i}.$$

Multiply by u^k and use $u^k u_{k;i} = 0$ to get $u^i \phi_{,i} = 0$. Hence the above equation becomes

$$\phi u^i u_{k;i} = \phi_{,k}.$$

The result to be proved follows.

Problem. If ξ^i is a Killing vector field and T^{ik} is a symmetric tensor satisfying the condition $T^{ik}_{;k} = 0$, then the vector $p^i = T^{ik}\xi_k$ has zero divergence.

Solution. We have that $p^i_{;i} = T^{ik}_{;i}\xi_k + T^{ik}\xi_{k;i} = T^{ik}\xi_{k;i} = \frac{1}{2}(\xi_{k;i} + \xi_{i;k})T^{ik} = 0$ by virtue of the symmetry of T^{ik} and the Killing equations. We have used the property that $T^{ik}A_{ik} \equiv 0$ if T^{ik} is symmetric and A_{ik} is antisymmetric.

Finite displacement. The above analysis of Killing vectors relates to infinitesimal displacement. In a special case it is possible to talk of a finite displacement. This is the case when all g_{ik} are independent of a particular coordinate, say x^0. Then direct substitution into (6.5) immediately shows that

$$\xi^i = (0, 0, 0, \epsilon), \tag{6.10}$$

where ϵ is an infinitesimal constant, is a solution. This means that a displacement of the form

$$x^0 \rightarrow x^0 + \epsilon, \qquad x^\mu \rightarrow x^\mu \tag{6.11}$$

leaves the spacetime invariant.

If x^0 is a timelike coordinate we say that the spacetime is *static*. When Equations (6.10) and (6.11) hold, we need not restrict ϵ to be infinitesimal. As is obvious, by a superposition of a series of infinitesimal displacements we can make up a finite displacement that leaves the spacetime invariant.

Relation to geodesics. If u^i is a tangent vector to a geodesic ζ and ξ_i is a Killing vector, then

$$\xi_i u^i = \text{constant along } \zeta. \tag{6.12}$$

The proof follows from the use of the geodesic equation (5.19) and the Killing equation (6.6):

$$u^i(\xi_k u^k)_{;i} = \xi_k u^i u^k_{;i} + u^k u^i \xi_{k;i} = 0.$$

We shall use this result in later work. Equation (6.12) represents a first integral of the geodesic equations.

Example 6.3.2 *Problem.* In the line element

$$ds^2 = e^\nu \, dT^2 - e^\lambda \, dR^2 - R^2(d\theta^2 + \sin^2\theta \, d\phi^2)$$

show that the timelike geodesic has the first integrals

$$e^\nu \frac{dT}{ds} = \text{constant}; \qquad R^2 \sin\theta \frac{d\phi}{ds} = \text{constant}.$$

Here ν and λ are functions of R only.

Solution. Since the metric is independent of T and ϕ, we deduce that the spacetime has Killing vectors $\xi^{i(1)} = (1, 0, 0, 0)$ and $\xi^{i(2)} = (0, 0, 0, 1)$, where $x^i \equiv (T, R, \theta, \phi)$. Hence we have two first integrals of the geodesic equations:

$$\xi^{i(1)} u_i = \text{constant}; \qquad \xi^{i(2)} u_i = \text{constant}.$$

The first one gives $u_0 = \text{constant}$. Since $u^0 = dT/ds$, we get $e^\nu \, dT/ds = \text{constant}$. The second relation likewise gives $u_3 = \text{constant}$. With $u^3 = d\phi/ds$, we get $R^2 \sin\theta \, d\phi/ds = \text{constant}$.

6.4 Homogeneity and isotropy

The physicist often refers to the above two properties of space and time. Of these, homogeneity implies the fact that the physical quantity he measures is the same at any two points P and Q in spacetime. Isotropy at a given point P implies invariance with respect to a change of direction at P. With the help of Killing vectors it is possible to express these properties more formally and precisely than the above statements. Since we might not always want the entire spacetime $\mathcal{M}$ to be homogeneous and/or isotropic, I shall consider below these properties in a spacetime $\mathcal{M}_n$ of n dimensions that is a subspace of $\mathcal{M}$.

Homogeneity. The spacetime $\mathcal{M}_n$ is said to be homogeneous if there are infinitesimal isometries that carry a typical point P to any point P' in its immediate neighbourhood. This means that the Killing vectors at P can take all possible values, and we can choose, at P, n linearly independent Killing vectors. By a suitable choice, we can therefore have a basis of n Killing vector fields $\xi_i^{(k)}(X, P)$ at a general point X in the neighbourhood of P such that

$$\lim_{X \to P} \xi_i^{(k)}(X, P) = \delta_i^k, \, (k = 1, \ldots, n). \qquad (6.13)$$

Clearly, by a succession of infinitesimal displacements we can take P to any distant point P'.

Example 6.4.1 The surface of the unit sphere is homogeneous because at each point it has two linearly independent Killing vectors. As shown in Example 6.2.1, there are *three* Killing vector fields on this surface, so at a general point we can choose any two of them.

Isotropy. The spacetime $\mathcal{M}_n$ is said to be isotropic at a given point P if there are Killing vectors ξ_i in the neighbourhood of P such that $\xi_i(P) = 0$ and $\xi_{i;k}(P)$ span the space of antisymmetric second-rank tensors at P. Thus we need $\frac{1}{2}n(n-1)$ linearly independent $\xi_{i;k}$ at P. In an isotropic spacetime at P we can choose coordinates in the neighbourhood of P such that there are $\frac{1}{2}n(n-1)$ Killing vector fields $\xi_i^{[pq]}(X, P)$ with the properties

$$\xi_i^{[pq]}(X, P) = -\xi_i^{[qp]}(X, P),$$

$$\xi_i^{[pq]}(P, P) = 0, \qquad (6.14)$$

$$\xi_{i;k}^{[pq]}(P, P) \equiv [\xi_{i;k}^{[pq]}(X, P)]_{X=P} = \delta_i^p \delta_k^q - \delta_i^q \delta_k^p$$

$$(p, q = 1, \ldots; n).$$

Example 6.4.2 Consider the same example of the surface of the unit sphere. In this case $n = 2$, i.e., $n(n-1)/2 = 1$. Since we have seen that this surface is homogeneous, we can take P to be the pole ($\theta = 0$) without loss of generality. At this point the Killing vector field $\xi^\theta = 0$, $\xi^\phi = 1$ shows isotropy. The coordinates are $x^1 \cong \theta \sin \phi$, $x^2 \cong \theta \cos \phi$ near the pole (where $\sin \theta \cong \theta$). In these coordinates the Killing vector field has covariant components

$$\xi_1^{[12]} = x^2, \qquad \xi_2^{[12]} = -x^1.$$

We may define $\xi_i^{[21]} = -\xi_i^{[12]}$. Then Equation (6.14) follows.

Theorem. Any $\mathcal{M}_n$ that is isotropic about every point is also homogeneous.

Consider Killing vectors $\xi_i^{[pq]}(X, P)$ and $\xi_i^{[pq]}(X, Q)$ that satisfy (6.14) at two neighbouring points P and Q, respectively. At point

X, $\xi_i^{[pq]}(X, Q) - \xi_i^{[pq]}(X, P)$ is also a Killing vector. On writing the coordinates of P and Q, respectively, as x_P^i and $x_P^i + \delta x_P^i$, we see that

$$\lim_{\delta x_P^k \to 0} \frac{1}{\delta x_P^k} \{\xi_i^{[pq]}(X, Q) - \xi_i^{[pq]}(X, P)\} = \frac{\partial \xi_i^{[pq]}(X, P)}{\partial x_P^k} \qquad (6.15)$$

is also a Killing vector at X. However, we also have, from Equation (6.14),

$$\xi_{i;k}^{[pq]}(P, P) = \lim_{\delta x_P^k \to 0} \frac{1}{\delta x_P^k} \xi_i^{[pq]}(Q, P)$$

$$= \delta_i^p \delta_k^q - \delta_i^q \delta_k^p.$$

On putting X = Q in (6.15) we get as Q → P

$$\left. \frac{\partial \xi_i^{[pq]}(X, P)}{\partial x_P^k} \right|_{X=P} = -\delta_i^p \delta_k^q + \delta_i^q \delta_k^p. \qquad (6.16)$$

The Killing vectors (6.16) obviously span the space of vectors at P. For, if α_i is any arbitrary vector at P, we can construct a general vector field in the neighbourhood of P:

$$\xi_i(X) = \frac{\alpha_l}{n-1} \frac{\partial \xi_i^{[lk]}(X, P)}{\partial x_p^k},$$

which is such that $\xi_i(P) = \alpha_i$ (arbitrary constants). This follows from Equation (6.16):

$$\xi_i(P) = \frac{\alpha_l}{n-1}(\delta_i^l \delta_k^k - \delta_i^k \delta_k^l) = \alpha_i.$$

Thus any vector at any arbitrary point P can be expressed in terms of the Killing vector fields at P. This proves the result.

Maximally symmetric spacetime. If $\mathcal{M}_n$ is homogeneous and isotropic, it is said to be *maximally symmetric*. From the above theorem, if $\mathcal{M}_n$ is isotropic at every point, it is maximally symmetric.

A maximally symmetric space has $\frac{1}{2}n(n+1)$ different Killing vector fields. To see this we consider the set of vector fields $\xi_i^{(k)}(X, P)$ and $\xi_i^{[pq]}(X, P)$. Suppose they satisfy a linear relation

$$\alpha_k \xi_i^{(k)}(X, P) + \beta_{[pq]} \xi_i^{[pq]}(X, P) \equiv 0,$$

at all points X where the αs and βs are constants. On setting X = P and using Equation (6.14) we get

$$\alpha_k \delta_i^k = \alpha_i = 0.$$

Next, by differentiating with respect to x^k and setting X = P, we get

$$\beta_{[pq]}(\delta_i^p \delta_k^q - \delta_i^q \delta_k^p) = 2\beta_{[ik]} = 0.$$

Hence these Killing vectors are linearly independent and the result follows. The maximally symmetric space has n Killing vectors for homogeneity and $\frac{1}{2}n(n-1)$ Killing vectors for isotropy.

6.5 Spacetime of constant curvature

We now obtain an important result for maximally symmetric spaces, which makes their explicit determination possible. In Equation (6.9) we have the general integrability condition. On applying it to the vectors $\xi^{[pq]}_{\;\;\;i}(X, P)$ at P we get, using Equation (6.14),

$$(\delta^p_i\delta^q_k - \delta^q_i\delta^p_k)[\delta^k_r\delta^i_l R^l_{\;tmn} - \delta^i_l\delta^k_t R^l_{\;rmn} - \delta^i_l\delta^k_n R^l_{\;mrt} - \delta^i_l\delta^k_m R^l_{\;nrt}] = 0.$$

The above equation can be simplified further. We get

$$R^p_{\;tmn}\delta^q_r - R^p_{\;rmn}\delta^q_t - R^p_{\;mrt}\delta^q_n - R^q_{\;nrt}\delta^p_m$$
$$= R^q_{\;tmn}\delta^p_r - R^q_{\;rmn}\delta^p_t - R^q_{\;mrt}\delta^p_n - R^p_{\;nrt}\delta^q_m. \qquad (6.17)$$

If we now consider Equation (6.9) for the vectors $\xi^i(X, P)$ and use the above relation at X = P, Equation (6.9) reduces to

$$R^i_{\;qmn;p} = R^i_{\;pmn;q}. \qquad (6.18)$$

On putting $q = r$ in Equation (6.17) and using the symmetry properties of R_{iklm}, we get

$$R_{ptmn} = \frac{1}{n-1}(R_{nt}g_{pm} - R_{mt}g_{pn}). \qquad (6.19)$$

On multiplying further by g^{tn} we get

$$R_{pm} = \frac{R}{n}g_{pm}. \qquad (6.20)$$

(In these reductions we have to use the relation $g^m_m = n$. For the four-dimensional spacetime we have $n = 4$.)

A spacetime satisfying Equation (6.20) is called an *Einstein space*. The maximally symmetric space, on the other hand, has more symmetries than in the Einstein space. Substitution of Equation (6.20) into Equation (6.19) gives

$$R_{ptmn} = \frac{R}{n(n-1)}(g_{nt}g_{pm} - g_{mt}g_{pn}). \qquad (6.21)$$

By taking the divergence of Equation (6.21) and using Equation (5.13) (for n-dimensional spacetime), we see that for $n \geq 3$

$$R = \text{constant} = n(n-1)K \text{ (say)}, \qquad (6.22)$$

where K is a constant. Equation (6.22) then tells us that the spacetime Riemann tensor has the form

$$R_{ptmn} = K(g_{nt}g_{pm} - g_{mt}g_{pn}).$$ (6.23)

In differential geometry this is known as the curvature tensor for a space of *constant* curvature K.

In the case $n = 2$, Equation (6.19) can be used to arrive at the same conclusion.

It can be shown that spaces of constant curvature are essentially unique. In other words, if we have two spacetimes $\mathcal{M}_n$ and $\mathcal{M}'_n$, with Equation (6.23) holding in $\mathcal{M}_n$ and

$$R'_{ptmn} = K(g'_{nt}g'_{pm} - g'_{mt}g'_{pn})$$ (6.24)

holding in $\mathcal{M}'_n$ with the metric tensor g'_{ik}, then there exists a coordinate transformation $x^i \rightarrow x'^i$ for $\mathcal{M}_n \rightarrow \mathcal{M}'_n$ that will take g_{ik} to g'_{ik} in the usual manner of tensor transformations.

The proof of this result will not be given here, for want of space (see Reference [7] for details). Using this result, however, it becomes easy to identify maximally symmetric spacetimes in n dimensions. The essential difference is in the sign of K, since the magnitude of K can be scaled by a suitable scale transformation of ds. We shall, later on, need the cases for which all the n dimensions are spacelike. In this case we have the following three line elements:

$$ds^2 = -K^{-1}\left\{(d\mathbf{x})^2 + \frac{(\mathbf{x} \cdot d\mathbf{x})^2}{1 - x^2}\right\} \quad (K > 0),$$ (6.25)

$$ds^2 = +K^{-1}\left\{(d\mathbf{x})^2 + \frac{(\mathbf{x} \cdot d\mathbf{x})^2}{1 + x^2}\right\} \quad (K < 0),$$ (6.26)

$$ds^2 = -(d\mathbf{x})^2 \quad (K = 0).$$ (6.27)

Here $\mathbf{x} = (x^1, \ldots, x^n)$ is the coordinate vector and x^2 is the square of its magnitude. It can be verified that these spaces do satisfy Equation (6.23) so that, by virtue of the uniqueness theorem, they contain all the required information about homogeneous and isotropic spaces.

Example 6.5.1 In two dimensions, for $K > 0$ we have

$$ds^2 = -\frac{1}{K}\left\{(dx^1)^2 + (dx^2)^2 + \frac{(x^1\,dx^1 + x^2\,dx^2)^2}{1 - (x^1)^2 - (x^2)^2}\right\}.$$

Put $x^1 = \sin\theta\cos\phi$, $x^2 = \sin\theta\sin\phi$. Then this becomes

$$ds^2 = -\frac{1}{K}[d\theta^2 + \sin^2\theta\,d\phi^2].$$

This is the surface of a sphere of radius $K^{-1/2}$.

6.6 Symmetric subspaces

In general the entire spacetime might not have many symmetries, but it may have subspaces with more symmetries. In particular it may have maximally symmetric subspaces. Although our eventual application will be to the $(3+1)$-dimensional spacetime we will continue to discuss m-dimensional subspaces in an n-dimensional $(n \geq m)$ spacetime $\mathcal{M}_n$.

Suppose $\{\varphi_m\}$ is a collection of subspaces within $\mathcal{M}_n$. We will choose a coordinate system x^i such that $x^1, \ldots, x^m$ denote different points on the same φ_m, while the remaining coordinates $x^{m+1}, \ldots, x^n$ for these points are the same. In other words, the variation of $(x^{m+1}, \ldots, x^n)$ denotes different numbers of $\{\varphi_m\}$ while the variation of $x^1, \ldots, x^m$ represents the variation on a given φ_m.

We say that the spaces φ_m are homogeneous in $\mathcal{M}_n$ if there exist at least m linearly independent Killing vectors that take any point P on φ_m into any other given point P' on φ_m while leaving φ_m as a whole invariant.

Example 6.6.1 The rotation of the 2-sphere about its centre in the $(3+1)$-dimensional spacetime of special relativity. Here ϕ_2 is the surface of the sphere.

A similar definition can be given for isotropy of φ_m. Of particular interest is the case in which φ_m is maximally symmetric. In this case there exist $\frac{1}{2}m(m+1)$ independent Killing vectors, each with the following property. For an infinitesimal displacement of the type

$$
\begin{aligned}
x^i &\to x^i + \xi^i, \quad i = 1, \ldots, m, \\
x^i &\to x^i, \quad i > m,
\end{aligned}
\tag{6.28}
$$

the whole space $\mathcal{M}_n$ is unchanged. The ξ^i therefore have zero components for $i > m$, although they can be functions of all x^i. The linear independence of all the $\frac{1}{2}m(m+1)$ different ξ^i implies therefore that there is no linear relation among them with coefficients depending on $x^1, \ldots, x^m$.

It can be then be shown (see Reference [7] for proof) that the line element of $\mathcal{M}_n$ can be written down in the form

$$
\begin{aligned}
\mathrm{d}s^2 = {} & f(x^{m+1}, \ldots, x^n) \sum_{i,k \leq m} h_{ik}(x^1, \ldots, x^m) \mathrm{d}x^i \, \mathrm{d}x^k \\
& + \sum_{i,k > m} g_{ik}(x^{m+1}, \ldots, x^n) \mathrm{d}x^i \, \mathrm{d}x^k.
\end{aligned}
\tag{6.29}
$$

We will consider two special cases of the above result applicable to the $(3+1)$-dimensional spacetime.

Spherically symmetric spacetime. In this case there are two-dimensional surfaces φ_2 of constant positive curvature, concentric about a fixed point O at all times. We may choose $x^1 = \theta, x^2 = \phi$ to denote the coordinates on φ_2, and $x^3 = r$, $x^4 = t$, to denote the variation among the $\{\varphi_2\}$ family. Then, from the above result (6.29), the line element has the form

$$ds^2 = A(r, t)dt^2 + 2H(r, t)dt\,dr + B(r, t)dr^2$$
$$+ F(r, t)\{d\theta^2 + \sin^2\theta\,d\phi^2\}, \tag{6.30}$$

where A, H, B and F are general functions. We shall need this spacetime in later work to describe the gravitational field of a spherically symmetric distribution of matter and energy.

Cosmological spacetimes. In this situation, there is a family of three-dimensional maximally symmetric spacelike subspaces $\{\varphi_3\}$. We choose $x^0 = t$ and use x^1, x^2, x^3 to denote points on any φ_3. On φ_3 we use the metric (6.25)–(6.27) depending on the sign of K. All three cases can be represented by a compact line element:

$$ds^2 = dt^2 - S^2(t)\left[\frac{dr^2}{1 - kr^2} + r^2(d\theta^2 + \sin^2\theta\,d\phi^2)\right], \quad k = 0, +1. \tag{6.31}$$

Note that the function g_{00} in Equation (6.31) can be made unity in this case because it can be absorbed in a pure time transformation $t \to t'$. Thus, if we start with t', then we can choose t such that

$$g_{00}(t')dt'^2 = dt^2.$$

The parameter k represents the sign of the curvature. A comparison with Equations (6.25)–(6.27) shows that

$$K = \frac{k}{S^2(t)}. \tag{6.32}$$

In applications to cosmology we will discuss the physical significance of the time coordinate t and of the subspaces $t = $ constant.

Example 6.6.2 *Problem.* Show that the de Sitter line element

$$ds^2 = dt^2 - e^{2Ht}(dr^2 + r^2\,d\theta^2 + r^2\sin^2\theta\,d\phi^2)$$

has a timelike Killing vector.

Solution. Consider the Killing equations

$$\xi^l g_{mn,l} + \xi^l{}_{,m} g_{ln} + \xi^l{}_{,n} g_{lm} = 0.$$

We will use coordinates $x^i \equiv (t, r, \theta, \phi)$ and will look for a solution in which $\xi^2 = 0, \xi^3 = 0$. Then we get the following equations:

$$\frac{\partial \xi^0}{\partial t} = 0, \qquad \frac{\partial \xi^0}{\partial r} = \frac{\partial \xi^1}{\partial t} e^{2Ht}, \qquad \frac{\partial \xi^1}{\partial r} + H\xi^0 = 0.$$

It is easy to verify that a solution to these equations is given by

$$\xi^0 = 1, \qquad \xi^1 = -Hr.$$

The fact that $\xi^0 \neq 0$ suggests that a symmetry with translation in the time direction is possible. The vector ξ^i is timelike for $H^2 r^2 e^{2Ht} < 1$.

Exercises

1. Show that the Gödel universe given by the line element

$$ds^2 = (dx^0)^2 + 2e^{x^1} dx^0 dx^2 - (dx^1)^2 + \frac{1}{2} e^{2x^1} (dx^2)^2 - (dx^3)^2$$

has the following Killing vector fields ξ^i:

$$(1, 0, 0, 0), \quad (0, 0, 0, 1), \quad (0, 0, 1, 0), \quad (0, 1, -x^2, 0),$$
$$(-4e^{-x^1}, 2x^2, -(x^2)^2 + 2e^{-2x^1}, 0).$$

Is this spacetime (i) homogeneous, (ii) isotropic and (iii) stationary?

2. Show that the subspaces $t =$ constant of the Heckmann–Schücking spacetime

$$ds^2 = dt^2 + 2e^{x^1} dt dx^2 - c_{11}(t)\{(dx^1)^2 + \alpha e^{2x^1} (dx^2)^2\}$$
$$- 2c_{12}(t)e^{x^1} dx^1 dx^2 - c_{33}(t)(dx^3)^2$$

are its homogeneous subspaces.

3. Show that the integrability condition (6.9) can be written in the form

$$\xi^m R_{ijkl;m} = \xi_{m;l} R^m{}_{kij} - \xi_{m;k} R^m{}_{lij} - \xi_{m;i} R^m{}_{jkl} + \xi_{m;j} R^m{}_{ikl}.$$

Deduce, by multiplication by g^{ik} or otherwise, that

$$\xi^m R_{il;m} = -\xi_{m;l} R_i{}^m + \xi_{m;i} R_l{}^m.$$

4. Find ten independent Killing vectors for the Minkowski spacetime.

5. Show that any Killing vector ξ^i satisfies the equation

$$\Box \xi^i + R^i{}_k \xi^k = 0.$$

6. If T^{ik} is the energy momentum tensor and ξ^i is a timelike Killing vector, show that the integral

$$\int T^i{}_k \xi^k \, d\Sigma$$

over the whole spacelike hypersurface is independent of the choice of the hypersurface.

7. Show that the line element (6.31) for $k = +1$ is manifestly conformal to the Minkowski line element through the following series of transformations (due to

L. Infeld and A. Schild):

$$r = \sin R, \qquad T = \int^t \frac{du}{S(u)}; \qquad \zeta = \frac{1}{2}(T + R), \qquad \eta = \frac{1}{2}(T - R);$$

$$X = \tan \zeta, \qquad Y = \tan \eta; \qquad \tau = \frac{1}{2}(X + Y), \qquad \rho = \frac{1}{2}(X - Y).$$

What transformations will do the same for the case $k = -1$?

Chapter 7
Physics in curved spacetime

7.1 Introduction

Having acquainted ourselves with the trials and tribulations of working in non-Euclidean spacetimes we are now prepared for the next step, that of describing physics in such curved spacetimes. For we recall from Chapter 2 that the Einstein programme for general relativity consists of replacing the Newtonian perception of gravitation as a force by the notion that its effect makes the geometry of spacetime 'suitably non-Euclidean'. What we mean by 'suitably' will be clear in the next two chapters. But *given that the geometry is non-Euclidean* we first need to know how the rest of physics is described in it.

For example, how do we describe the motion of a particle under a *non-gravitational* force? How do we write Maxwell's equations? What is the role of energy-momentum tensors? Can a dynamical action principle be written in curved spacetime? Such questions need our attention before we turn to the basic issue of how gravity actually leads to curved spacetime.

To this end we will introduce a concept that Einstein took as a basic principle in formulating general relativity. It is known as the *principle of equivalence*.

7.2 The principle of equivalence

Let us go back to the purely mathematical result embodied in the relations shown in Section 4.6 and attempt to describe their physical meaning. These relations tell us that special (locally inertial) coordinates that behave like the coordinates (t, x, y, z) of special relativity exist in the neighbourhood of any point P in spacetime. Physically, these coordinates imply a frame of reference in which a momentary illusion is created at P and in a small neighbourhood of P that the geometry is of special relativity. The illusion is momentary and local to P because we have seen that the relations of (4.15) cannot be made to hold everywhere and at all times.

In view of the assertion made in Section 2.1 that gravitation manifests itself as non-Euclidean geometry, we would have to argue that in the above locally inertial frame gravitation has been transformed away momentarily and in a small neighbourhood of P. How does this happen in practice? Consider Einstein's celebrated example of the freely falling lift. A person inside such a lift feels weightless. The accelerated frame of reference of the lift provides the locally inertial frame in the small neighbourhood of the falling person. Similarly, a spacecraft circling around Earth is in fact freely falling in the Earth's gravity, and the astronauts inside it feel weightless. (See Figure 7.1 showing an astronaut floating in space.)

Fig. 7.1. A floating astronaut in the micro-gravity environment of a space shuttle. Photograph by courtesy of NASA.

It should be emphasized that this feeling of weightlessness in a falling lift or a spacecraft is limited to local regions: there is no universal frame that transforms away Earth's gravity everywhere, at all times. If we demand that the relations of (4.15) hold at all points of spacetime, we would need to have $\partial \Gamma^i_{kl} / \partial x^m = 0$ everywhere, leading to $R^i_{klm} = 0$, that is, to a flat spacetime. Thus a curved spacetime with a non-vanishing Riemann tensor is necessary to describe genuine effects of gravitation.

The *weak principle of equivalence* states that effects of gravitation can be transformed away locally and over small intervals of time by using suitably accelerated frames of reference. Thus it is the physical statement of the mathematical relations given by (4.15). It is possible, however, to go from here to a much stronger statement, the so-called *strong principle of equivalence*, which states that any physical interaction (other than gravitation, which has now been identified with geometry) behaves in a locally inertial frame as if gravitation were absent. For example, Maxwell's equations will have their familiar form in a locally inertial frame. Thus an observer performing a local experiment in a freely falling lift would measure the speed of light to be c.

The strong principle of equivalence enables us to extend any physical law that is expressed in the covariant language of special relativity to the more general form it would have in the presence of gravitation. The law is usually expressed in terms of vectors, tensors, or spinors in the Minkowski spacetime of special relativity. All we have to do is to write it in terms of the corresponding entities which are covariant in curved spacetime. Thus, in the flat spacetime of special relativity, the Maxwell electromagnetic field tensor F^{ik} is related to the current vector j^k by

$$F^{ik}_{\ ,i} = 4\pi j^k. \tag{7.1}$$

In curved spacetime the ordinary derivative is replaced by the covariant derivative:

$$F^{ik}_{\ ;i} = 4\pi j^k. \tag{7.2}$$

Notice that the effect of gravitation enters through the Γ^i_{kl} terms that are present in (7.2). This generalization of (7.1) to (7.2) is called the *minimal coupling* of the field with gravitation, since it is the simplest one possible.

So, in order to describe how other interactions behave in the presence of gravitation, we use the covariance under the general coordinate transformation as the criterion to be satisfied by their underlying equations. Thus, it is immediately clear from the example of the electromagnetic field that a light ray describes a null geodesic.

In the same vein we can now describe a moving object that is acted on by no other interaction except gravitation – for example, a probe moving in the gravitational field of the Earth. *In the absence of gravity,* this object would move in a straight line with uniform velocity; that is, with the equation of motion

$$\frac{du^i}{ds} = 0,\qquad(7.3)$$

where u^i is the 4-velocity. In the presence of gravity, (7.3) is modified to our geodesic equation (5.19).

7.3 A uniformly accelerated frame

We now describe another example that provides a clue about how gravitational effects show up in spacetime geometry according to general relativity. Consider the Minkowski spacetime with the standard line element

$$ds^2 = c^2\,dt^2 - dx^2 - dy^2 - dz^2.\qquad(7.4)$$

If we make the coordinate transformation for a constant g,

$$x = \frac{c^2}{g}\left(\cosh\left(\frac{gt'}{c}\right) - 1\right) + x'\cosh\left(\frac{gt'}{c}\right),\qquad(7.5)$$

where

$$y = y',$$
$$z = z',$$
$$t = \frac{c}{g}\sinh\left(\frac{gt'}{c}\right) + \frac{x'}{c}\sinh\left(\frac{gt'}{c}\right).$$

This leads to the line element

$$ds^2 = \left(1 + \frac{gx'}{c^2}\right)^2 dt'^2 - dx'^2 - dy'^2 - dz'^2.\qquad(7.6)$$

What interpretation can we give to (7.6)? The origin of the (x', y', z') system has a world line whose parametric form in the old coordinates is given by

$$x = \frac{c^2}{g}\left(\cosh\left(\frac{gt'}{c}\right) - 1\right),\quad y = 0,\quad z = 0,\quad t = \frac{c}{g}\sinh\left(\frac{gt'}{c}\right).\qquad(7.7)$$

Using the kinematics of special relativity described in Section 1.8 of Chapter 1, it can be easily seen that (7.7) describes the motion of a point that has a uniform acceleration g in the x direction, a point that is momentarily at rest at the origin of (x, y, z) at $t = 0$. We may interpret the line element (7.6) and the new coordinate system as describing the spacetime in the rest frame of the uniformly accelerated observer.

Direct calculation shows that not all Γ^i_{kl} are zero in (7.6) at $x' = 0$, $y' = 0$, $z' = 0$. The frame is therefore non-inertial. For the neighbourhood of the origin, the metric component

$$g_{00} \cong 1 + \frac{2gx'}{c^2} = 1 + \frac{2\phi}{c^2}, \tag{7.8}$$

where ϕ is the Newtonian gravitational potential for a uniform gravitational field that induces an acceleration due to gravity of $-g$. We have here the reverse situation to that of the falling lift: we seem to have generated a pseudo-gravitational field by choosing a suitably accelerated observer. The prefix 'pseudo-' is used because the gravitational field is not real – it is an illusory effect arising from the choice of coordinates. The Riemann tensor for the metric is zero, thus confirming the above statement.

An example of an accelerated frame is provided by a bus or an aircraft starting off from rest. All passengers facing in the forward direction feel a force pressing their backs to their seats. This force 'attracting' them to the seats is illusory and momentary, lasting only so long as the acceleration persists. Astronauts taking off in rocket-driven spaceships feel their weight increase several times at the time of lift off, again because of the initial acceleration. All these examples tell us how intimately related the accelerated frames are to gravity.

Nevertheless the relation (7.8) is also suggestive of the real gravitational field, as we shall see in the following example and later in Chapter 8.

Example 7.3.1 Consider a particle held at rest at the origin $x = 0, y = 0, z = 0$ in the manifestly Minkowski frame (7.4). What is its trajectory in the uniformly accelerated frame (7.6)?

On setting $x = 0$ in Equation (7.5), we get,

$$x' = \frac{c^2}{g} \left(\mathrm{sech} \left(\frac{gt'}{c} \right) - 1 \right),$$

which, for small t', i.e., for $t' \ll c/g$, approximates to

$$x' = -\frac{1}{2}gt'^2.$$

Thus to an observer at rest in the accelerated frame, the particle will appear to have a 'free fall' in the negative x' direction, and the observer will ascribe this to gravity in that direction.

Example 7.3.2 *Problem.* A particle of unit rest mass is uniformly accelerated as described in Section 7.3. Show that, at time t, its energy has grown to $\gamma(t)c^2$, where

$$\gamma(t) = \left(1 + \frac{g^2 t^2}{c^2}\right)^{1/2}.$$

Solution. Without loss of generality we take the particle at $x' = 0$. Using (7.7) we get $dt/ds = \cosh(gt'/c)$. We may identify dt/ds with the energy of motion per unit mass. That is, $\gamma(t) = \cosh(gt'/c)$. Using (7.7) again to relate t' to t, we get

$$\cosh^2\left(\frac{gt'}{c}\right) = 1 + \sinh^2\left(\frac{gt'}{c}\right) = 1 + \left(\frac{gt}{c}\right)^2.$$

From this we get the required result

$$\gamma(t) = \left(1 + \frac{g^2 t^2}{c^2}\right)^{1/2}.$$

7.4 The action principle and the energy-momentum tensors

Let us now see how we can write the laws of physics in the covariant language in a Riemannian spacetime using the strong principle of equivalence. We take the familiar example of charged particles interacting with the electromagnetic field. The physical laws can be derived from an action principle. First we write the action in Minkowski spacetime:

$$A = -\sum_a cm_a \int ds_a - \frac{1}{16\pi c}\int F_{ik}F^{ik}\, d^4x - \sum_a \frac{e_a}{c}\int A_i\, dx_a^{\,i}. \tag{7.9}$$

Here we assume that the action describes physics in a volume $\mathcal{V}$ of spacetime bound by surface Σ. All variations of physical quantities are supposed to vanish on Σ. A_i are the components of the 4-potential, which are related to the field tensor F_{ik} by

$$A_{k,i} - A_{i,k} = F_{ik}, \tag{7.10}$$

while e_a and m_a are the charge and rest mass of particle a, whose coordinates are given by x_a^i and the proper time by s_a with

$$ds_a^2 = \eta_{ik}\, dx_a^{\,i}\, dx_a^{\,k}. \tag{7.11}$$

How do we generalize (7.9) to Riemannian spacetime? First, we note that η_{ik} in (7.11) are replaced by g_{ik}. Next, starting from the covariant vector A_i, we generate F_{ik} by the covariant generalization of (7.10):

$$A_{k;i} - A_{i;k} = F_{ik}. \tag{7.12}$$

However, since the expression (7.12) is antisymmetric in (i, k), the extra terms involving the Christoffel symbols cancel out and we are back to (7.10)! The volume integral in (7.9) is modified to

$$\int F_{ik} F^{ik} \sqrt{-g}\, \mathrm{d}^4 x. \tag{7.13}$$

The extra factor $\sqrt{-g}$ has crept in because the combination

$$\sqrt{-g}\, \mathrm{d}x^1\, \mathrm{d}x^2\, \mathrm{d}x^3\, \mathrm{d}x^0 = \frac{1}{24} e_{ijkl}\, \mathrm{d}x^i\, \mathrm{d}x^j\, \mathrm{d}x^k\, \mathrm{d}x^l \tag{7.14}$$

acts as a scalar. (Refer to Example 3.3.4.) We therefore have the following generalized form of (7.9):

$$\mathcal{A} = -\sum_a cm_a \int \mathrm{d}s_a - \frac{1}{16\pi c} \int F_{ik} F^{ik} \sqrt{-g}\, \mathrm{d}^4 x - \sum \frac{e_a}{c} \int A_i\, \mathrm{d}x_a^i. \tag{7.15}$$

The variation of the world line of particle a gives its equation of motion

$$\frac{\mathrm{d}^2 x_a^i}{\mathrm{d}s_a^2} + \Gamma_{kl}^i \frac{\mathrm{d}x_a^k}{\mathrm{d}s_a} \frac{\mathrm{d}x_a^l}{\mathrm{d}s_a} = \frac{e_a}{m_a} F^i{}_l \frac{\mathrm{d}a^l}{\mathrm{d}s_a}, \tag{7.16}$$

while the variation of A_i gives the field equations (7.2).

We summarize the situation by stating a general rule. Whatever variables we introduce to specify the dynamics of the observed situation, we apply the principle of stationarity of action for small variations of these variables so that we end up knowing the 'equations of motion' that specify how these variables change over space and time.

7.5 Variation of the metric tensor

The transition from (7.9) to (7.15) has, however, introduced an additional independent feature into the action, besides the particle world lines and the potential vector. The new feature is the spacetime geometry typified by the metric tensor g_{ik}. We argue, quite plausibly, that the entire problem is specified not just by the dynamical and field variables, but also by the spacetime geometry. What will happen if we demand that the g_{ik} are also dynamical variables and that the action $\mathcal{A}$ remains stationary for small variations of the type

$$g_{ik} \to g_{ik} + \delta g_{ik}? \tag{7.17}$$

From the generalized action principle, should we not expect to get the equations that determine the g_{ik}, and through them the spacetime geometry? Let us investigate.

A glance at the action (7.15) shows that the last term does not contribute anything under (7.17) if we keep the world lines and A_i fixed in spacetime. The first two terms, however, do make contributions. Let us consider them in that order. First note that

$$\delta(ds_a^2) = \delta g_{ik}\, dx_a^{\ i}\, dx_a^{\ k},$$

that is,

$$\delta(ds_a) = \frac{1}{2}\, \delta g_{ik}\, \frac{dx_a^{\ i}}{ds_a}\, \frac{dx_a^{\ k}}{ds_a}\, ds_a.$$

Therefore,

$$\delta \sum_a cm_a \int ds_a = \frac{1}{2}\sum_a c \int m_a \frac{dx_a^{\ i}}{ds_a}\frac{dx_a^{\ k}}{ds_a}\, ds_a\, \delta g_{ik}. \qquad (7.18)$$

Let us consider this variation in a small 4-volume $\mathcal{V}$ near a point P. If we look at a locally inertial coordinate system near P we can identify the above expression in a more familiar form. Let us first identify

$$p_{(a)}^i = cm_a \frac{dx_a^{\ i}}{ds_a}$$

as the 4-momentum of particle a. Then $cp_{(a)}^0 = E_a$ is the energy of the particle, and we get

$$\frac{1}{2}cm_a \frac{dx_a^{\ i}}{ds_a}\frac{dx_a^{\ k}}{ds_a}\, ds_a = \frac{c^2}{2E_a}p_{(a)}^i p_{(a)}^k\, dt_a = \frac{c}{2E_a}p_{(a)}^i p_{(a)}^k\, dx_a^0.$$

Figure 7.2 shows the volume $\mathcal{V}$ as a shaded region in the neighbourhood of P, t being the local time coordinate and x^μ ($\mu = 1, 2, 3$) the local rectangular space coordinates. We will shortly discuss the various cases described in Figure 7.2. The expression (7.18) can then be looked upon as a volume integral over $\mathcal{V}$ of the form

$$\delta \sum_a cm_a \int ds_a = \frac{1}{2c}\int_{\mathcal{V}} \delta g_{ik} T_{(m)}^{ik}\, d^4x, \qquad (7.19)$$

where $T_{(m)}^{ik}$ is the sum of the expressions like

$$\frac{c^2}{E_a}p_{(a)}^i p_{(a)}^k$$

for each particle a that crosses a unit volume of the shaded region near P. We now interpret this sum under various conditions. In each case the trick is to look at the problem in the locally inertial frame and then transform to a general frame.

7.5.1 The energy tensor of matter

This expression for T_{ik} is none other than the usual expression for the energy tensor of matter (also called the *energy-momentum tensor* or the

Fig. 7.2. In (a) we have matter particles moving along parallel worldlines, with no collisions and very little relative motion. In (b) we see particles moving relativistically in random directions, while in (c) we have an intermediate situation wherein particles have small, random, relative motions.

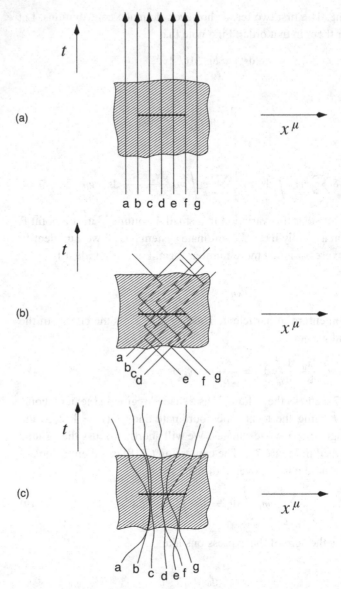

stress energy tensor). Since we will need this tensor frequently, it is derived below for three different types of matter.

Dust

This is the simplest situation, in which all of the particle world lines going through the shaded region in Figure 7.2(a) are more or less parallel, indicating that the particles of matter are moving without any relative motion in the neighbourhood of P. Writing the typical 4-velocity as u^i and using a Lorentz transformation to make $u^i = (1, 0, 0, 0)$ (that is, transforming to the rest frame of the dust), the only non-zero component of the energy tensor is

$$T^{00} = \sum_a m_a c^2 = \rho_0 c^2,$$

where the summation is over a unit volume in the neighbourhood of P. Here ρ_0 is the rest mass density of dust. In any other Lorentz frame we get

$$\underset{(m)}{T^{ik}} = \rho_0 c^2 u^i u^k, \tag{7.20}$$

an expression that is easily generalized to any (non-Lorentzian) reference system.

Relativistic particles

This situation, described in Figure 7.2(b), represents the opposite extreme. Here we have highly relativistic particles moving at random through $\mathcal{V}$. The 4-momentum of a typical particle is then approximated to the form

$$p^i = \left(\frac{E}{c}, P\right), \qquad E^2 = c^2 |\mathbf{P}|^2 + m^2 c^4 \cong c^2 P^2, \qquad P = |\mathbf{P}|.$$

Here m is the rest mass of a typical particle. In the highly relativistic approximation we have $|\mathbf{P}| \gg mc$.

Using the fact that the particles are moving randomly, we find that the energy tensor has pressure components also:

$$T^{00} = \sum_a E_a = \epsilon,$$

$$T^{11} = T^{22} = T^{33} = \sum_a \frac{P^2 c^2}{3 E_a} \cong \sum_a \frac{1}{3} E_a. \tag{7.21}$$

The factor 1/3 comes from randomizing in all directions. These are the only non-zero pressure components. Here ϵ is the energy density. Thus for highly relativistic particles we get

$$\underset{(m)}{T^{ik}} = \text{diag}(\epsilon, \epsilon/3, \epsilon/3, \epsilon/3). \tag{7.22}$$

This form is applicable to randomly moving neutrinos or photons.

Fluid

This situation is illustrated in Figure 7.2(c) and consists of a collection of particles with small (non-relativistic) random motions. If we choose the locally inertial frame in which the fluid as a whole is at rest as the frame of reference, we can evaluate the components of $T^{ik}_{(m)}$ as follows. Let a typical particle have the 4-momentum vector given by

$$p^0 = \frac{mc^2}{\sqrt{1 - \dfrac{v^2}{c^2}}}, \qquad p^\mu = \frac{m\mathbf{v}}{\sqrt{1 - \dfrac{v^2}{c^2}}} \qquad (\mu = 1, 2, 3). \tag{7.23}$$

Then

$$T^{00} = \sum mc^2 \left(1 - \frac{v^2}{c^2}\right)^{-1/2} \cong \sum mc^2 \left(1 + \frac{v^2}{2c^2}\right) = \rho c^2,$$

$$T^{11} = T^{22} = T^{33} = \frac{1}{3} \sum mv^2 \left(1 - \frac{v^2}{c^2}\right)^{-1/2} \cong p. \tag{7.24}$$

Here ρ and p are the density and pressure of the fluid. In a frame of reference in which the fluid as a whole has a 4-velocity u^i, the energy tensor becomes

$$T^{ik}_{(m)} = (p + \rho c^2)u^i u^k - p\eta^{ik}. \tag{7.25}$$

The generally covariant form of (7.25) is obviously

$$T^{ik}_{(m)} = (p + \rho c^2)u^i u^k - pg^{ik}. \tag{7.26}$$

Note that ρ is not just the rest-mass density, but also includes the energy density of internal motion, as seen in (7.24).

We may now relax our restriction to the locally inertial coordinate system at P. The generalized form of (7.19) is then

$$\delta \sum_a cm_a \int ds_a = \frac{1}{2c} \int T^{ik}_{(m)} \sqrt{-g}\, \delta g_{ik}\, d^4x. \tag{7.27}$$

7.5.2 The energy tensor of the electromagnetic field

We next consider the variation of the second term of (7.9). If we keep A_i fixed, the F_{ik}, as given by (7.12) or (7.10), remain unchanged under the variation of g_{ik}. Hence

$$\delta(F_{ik}F^{ik}\sqrt{-g}) = F_{ik}F_{lm}\, \delta(g^{il}g^{km}\sqrt{-g}).$$

From the basic definition we get

$$\delta g^{ik} g_{kl} = -g^{ik}\, \delta g_{kl},$$

that is,

$$\delta g^{ik} = -g^{im}g^{kn}\, \delta g_{mn}. \tag{7.28}$$

Also, from (4.12) we have

$$\delta\sqrt{-g} = \frac{1}{2}g^{ik}\sqrt{-g}\,\delta g_{ik}. \tag{7.29}$$

Substituting these expressions into the variation of the second term of the action gives

$$\delta\frac{1}{16\pi c}\int_V F_{ik}F^{ik}\sqrt{-g}\,\mathrm{d}^4x = \frac{1}{2c}\int_V T_{(em)}^{\ \ ik}\sqrt{-g}\,\delta g_{ik}\,\mathrm{d}^4x, \tag{7.30}$$

with the electromagnetic energy tensor given by

$$T_{(m)}^{\ \ ik} = \frac{1}{4\pi}\left(\frac{1}{4}F_{mn}F^{mn}g^{ik} - F^i_{\ l}F^{lk}\right). \tag{7.31}$$

The above two examples can be generalized to any field Λ that is described by an action

$$\mathcal{A}_\Lambda = \int L_\Lambda\sqrt{-g}\,\mathrm{d}^4x. \tag{7.32}$$

Here L_Λ is the *Lagrangian density* of Λ. The variation of $\mathcal{A}_\Lambda$ may be written as

$$\delta\mathcal{A}_\Lambda = -\frac{1}{2c}\int T_{(\Lambda)}^{\ \ ik}\sqrt{-g}\,\delta g_{ik}\,\mathrm{d}^4x. \tag{7.33}$$

This may be taken as a formal definition of T^{ik}, the energy-momentum tensor.

In theories defined only in Minkowski spacetime the appearance of energy tensors is somewhat *ad hoc*. They do not enter explicitly into any dynamic or field equations. They appear only through their divergences, the typical rule for conservation of energy and momentum being given by $T^{ik}_{\ \ ,k} = 0$. In our curved spacetime framework the T^{ik} find a natural expression through the variation of g_{ik}. Moreover, as we shall show next, the above derivation of the T^{ik} leads to the zero-divergence equation as an automatic consequence.

7.5.3 Conservation of energy and momentum

We begin with the observation that L_Λ in the action leading to (7.32) is a scalar quantity, so any change of coordinates does not change it. Using this result, we make an infinitesimal change of coordinates:

$$x'^i = x^i + \xi^i, \tag{7.34}$$

where the ξs are infinitesimally small. Clearly, for such a coordinate change, the change $\delta\mathcal{A}_\Lambda$ in the action will be zero. But we can express the change in another way. The coordinate change introduces a change of metric tensor implying that for the same geometry there will in general be a non-zero δg_{ik}. So we will get the change of action as in (7.33). We will therefore evaluate it first. For brevity we will denote $\partial \xi^l / \partial x^k$ by $\xi^l_{,k}$.

Tensor transformation law will give

$$\frac{\partial x^l}{\partial x'^i} \approx \delta^l_i - \xi^l_{,i},$$

so we get

$$g'_{ik}(x^i + \xi^i) = (\delta^l_i - \xi^l_{,i})(\delta^m_k - \xi^m_{,k}) \times g_{lm}.$$

Expand the left-hand side by Taylor expansion around x^i retaining only up to the first-order term in ξ^i. On the right-hand side likewise retain terms of that order only to get

$$\delta g_{ik} \equiv g'_{ik}(x^i) - g_{ik}(x^i) = -g_{im}\xi^m_{,k} - g_{lk}\xi^l_{,i} - g_{ik,l}\xi^l.$$

Now convert the ordinary derivatives of ξ into covariant derivatives by adding the terms with Christoffel symbols and use the identities of Section 4.5 to express the derivative $g_{ik,l}$ in terms of the same symbols. A simple manipulation along these lines leads to a result somewhat similar to that of Chapter 6 (see Equation (6.6) there):

$$\delta g_{ik} = -[\xi_{i;k} + \xi_{k;i}]. \tag{7.35}$$

We therefore get the change in action, using Equation (7.33), as

$$\delta \mathcal{A}_\Lambda = \frac{1}{2c} \int_{(\Lambda)} T^{ik} \sqrt{-g} \, \delta g_{ik} \, d^4x \tag{7.36}$$

$$= -\frac{1}{2c} \int_{(\Lambda)} T^{ik} \sqrt{-g} \, [\xi_{i;k} + \xi_{k;i}] d^4x. \tag{7.37}$$

Since T_{ik} is a symmetric tensor, this expression can be further simplified and rewritten (after suppressing the suffix Λ) as

$$\delta \mathcal{A}_\Lambda = -\frac{1}{c} \int [(T^{ik}\xi_i)_{;k} - \xi_i T^{ik}_{;k}] \sqrt{-g} \, d^4x. \tag{7.38}$$

Of the two terms inside the square brackets, the first gets transformed to a surface integral by Green's theorem and, since in the variational process changes like ξ_i are supposed to vanish on the boundary, we are left with the second term only. Since we expect $\delta \mathcal{A}_\Lambda$ to vanish for arbitrary ξ_i, we conclude that

$$T^{ik}_{;k} \equiv 0, \tag{7.39}$$

i.e., the energy-momentum tensor is conserved.

Notice that this result was deduced from the scalar property of the action, that is, from its invariance with respect to coordinate transformation. This is a 'symmetry' property of the action and the above result may be seen as an example of the general theorem due to Emily Noether, which states that for every symmetry of the action there is a conservation law. We encounter several examples of Noether's theorem in theoretical physics.

Example 7.5.1 *Problem.* For a scalar field with Lagrangian density

$$L = \frac{1}{2}\phi_{,i}\phi_{,k}g^{ik}$$

derive the energy-momentum tensor.

Solution. By performing the variation of g_{ik}, g^{ik}, etc. we get

$$\delta A_\phi = \delta \int \frac{1}{2}\phi_{,i}\phi_{,k}g^{ik}\sqrt{-g}\, d^4x = \frac{1}{2}\int \phi_{,i}\phi_{,k}\,\delta(g^{ik}\sqrt{-g})\, d^4x.$$

Using the result

$$\delta(g^{ik}\sqrt{-g}) = \delta g^{ik}\sqrt{-g} + g^{ik}\,\delta\sqrt{-g}$$

$$= \delta g^{ik}\sqrt{-g} - \frac{1}{2}\sqrt{-g}g_{pq}\,\delta g^{pq}\,g^{ik}$$

we get

$$\delta A_\phi = \frac{1}{2}\int \phi_{,i}\phi_{,k}\left[\delta g^{ik}\sqrt{-g} - \frac{1}{2}\sqrt{-g}g_{pq}\,\delta g^{pq}\,g^{ik}\right] d^4x$$

$$= \frac{1}{2}\int T_{ik}\,\delta g^{ik}\sqrt{-g}d^4x,$$

where

$$T_{ik} = \phi_{,i}\phi_{,k} - \frac{1}{2}g_{ik}\phi^{,l}\phi_{,l}.$$

This is the required energy-momentum tensor for the scalar field.

Problem. Show that, if the Lagrangian density L of a field explicitly depends on g_{ik} and $g_{ik,l}$, then the corresponding energy tensor is given by

$$T^{ik} = 2\left[\left(\frac{\partial L}{\partial g_{ik,l}}\right)_{,l} + \frac{1}{2}\frac{\partial L}{\partial g_{ik,l}}g^{mn}g_{mn,l} - \frac{\partial L}{\partial g_{ik}} - \frac{1}{2}Lg^{ik}\right].$$

Solution. We have, from (7.33) and (7.29),

$$\delta(L\sqrt{-g}) = \left[\frac{\partial L}{\partial g_{ik}}\delta g_{ik} + \frac{\partial L}{\partial g_{ik,l}}\delta g_{ik,l}\right]\sqrt{-g} + \frac{1}{2}L\sqrt{-g}g^{ik}\,\delta g_{ik}.$$

However, $\delta g_{ik,l}$ may be written as $(\delta g_{ik})_{,l}$ and one can use the divergence theorem to get

$$\int_V \frac{\partial L}{\partial g_{ik,l}}\delta g_{ik,l}\sqrt{-g}\, d^4x = -\int_\vartheta \left\{\frac{\partial L}{\partial g_{ik,l}}\sqrt{-g}\right\}_{,l}\delta g_{ik}\, d^4x.$$

Here we have used the fact that the variations δg_{ik} vanish on the boundary of $\mathcal{V}$, so that the surface integral on the right-hand side vanishes. Hence from (7.33) we get

$$T^{ik} = 2\left[\frac{1}{\sqrt{-g}}\left\{\frac{\partial L}{\partial g_{ik,l}}\sqrt{-g}\right\}_{,l} - \frac{\partial L}{\partial g_{ik}} - \frac{1}{2}Lg^{ik}\right].$$

The stated answer follows when we recall that the first term on the right-hand side contains $\sqrt{-g}$, and it gives $(-g)_{,l} = \frac{1}{2}\sqrt{-g}g^{mn}g_{mn,l}$.

It was this variation of the metric tensor that led Hilbert to derive the field equations of general relativity shortly after Einstein had proposed them from heuristic considerations. We now turn our attention to this topic in the following chapter.

Exercises

1. Calculate the energy-momentum tensor for the scalar field ϕ given by the action integral

$$\int (\phi_{;i}\phi_{;k}g^{ik} + m^2\phi^2)\sqrt{-g}\,\mathrm{d}^4x,$$

where $m = $ constant. (m is usually identified with the mass of the field.)

2. Fluid with isotropic pressure p and density ρ fills a spherically symmetric region with the line element

$$\mathrm{d}s^2 = \mathrm{e}^{\nu}\,\mathrm{d}t^2 - \mathrm{e}^{\lambda}\,\mathrm{d}r^2 - \mathrm{e}^{\mu}(\mathrm{d}\theta^2 + \sin^2\theta\,\mathrm{d}\phi^2),$$

where λ, μ and ν depend on r and t only. From the conservation law deduce the relations

$$\frac{\partial}{\partial t}(\lambda + 2\mu) = -\frac{2}{p + \rho}\frac{\partial\rho}{\partial t}; \qquad \frac{\partial\nu}{\partial r} = -\frac{2}{p + \rho}\frac{\partial p}{\partial r}.$$

(The velocity of light is unity.)

3. Dust of density $\rho(t)$ and radiation of density $u(t)$ fill the spacetime given by the line element

$$\mathrm{d}s^2 = \mathrm{d}t^2 - S^2(t)\left[\frac{\mathrm{d}r^2}{1 - kr^2} + r^2(\mathrm{d}\theta^2 + \sin^2\theta\,\mathrm{d}\phi^2)\right],$$

where $k = 1$, 0 or -1. From the conservation law deduce that

$$\frac{1}{S^3}\frac{\partial}{\partial t}(\rho S^3) + \frac{1}{S^4}\frac{\partial}{\partial t}(uS^4) = 0.$$

(The velocity of light is unity.)

4. Verify by direct calculation that the divergence of the electromagnetic energy-momentum tensor vanishes everywhere except at the location of the charged

particles. By a suitable limiting process deduce the equations of motion of the electric charge by evaluating

$$T^{ik}_{;k(m)} + T^{ik}_{;k(em)}$$

at the particle.

5. The action $\mathcal{A}_\Lambda$ is *conformally invariant*, i.e., it does not change when the spacetime metric g_{ik} is changed to $\Omega^2 g_{ik}$, where Ω is a well-behaved function of x^i and $0 < \Omega < \infty$. Show that the trace of $T^{ik}_{(\Lambda)}$ vanishes identically. (The trace of a tensor A^{ik} is $g_{ik}A^{ik}$.)

6. Show that for dust $T^{ik} = \rho u^i u^k$ conservation means that u^i follows a geodesic.

7. Suppose that in a specific coordinate system the metric g_{ik} is independent of x^1. Show that the conservation law $T^i_{1;i} = 0$ for the energy-momentum tensor becomes expressible as

$$\frac{1}{\sqrt{-g}} \frac{\partial}{\partial x^i}(\sqrt{-g}T^i_1) = 0.$$

(Both T^i_k and g_{ik} are assumed diagonal.)

8. Show by direct calculation that Maxwell's equations are conformally invariant. Work out how masses of electric charges must transform if the Maxwell–Lorentz equation of motion also is conformally invariant.

9. Calculate the form of the energy-momentum tensor for a plane electromagnetic wave in Minkowski spacetime.

10. Write down Poynting's theorem in the older three-dimensional form of the electromagnetic theory. Work out its form in the four-dimensional notation of special relativity and generalize it to curved spacetime.

Chapter 8
Einstein's equations

8.1 A heuristic approach

The preceding chapter showed that the variation of the action $\mathcal{A}$ with respect to g_{ik} leads us to the energy tensor of various interactions. We still do not have dynamical equations that tell us how to determine the g_{ik} in terms of the distribution of matter and energy. It was Einstein's conjecture that the energy tensors should act as the 'sources' of gravity. Thus what we have so far achieved is identification of sources of gravitation. But we further need the basic variables *whose sources* are these T_{ik}. Einstein felt that the variables are not to be found in physics *but in the geometry of spacetime*. We have already seen that the basic measurements of the geometry are carried out through the g_{ik}, the concept of parallelism is expressed through the $\Gamma^i{}_{kl}$ while spacetime curvature appears through the Riemann tensor R_{iklm}. Einstein reasoned in a heuristic way to arrive at equations linking these quantities to the energy-momentum tensors. Below we capture the reasoning he used.

Following the general trend of nineteenth-century physics, especially the Maxwell equations, Einstein looked for an expression that would act like a wave equation for g_{ik}, with T_{ik} as the source. It is immediately clear that the standard wave equation in the covariant form

$$g^{mn} g_{ik;mn} = \kappa T_{ik}, \tag{8.1}$$

where κ is a constant, *will not do*, for the left-hand side vanishes identically. In fact any covariant linear combination of the first and second derivatives of the metric tensor will be expressed in terms of their covariant derivatives and will vanish because of the identity $g_{ik;l} \equiv 0$. However,

116

if we go to covariant non-linear expressions involving ordinary derivatives, this need not be so.

Is there a second-rank tensor symmetric in its indices (like the T_{ik}) that involves second derivatives of g_{ik} in a *non-linear* form? Does such a tensor appear naturally when one studies the geometry of spacetime? Clearly, if the tensor is to bring out the special feature of curvature of spacetime, it must be related to the Riemann tensor. Einstein first tried R_{ik}, writing his equations as

$$R_{ik} = \text{constant} \times T_{ik}. \tag{8.2}$$

In order to ensure that energy and momentum are conserved, he had to impose the additional requirement that the right-hand side of these equations have a zero divergence. However, after some trial and error he improved on this conjecture, finally arriving at the tensor G_{ik} of (5.11). His field equations of general relativity, published in 1915 (see Reference [8]), took the form

$$R_{ik} - \frac{1}{2} g_{ik} R \equiv G_{ik} = -\kappa T_{ik}. \tag{8.3}$$

The constant κ is to be determined by the requirement that the above equations resemble Newton's when describing slow motion ($v \ll c$) in a weak gravitational field. We will return to this problem in Section 8.3.

These equations have the added advantage that in view of the Bianchi identities in (5.13) all solutions of these equations must satisfy the condition

$$T^{ik}_{\;;k} \equiv 0. \tag{8.4}$$

That is, the law of conservation of energy and momentum follows naturally from (8.3).

Although there are ten Einstein equations for ten unknown g_{ik}, the divergence condition of (8.4) reduces the number of independent equations to six. This underdeterminacy of the problem can be related to the general covariance of the theory: if g_{ik} is a solution, then so is any tensor transform of g_{ik} obtained through a change of coordinates. In short, there is a degeneracy of solutions: several apparently different solutions represent the same physical reality. One solution in this set can be obtained from another by a suitable coordinate transformation.

The expression (8.4) follows for any T^{ik} obtained from an action principle by the variation of g_{ik} as found in the last chapter. As mentioned there, this result is an example of Noether's theorem, which relates a conservation law to a basic symmetry. In this particular case the symmetry is that of coordinate invariance. It is therefore pertinent to ask whether the Einstein tensor can also be obtained naturally by deriving

Equations (8.3) from an action principle. This problem was solved by Hilbert [9] soon after Einstein proposed his equations of gravitation.

8.2 The Hilbert action principle

If we wish to derive the Einstein tensor from an action principle, we naturally look for a scalar of geometrical origin that contains up to first derivatives of g_{ik}. No such scalar exists! However, if we go to second derivatives, then the simplest scalar is R. It was therefore taken as the starting point by Hilbert for his action principle.

Hilbert's problem can be posed as follows. Consider the variation of the term defined over a spacetime volume $\mathcal{V}$,

$$\int_{\mathcal{V}} R\sqrt{-g}\, \mathrm{d}^4 x \tag{8.5}$$

for $g^{ik} \to g^{ik} + \delta g^{ik}$ with the restriction that δg^{ik} and $\delta g^{ik}{}_{,l}$ vanish on the boundary of $\mathcal{V}$. We now show that

$$\delta \int_{\mathcal{V}} R\sqrt{-g}\, \mathrm{d}^4 x = \int_{\mathcal{V}} \delta g^{ik}\left(R_{ik} - \frac{1}{2}g_{ik}R\right)\sqrt{-g}\, \mathrm{d}^4 x$$

$$= -\int_{\mathcal{V}} \delta g_{ik}\left(R^{ik} - \frac{1}{2}g^{ik}R\right)\sqrt{-g}\, \mathrm{d}^4 x. \tag{8.6}$$

To show this, first note that, under the variation $g_{ik} \to g_{ik} + \delta g_{ik}$, the variation $\delta\Gamma^i{}_{kl}$ transforms as a tensor. This follows on applying the transformation formula (4.5) to both $\Gamma^i{}_{kl}$ and $\Gamma^i{}_{kl} + \delta\Gamma^i{}_{kl}$ and taking the difference. The second 'gamma-independent' terms on the right-hand side of both the formulae are cancelled out, leaving a pure tensor transformation law.

Coming now to the main result, we write

$$R = R_{ik}g^{ik}$$

so that

$$\delta R = \delta R_{ik} + R_{ik}\,\delta g^{ik}.$$

Thus we can deduce (8.6) provided we can show that $\int_{\mathcal{V}} \delta R_{ik}\, g^{ik}\sqrt{-g}\, \mathrm{d}^4 x = 0$. To prove this result, use a locally inertial coordinate system to deduce that Equation (5.9) leads to

$$\sqrt{-g}g^{ik}\,\delta R_{ik} = -\sqrt{-g}[(g^{ik}\,\delta\Gamma^l{}_{ik})_{,l} - (g^{il}\,\delta\Gamma^k{}_{ik})_{,l}] = \sqrt{-g}w^l{}_{,l},$$

where we write

$$w^l = g^{ik}\,\delta\Gamma^l{}_{ik} - g^{il}\,\delta\Gamma^k{}_{ik}.$$

Here w^k is seen to be a vector since it involves terms like $\delta \Gamma^i_{kl}$ that are tensors. Then use Green's theorem over the specified volume to show that, since the variations of gammas are supposed to vanish on the boundary, the variation part δR_{ik} in the above expression leads to zero. Note that, because locally the Γ symbols are zero, we can write $w^l_{,l} = w^l_{;l}$.

Caution. There has been a subtle departure from the usual variational procedure here! Normally the Lagrangian in the action is limited to first derivatives of any dynamical variable, so the variations of that variable (and not its derivative) are assumed to vanish on the boundary. Here the Lagrangian contains second derivatives of the metric tensor, so we need both $\delta g_{ik,l}$ and δg_{ik} to vanish on the boundary. In short, we are dealing here with a variational problem involving derivatives one level higher than in the standard Euler–Lagrange variational problem.

One way of avoiding having second derivatives in the action is to replace R by

$$g^{il}(\Gamma^k_{il}\Gamma^n_{kn} - \Gamma^n_{im}\Gamma^m_{ln})$$

in the integral (8.5). One can show that the modified action also leads to the same Einstein equations. However, the modified integrand suffers from one defect: it is not a scalar!

Ignoring these hiccups, it follows that Einstein's equations can be derived from an action principle if we add to $\mathcal{A}$ the term

$$\mathcal{A}_{\text{gravitation}} = \frac{1}{2\kappa c} \int_{\mathcal{V}} R \sqrt{-g}\, \mathrm{d}^4 x. \tag{8.7}$$

Further, if to the scalar R we add a constant (2λ, say) that is trivially a scalar, we get a modified set of field equations:

$$R_{ik} - \frac{1}{2} g_{ik} R + \lambda g_{ik} = -\kappa T_{ik}. \tag{8.8}$$

We may consider this equation as representing the variation of action (8.7) in a spacetime region of prescribed volume, with λ playing the role of a Lagrangian undetermined multiplier. We will consider these equations only when we discuss cosmology, since the extra term (the λ-term) has cosmological significance. λ is often referred to as the 'cosmological constant'. For the time being we move on to (8.3) and relate κ to known physical constants.

8.3 The Newtonian approximation

The important question of the magnitude of κ can be settled by examining the relationship between general relativity and Newtonian

gravitation. The first hint of a connection between Newtonian gravitation and the present theory was provided by (7.8), where we saw
that, provided g_{00} did not differ significantly from unity, the difference
$(g_{00} - 1)$ is proportional to the Newtonian gravitation potential. We
now seek to formalize this relationship and thereby determine κ. We
will show that in the so-called *slow-motion + weak-field approximation*,
general relativity reduces to Newtonian gravitation.

This approximation is quantified by the following assumptions.

1. The motions of particles are non-relativistic: $v \ll c$. In this case we are back
   to Newtonian mechanics.
2. The gravitational fields are weak in the sense that

$$g_{ik} = \eta_{ik} + h_{ik}, \qquad |h_{ik}| \ll 1. \tag{8.9}$$

The inequality suggests that we ignore powers of $|h_{ik}|$ higher than 2 in the
action principle and higher than 1 in the field equations. We expect this to
lead to a spacetime geometry not very different from Euclid's.

3. The fields change slowly with time. This means we ignore time derivatives
   in comparison with space derivatives. This assumption asks us to ignore
   possible effects of gravitational waves. In a sense, this approximation brings
   us back to the Newtonian concept of instantaneous action at a distance.

Let us now see how the action is simplified under these approximations. First note that, with $x^0 = ct$,

$$ds^2 = (\eta_{ik} + h_{ik})dx^i \, dx^k \approx (1 + h_{00})c^2 \, dt^2 - v^2 \, dt^2,$$

that is,

$$ds \approx \left(\sqrt{1 + h_{00} - \frac{v^2}{c^2}} \right) c \, dt \approx \left(1 + \frac{1}{2} h_{00} - \frac{v^2}{2c^2} \right) c \, dt. \tag{8.10}$$

We next look at the term involving the scalar curvature. The linearized
expression for the Riemann tensor (see (5.5)) is

$$R_{iklm} \approx \frac{1}{2} (h_{kl,im} + h_{im,kl} - h_{km,il} - h_{il,km}). \tag{8.11}$$

The corresponding values of R_{ik} and R can also be calculated. However,
care is needed if we are to look at the action principle rather than the field
equations in this approximation, for we expect quadratic expressions in
the h_{ik} to appear in the geometrical term (8.7).

Item 3 above eliminates time derivatives altogether. Further, the
ratios of typical space and time displacements are $\delta x^\mu / \delta x^0 = v^\mu / c$,
where v^μ are typical Newtonian velocities. Thus h_{00} is more important

than any other h_{ik}, at least by the factor (c/v). We will henceforth ignore all other h_{ik} in comparison with h_{00}. We then get

$$g^{00} \approx 1 - h_{00}, \tag{8.12}$$

$$\sqrt{-g} \approx 1 + \frac{1}{2} h_{00} \tag{8.13}$$

and

$$R \sqrt{-g} \approx - \left(1 - \frac{1}{2} h_{00} \right) \nabla^2 h_{00}. \tag{8.14}$$

Using these relations, we finally get the approximate action as

$$\mathcal{A} \approx -\frac{1}{2\kappa} \int \int \left(1 - \frac{1}{2} h_{00} \right) \nabla^2 h_{00} \, d^3x \, dt - \sum \frac{1}{2} mc^2 \int h_{00} \, dt$$
$$+ \sum \frac{1}{2} m \int v^2 \, dt + \text{constant}. \tag{8.15}$$

The constant represents path-independent terms that can be ignored in a variational problem. Here we have switched over to Newtonian three-dimensional notation, dropping particle labels a, b, ... and using the 3-vector $\mathbf{x}$ to denote x^μ ($\mu = 1, 2, 3$). We can use Green's theorem and ignore surface terms. Thus, in the three-dimensional spatial volume, we get

$$\int_{\text{3-volume}} \left(1 - \frac{1}{2} h_{00} \right) \nabla^2 h_{00} \, d^3x = \int_{\text{2-surface}} \left(1 - \frac{1}{2} h_{00} \right) \nabla h_{00} \, d\mathbf{S}$$

$$= \frac{1}{2} \int_{\text{3-volume}} (\nabla h_{00})^2 \, d^3x.$$

Since we are dynamically interested only in the 3-volume term, we ignore the surface term. Hence

$$\mathcal{A} \approx -\frac{1}{4\kappa} \int \int (\nabla h_{oo})^2 \, d^3x \, dt - \sum \frac{1}{2} mc^2 \int h_{00} \, dt + \sum \frac{1}{2} m \int v^2 \, dt. \tag{8.16}$$

Now compare this with the Newtonian action

$$\mathcal{A}_N \approx -\frac{1}{8\pi G} \int \int (\nabla \phi)^2 \, d^3x \, dt - \sum m \int \phi \, dt + \sum \frac{1}{2} m \int v^2 \, dt, \tag{8.17}$$

with ϕ as the gravitational potential. Clearly, (8.16) becomes the same as (8.17) if we put

$$\phi = \frac{1}{2} c^2 h_{00}, \qquad \kappa = \frac{8\pi G}{c^4}. \tag{8.18}$$

Example 8.3.1 *Problem.* Show by a direct Newtonian approximation of Einstein's field equations and geodesic equations that the result of (8.18) follows.

Solution. First note that the contraction of field equation gives $R = \kappa T$, where $T = g_{ik} T^{ik}$. Hence the equations may be rewritten as

$$R_{ik} = -\kappa \left[T_{ik} - \frac{1}{2} g_{ik} T \right].$$

From (8.11) we get, with $h = \eta^{ik} h_{ik}$,

$$R_{ik} \cong \frac{1}{2} \Box h_{ik} + \frac{1}{2} \{ h_{,ik} - h^l{}_{i,lk} - h^l{}_{k,li} \}.$$

Again ignoring time derivatives and retaining only h_{00} from all h_{ik}, we get

$$R_{00} \cong -\frac{1}{2} \nabla^2 h_{00}.$$

Likewise, for dust of density ρ,

$$-\kappa \left[T_{00} - \frac{1}{2} g_{00} T \right] = -\kappa \rho/2.$$

Therefore the (00) component of the field equations gives

$$\nabla^2 h_{00} = \kappa \rho. \tag{A}$$

Next consider the μ components of the geodesic equations. We write in the present approximation

$$\frac{dx^i}{ds} \cong (1, \mathbf{v})$$

with the 3-velocity $\mathbf{v}$ of a particle being small in magnitude compared with $c = 1$. The only relevant Γ^i_{kl} is

$$\Gamma^\mu_{00} \cong \frac{1}{2} h_{00,\mu},$$

so we get, from the geodesic equations

$$\frac{d^2 x^\mu}{ds^2} + \Gamma^\mu_{kl} \frac{dx^k}{ds} \frac{dx^l}{ds} = 0,$$

the 'Newtonian' equations of motion

$$\frac{d\mathbf{v}}{dt} = -\frac{1}{2} \nabla h_{00}. \tag{B}$$

These will exactly correspond to the Newtonian equations if we define the potential ϕ by $h_{00} = 2\phi/c^2 = 2\phi$ for $c = 1$. Equation (A) then becomes the familiar Poisson equation

$$\nabla^2 \phi = 4\pi G \rho,$$

provided that we define $\kappa = 8\pi G \, (= 8\pi G/c^4)$. Thus the match with Newtonian physics is complete.

Problem. Show that, for a spacetime of constant curvature K satisfying Einstein's field equations with energy-momentum tensor T_{ik},

$$K = -\frac{2\pi G}{3}T,$$

where T is the trace of T_{ik}.

Solution. We have for the given spacetime

$$R_{iklm} = K[g_{il}g_{km} - g_{im}g_{kl}].$$

This leads to $R_{kl} = -3Kg_{kl}$ and $R = -12K$. Therefore

$$R_{ik} - \frac{1}{2}g_{ik}R = -3Kg_{ik} + 6Kg_{ik} = 3Kg_{ik}.$$

By equating this to $-8\pi G T_{ik}$ $(c = 1)$, we get

$$3Kg_{ik} = -8\pi GT_{ik}.$$

Hence on multiplication by g^{ik} we get

$$12K = -8\pi GT,$$

from which the result follows.

Problem. In a spacetime containing pure isotropic radiation, show that a positive cosmological constant is needed in order to have a positive scalar curvature for the spacetime.

Solution. The field equations with λ are

$$R_{ik} - \frac{1}{2}g_{ik}R + \lambda g_{ik} = -\kappa T_{ik}.$$

Take the trace of these equations, recalling that, for pure isotropic radiation, $T = 0$. Hence we get

$$R - 2R + 4\lambda = -\kappa T = 0,$$

i.e.,

$$\lambda = \frac{1}{4}R.$$

Thus, for $R > 0$, we need $\lambda > 0$.

Thus we have completed our project of evaluating κ and relating the relativistic framework to Newtonian gravitation. Assumptions 1 to 3 above are known as the *Newtonian approximation*. It leads to the linear gravitation theory of Newton, which has wide applications, ranging from the tidal phenomenon of the Earth's oceans to motions of planets of the Solar System and of stars and galaxies in clusters. Provided that these three assumptions hold, general relativity does not add anything new. If assumption 3 is dropped but assumptions 1 and 2 are retained, we are in the domain of the weak-field theory of *gravitational radiation*. For, in the weak-field limit, it is seen that spacetime-curvature effects propagate as waves with the speed of light. We shall discuss this intriguing phenomenon in detail in Chapter 11. To get the full effects of general relativity, however, we must drop all three assumptions and face the non-linear equations of (8.3) in their most general form. Naturally this is a complicated task, and after nearly a century of this theory there are only a handful of exact solutions of direct physical relevance. We will discuss the earliest, simplest and most important of these solutions in the following chapter.

Exercises

1. Assume that Γ^i_{kl} are not explicitly related to g_{ik} in the expression for R, which is as given in Chapter 5. Determine the form of Γ^i_{kl} by requiring that

$$\delta \int R\sqrt{-g}\, \mathrm{d}^4x = 0$$

for $\Gamma^i_{kl} \to \Gamma^i_{kl} + \delta\Gamma^i_{kl}$ while the coordinates and the metric tensor remain unchanged. Show that this method, known as the *Palatini method*, leads to the familiar Riemannian affine connection.

2. Verify that Einstein's equations can be obtained if instead of the term $\int R\sqrt{-g}\, \mathrm{d}^4x$ we have the following term in the action:

$$\int g^{il}(\Gamma^k_{il}\Gamma^n_{kn} - \Gamma^n_{im}\Gamma^m_{ln})\sqrt{-g}\, \mathrm{d}^4x.$$

Notice that this term does not contain second derivatives of g_{ik}. However, it is *not* an invariant.

3. Show that, if the gravitational equations are obtained from an action principle, subject to the restriction that the 4-volume of the region $\mathcal{V}$ in question,

$$\int_{\mathcal{V}} \sqrt{-g}\, \mathrm{d}^4x,$$

remains unchanged, the Einstein tensor is replaced by

$$R_{ik} - \frac{1}{2}g_{ik}R + \lambda g_{ik},$$

where λ is a Lagrange multiplier.

4. Show from the linearized form of R_{iklm} that G_{00} and $G_{0\mu}$ do not contain time derivatives (of any h_{ik}) of order two. The equations $G_{0i} = \kappa T_{0i}$ are called *constraint equations*, which must be satisfied by any initial data specified for solving the problem.

5. Derive the Newtonian approximation of Einstein's field equations with the cosmological constant.

6. Show that, in Newtonian gravitation *with* the cosmological term, two masses can stay in equilibrium at a specific distance. Is this equilibrium stable, unstable or neutral?

7. In a given volume $\mathcal{V}$ of spacetime the Ricci scalar R is expected to be positive. Why?

8. To avoid having to demand the surface condition $\delta g_{ik,l} = 0$ in the Hilbert action problem, Gibbons and Hawking suggested adding an extra term to the action in the form of a surface integral:

$$A_{\mathrm{GH}} = \frac{1}{8\pi G} \int_{\partial \mathcal{V}} n_{i;k}(g^{ik} - en^i n^k)\sqrt{-h}\, \mathrm{d}^3 x,$$

where $\partial \mathcal{V}$ is the surface of the volume $\mathcal{V}$ over which the Hilbert term was defined, n_i is the unit normal to $\partial \mathcal{V}$ and $h^i{}_k = g^i{}_k - en^i n_k$. The quantity $e = +1$ for timelike n_i and $e = -1$ for spacelike n_i, with $n_i n^i = e$. Show that the surface variational term in the Hilbert action is now cancelled out by the variation of the above surface integral.

Chapter 9
The Schwarzschild solution

9.1 The exterior solution

Shortly after Einstein published his equations of general relativity, Karl Schwarzschild solved them to find the geometry in the empty spacetime outside a spherical distribution of matter of mass M (see Reference [10]). As we know, this is the simplest finite source of matter that gives rise to gravitational effects. The corresponding problem in Newtonian gravitation yields the solution for the gravitational potential as

$$\phi = -\frac{GM}{r}, \tag{9.1}$$

r being the distance from the centre of the spherical distribution. Perhaps it is worth commenting that this 'simplest' problem took Newton many years to solve to his satisfaction. The above solution was seen as the correct one for a point mass. Yet, was it the same for a finite spherically symmetric distribution of matter? Since Newton wanted to apply his theory to planets and the Moon, all extended spherical objects, he wanted to be clear on this issue. For example, the solution, if correct, does not carry any information about the size or radial inhomogeneity of the source. For an inverse square law of force, this happens to be correct, as Newton eventually proved to himself. Today we can prove this result by solving the Laplace equation for a finite spherically symmetric source.

Let us now look at the relativistic counterpart of this solution. We have to determine the spacetime metric for the *non-Euclidean geometry* outside the source. At a large distance from the centre, we expect the gravitational field to be weak. So under the Newtonian approximation we expect

$$g_{00} \sim 1 - \frac{2GM}{c^2 r}. \tag{9.2}$$

We will now show how the Schwarzschild solution is obtained and how this exact solution relates to the above asymptotic form.

The problem is simplified by making use of symmetry arguments. If the spacetime outside such a spherical distribution is empty, then its geometry should be spherically symmetric about the centre O of the distribution. So we start with the most general form of the line element that satisfies this requirement of spherical symmetry.

It can be shown, using arguments from Chapter 6, that the most general form of such a line element is given by Equation (6.30). We recall the form of (6.30) here as

$$ds^2 = A(r, t)dt^2 + 2H(r, t)dt\,dr + B(r, t)dr^2 + F(r, t)(d\theta^2 + \sin^2\theta\,d\phi^2). \tag{9.3}$$

We now redefine the radial coordinate as r' by setting $r'^2 = F(r, t)$. This would lead to changes in the forms of A, B and H. However, as can easily be verified, the choice of a new time coordinate t' can be made such that the cross-product $dt'\,dr'$ disappears from the expression for ds^2. Writing the coefficients of $c^2\,dt'^2$ and dr'^2 as e^ν and e^λ, respectively, and dropping the primes on the coordinates t' and r', the line element may be rewritten as

$$ds^2 = e^\nu c^2\,dt^2 - e^\lambda\,dr^2 - r^2(d\theta^2 + \sin^2\theta\,d\phi^2), \tag{9.4}$$

where ν and λ are functions of r and t. The advantage of the exponential form is that, for real ν and λ, $g_{00} > 0$ and $g_{11} < 0$ as required by the timelike coordinate t and spacelike coordinate r. If $\nu = \lambda = 0$, we get the Minkowski line element in spherical polar space coordinates. The non-Euclidean effects are therefore contained in the functions λ and ν. Although in this case r ceases to measure the radial distance from O, it still has the meaning that the spherical surface $r = \text{constant} = r_0$ (for example) has the surface area $4\pi r_0^2$.

Given the line element (9.4), the next step is to calculate g^{ik}, $\sqrt{-g}$ and $\Gamma_{kl}{}^i$. We then calculate the components of R_{kl}, which are given by (5.9) and are expressible in the form

$$R_{kl} = -\frac{\partial \Gamma_{kl}^i}{\partial x^i} + \frac{\partial^2 (\ln\sqrt{-g})}{\partial x^k \partial x^l} + \Gamma_{kn}^m \Gamma_{lm}^n - \frac{\partial}{\partial x^n}(\ln\sqrt{-g})\Gamma_{kl}^n. \tag{9.5}$$

Since the space outside the distribution is empty, it has $T_{kl} = 0$. Therefore the contraction of the field equations (8.3) gives $R = 0$, and these equations reduce to

$$R_{kl} = 0. \tag{9.6}$$

We now proceed to carry out these steps of calculation. The non-zero components of g_{ik} and $\Gamma^i{}_{kl}$ are given below, using the coordinates defined by $x^0 = t, x^1 = r, x^2 = \theta, x^3 = \phi$:

$$g_{00} = e^\nu, \qquad g_{11} = -e^\lambda, \qquad g_{22} = -r^2, \qquad g_{33} = -r^2 \sin^2\theta,$$

$$g^{00} = e^{-\nu}, \qquad g^{11} = -e^{-\lambda}, \qquad g^{22} = -r^{-2}, \qquad g^{33} = -r^{-2}\operatorname{cosec}^2\theta,$$

$$\Gamma_{0|00} = \frac{1}{2}e^\nu \dot{\nu}, \qquad \Gamma_{1|00} = -\frac{1}{2}e^\nu \nu',$$

$$\Gamma_{0|01} = \frac{1}{2}e^\nu \nu',$$

$$\Gamma_{1|01} = -\frac{1}{2}e^\lambda \dot{\lambda}, \qquad \Gamma_{0|11} = \frac{1}{2}e^\lambda \dot{\lambda}, \qquad \Gamma_{1|11} = -\frac{1}{2}e^\lambda \lambda',$$

$$\Gamma_{2|12} = -r, \qquad \Gamma_{3|13} = -r\sin^2\theta,$$

$$\Gamma_{1|22} = r, \qquad \Gamma_{3|23} = -r^2 \sin\theta\cos\theta,$$

$$\Gamma_{1|33} = r\sin^2\theta, \qquad \Gamma_{2|33} = r^2 \sin\theta\cos\theta,$$

$$\Gamma^0{}_{00} = \frac{1}{2}\dot{\nu}, \qquad \Gamma^1{}_{00} = \frac{1}{2}e^{\nu-\lambda}\nu',$$

$$\Gamma^0{}_{01} = \frac{1}{2}\nu', \qquad \Gamma^1{}_{01} = \frac{1}{2}\dot{\lambda}, \qquad \Gamma^0{}_{11} = \frac{1}{2}e^{\lambda-\nu}\dot{\lambda}, \qquad \Gamma^1{}_{11} = \frac{1}{2}\lambda',$$

$$\Gamma^2{}_{12} = \Gamma^3{}_{13} = \frac{1}{r}, \qquad \Gamma^1{}_{22} = -re^{-\lambda},$$

$$\Gamma^3{}_{23} = \cot\theta, \qquad \Gamma^1{}_{33} = -re^{-\lambda}\sin^2\theta, \qquad \Gamma^2{}_{33} = -\sin\theta\cos\theta.$$

(Here a prime denotes differentiation with respect to r, and an overdot denotes differentiation with respect to t.) We next compute the various components of the Ricci tensor. The (00) and (11) components of (9.6) give, after some manipulation, the following equations:

$$e^{-\lambda}\left(\frac{\lambda'}{r} - \frac{1}{r^2}\right) + \frac{1}{r^2} = 0. \tag{9.7}$$

$$-e^{-\lambda}\left(\frac{\nu'}{r} + \frac{1}{r^2}\right) + \frac{1}{r^2} = 0. \tag{9.8}$$

From these we get, by subtracting (9.8) from (9.7),

$$\nu' + \lambda' = 0,$$

that is,

$$\nu + \lambda = f(t).$$

The arbitrary function $f(t)$ can, however, be set to equal zero since we still have an arbitrary time tranformation

$$t = g(\bar{t})$$

at our disposal, which changes ν to

$$\bar{\nu} = \nu + 2\ln\left(\frac{dg}{d\bar{t}}\right)$$

and preserves the form of the line element (9.4). Therefore we can take, without loss of generality,

$$\nu + \lambda = 0. \tag{9.9}$$

However, we also have, from $R_{01} = 0$,

$$\dot{\lambda} = 0. \tag{9.10}$$

Thus both λ and $\nu\,(=-\lambda)$ are functions of r only. Equations (9.7) and (9.8) then yield the solution

$$e^{\nu} = e^{-\lambda} = 1 - \frac{A}{r}, \qquad A = \text{constant}.$$

However, if we are given that the mass of the object is M, we may use the boundary condition (9.2) as $r \to \infty$ to set $A = 2GM/c^2$. Thus we get our required solution as the line element

$$ds^2 = \left(1 - \frac{2GM}{c^2 r}\right)c^2\,dt^2 - \left(1 - \frac{2GM}{c^2 r}\right)^{-1} dr^2 - r^2(d\theta^2 + \sin^2\theta\,d\phi^2).$$

$$\tag{9.11}$$

This is known as the *Schwarzschild line element*. It turns out that because of the symmetries of the problem the other field equations are automatically satisfied: we need only the (11), (00) and (01) components in order to arrive at the solution. One may notice that the metric behaves strangely for small r, namely for $r \leq R_s$, where

$$R_s = \frac{2GM}{c^2}. \tag{9.12}$$

This quantity is called the *Schwarzschild radius* of the mass. It is easy to verify that this is an extremely small 'radius' and most known objects have a radius exceeding it by a large factor. For the Sun, for example, the Schwarzschild radius is about 3 km, whereas its actual radius is nearly 700 000 km. Idealized objects whose radius equals their Schwarzschild radius are known as *black holes*. We shall return to a discussion of black holes in Chapter 13.

The Schwarzschild solution, as derived above, is manifestly static. Thus there is no scope for a dynamical solution such as one involving gravitational radiation, even if our spherical source is expanding, contracting, or oscillating. This remarkable result is known as *Birkhoff's theorem*.

Example 9.1.1 *Problem.* Find a coordinate transformation $r = f(R)$ that will transform the Schwarzschild exterior line element to a manifestly isotropic form:

$$ds^2 = e^\mu \, dt^2 - e^\sigma [dR^2 + R^2(d\theta^2 + \sin^2\theta \, d\phi^2)].$$

Find μ and σ.

Solution. On comparing the two line elements, we find

$$r^2 = R^2 e^\sigma \quad \text{and} \quad e^\lambda \, dr^2 = e^\sigma \, dR^2.$$

By eliminating the unknown σ we get

$$\frac{e^\lambda \, dr^2}{r^2} = \frac{dR^2}{R^2}, \quad \text{i.e.,} \quad \ln R = \int \left(1 - \frac{2GM}{r}\right)^{-1/2} \frac{dr}{r}.$$

This can easily be integrated to give $R = \frac{1}{2}[r - GM + \sqrt{r(r - 2GM)}]$. Some simple algebra then yields $r - GM = R + G^2 M^2/(4R)$. This corresponds to the transformation

$$r = R\left[1 + \frac{GM}{2R}\right]^2.$$

The corresponding e^μ and e^σ are

$$e^\mu = \left(\frac{1 - \dfrac{GM}{2R}}{1 + \dfrac{GM}{2R}}\right)^2, \quad e^\sigma = \left(1 + \frac{GM}{2R}\right)^4.$$

We end this section with another observation. Suppose we have a point mass in Newtonian gravitation and we wanted to solve the Laplace equation to determine the potential. We could do so by invoking the delta function $\delta(x)$ which vanishes for any non-zero x but has an integral equal to unity over any interval containing the point $x = 0$. In the three-dimensional case under consideration the point mass M is represented by the density

$$\rho = M \times \frac{\delta(r)}{4\pi r^2}. \tag{9.13}$$

The integration of the Laplace equation then leads us to the solution (9.1). In the relativistic version, however, we cannot do so! For we have a problem of singularity at $r = 0$. The metric diverges, so defining a point mass is not possible. What we have done therefore is to determine the constant in e^ν by appealing to the Newtonian limit at large distances.

We can, of course, avoid this issue by appealing to the finite size of any mass. So we next consider the extension of the Schwarzschild solution to a finite distribution of matter.

9.2 The interior solution

Let us assume that the source in the above problem is not a point mass but a spherical distribution of matter confined to a coordinate radius $r = R_0$. Let T_{ik} denote the energy-momentum tensor of the interior matter. Then Equations (9.7) and (9.8) get modified to

$$e^{-\lambda}\left(\frac{\lambda'}{r} - \frac{1}{r^2}\right) + \frac{1}{r^2} = 8\pi G T_0^{\,0}, \qquad (9.14)$$

$$-e^{-\lambda}\left(\frac{\nu'}{r} + \frac{1}{r^2}\right) + \frac{1}{r^2} = 8\pi G T_1^{\,1}. \qquad (9.15)$$

Schwarzschild had investigated a special solution in which the interior is an incompressible fluid of *constant* density ρ and (variable) pressure p. Thus the energy tensor was taken to be

$$T^{lk} = (\rho + p)u^i u^k - p g^{ik}, \qquad (9.16)$$

where $u^i \equiv (u^0, 0, 0, 0)$ is the flow vector of the fluid at rest. Since $u^i u_i = 1$, we get $(u^0)^2 e^\nu = 1$, i.e., $u^0 = e^{-\nu/2}$. Therefore, $T_0^{\,0} = \rho$, $T_1^{\,1} = -p$.

Equation (9.14) can easily be integrated. We have

$$e^{-\lambda} = 1 - \frac{8\pi G\rho}{3}r^2 = 1 - \alpha r^2, \qquad (9.17)$$

say, where $\alpha = 8\pi G\rho/3 = $ constant. We have set $\lambda = 0$ at $r = 0$.

Next consider the energy-conservation equation $T_i^{\,k}{}_{;k} = 0$, for $i = 1$. From (9.16) we get $T_0^{\,0} = \rho$ and $T_1^{\,1} = T_2^{\,2} = T_3^{\,3} = -p$, so

$$0 = T_1^{\,k}{}_{;k} = \frac{1}{\sqrt{-g}}\frac{\partial}{\partial x^k}(\sqrt{-g}\,T_1^{\,k}) - \Gamma^l_{1k}T_l^{\,k}$$

$$= \frac{1}{r^2}e^{-(\nu+\lambda)/2}\frac{\partial}{\partial r}\left(r^2 e^{(\lambda+\nu)/2}T_1^{\,1}\right) - \Gamma^0_{10}T_0^{\,0} - (\Gamma^1_{11} + \Gamma^2_{12} + \Gamma^3_{13})T_1^{\,1}$$

$$= -\left[\frac{2}{r} + \frac{1}{2}(\nu' + \lambda')\right]p - \frac{\partial p}{\partial r} - \frac{1}{2}\nu'\rho + \left(\frac{1}{2}\lambda' + \frac{2}{r}\right)p$$

$$= -\frac{1}{2}\nu'(p + \rho) - \frac{\partial p}{\partial r},$$

i.e.,

$$\frac{\partial p}{\partial r} = -\frac{1}{2}(p + \rho)\nu'. \qquad (9.18)$$

Returning to (9.15), we have v' given in terms of e^λ, r and p by

$$v' = -\frac{1}{r} + \frac{e^\lambda}{r} + 8\pi Gpre^\lambda. \qquad (9.19)$$

Substitute (9.19) for v' in (9.18), then use the definition of α and the expression (9.17) for $e^{-\lambda}$ to arrive at the following differential equation:

$$\frac{dp}{dr} = -\frac{4\pi G(p + \rho)r}{1 - \alpha r^2}\left(p + \frac{1}{3}\rho\right). \qquad (9.20)$$

Assuming that $p = 0$ at the surface $r = R_0$, we can integrate (9.20). A straightforward but tedious integration gives

$$p = \rho\left\{\frac{\sqrt{1 - \alpha r^2} - \sqrt{1 - \alpha R_0^2}}{3\sqrt{1 - \alpha R_0^2} - \sqrt{1 - \alpha r^2}}\right\}. \qquad (9.21)$$

Finally, the equation for e^v leads to

$$e^v = \frac{1}{4}\left\{3\sqrt{1 - \alpha R_0^2} - \sqrt{1 - \alpha r^2}\right\}^2. \qquad (9.22)$$

This is known as the *Schwarzschild interior solution*.

The pressure p can be arbitrarily high, even exceeding ρ. The largest value of p is at the centre ($r = 0$) and it reaches the limit $p \to \infty$ when $\alpha R_0^2 = 8/9$, i.e., when

$$R_0 = \sqrt{\frac{1}{3\pi G\rho}}. \qquad (9.23)$$

The interior solution should match the exterior solution obtained earlier in Section 9.1 across the boundary $r = R_0$. We see that, across this boundary, g_{22} and g_{33} are continuous. What about g_{11} and g_{00}? We expect that an observer moving across the boundary with a clock should not notice any discontinuity of time measurement. That is, e^v should be continuous. From (9.22) we have at $r = R_0$ that $e^v = 1 - \alpha R_0^2$. The exterior solution has $e^v = 1 - 2GM/R_0 = 1 - 2G \times (4\pi/3)R_0^2\rho = 1 - \alpha R_0^2$. Thus the continuity of e^v is maintained. What about e^λ? We find that it is continuous also. In general this need not be the case. For, in measuring ds in the radial direction, we are using the exterior solution for $r > R_0$ and the interior solution for $r < R_0$. So the prescription for measuring dr need not be the same in the two regions.

Example 9.2.1 *Problem*. From equations set up to describe a spherically symmetric situation show that in general $v + \lambda < 0$ at any finite r if the spacetime is asymptotically flat.

Solution. We have the two equations

$$e^{-\lambda}\left(\frac{\nu'}{r} + \frac{1}{r^2}\right) - \frac{1}{r^2} = -8\pi G T_1{}^1,$$

$$e^{-\lambda}\left(\frac{1}{r^2} - \frac{\lambda'}{r}\right) - \frac{1}{r^2} = -8\pi G T_0{}^0.$$

On subtracting the second equation from the first we get

$$\frac{e^{-\lambda}(\nu' + \lambda')}{r} = 8\pi G(T_0{}^0 - T_1{}^1).$$

Now, for most physically relevant $T_k{}^i$, we have $T_0{}^0 > 0$, $T_1{}^1 < 0$. For example, for a fluid $T_0{}^0 = \rho > 0$, $T_1{}^1 = -p < 0$. Thus we have, from the above equation, $\nu' + \lambda' > 0$, i.e., $(\nu + \lambda)$ at a finite value of $r = r_0$ equals $(\nu + \lambda)_\infty - \int_{r_0}^{\infty}(\nu' + \lambda')dr$.

But, assuming spacetime to be asymptotically flat, i.e., that of special relativity, $(\nu + \lambda)_\infty = 0$. Since the integral itself is positive, the above equation gives that $(\nu + \lambda)$ at $r = r_0$ is *negative*.

Problem. For a spherical object in equilibrium under gravitation and its fluid pressures, show that

$$\frac{dp}{dr} = -\frac{4\pi G r(p + \rho)}{1 - \frac{2Gm(r)}{r}} \cdot \left\{p + \frac{m(r)}{4\pi r^3}\right\},$$

where $m(r) = \int_0^r 4\pi r_1^2 \rho(r_1)dr_1$.

Solution. From Equations (9.14), (9.15) and (9.18) we get a series of results that lead to the desired answer as follows. On writing $T_0{}^0 = \rho$, we get from (9.14)

$$e^{-\lambda} = 1 - \frac{2Gm(r)}{r}, \quad m(r) \text{ as defined.} \tag{A}$$

From (9.18) we get

$$\frac{dp}{dr} = -\frac{1}{2}(p + \rho)\nu', \tag{B}$$

while from (9.15) we have

$$\nu' = 8\pi G p r e^\lambda + \frac{e^\lambda}{r} - \frac{1}{r}. \tag{C}$$

We substitute for e^λ from (A) in the above equation to get

$$\nu' = -\frac{1}{r} + \frac{1}{1 - \frac{2Gm}{r}}\left\{\frac{1}{r} + 8\pi G p r\right\}$$

$$= \frac{1}{1 - \frac{2Gm}{r}}\left\{\frac{1}{r} + 8\pi G p r - \frac{1}{r}\left(1 - \frac{2Gm}{r}\right)\right\}$$

$$= \frac{1}{1 - \frac{2Gm}{r}} \left\{ \frac{2Gm}{r^2} + 8\pi Gpr \right\}$$

$$= \frac{8\pi Gr}{1 - \frac{2Gm}{r}} \left\{ p + \frac{m}{4\pi r^3} \right\}.$$

Use this value v' in Equation (B) to get the required result.

We will discuss in Chapter 12 situations more general than the somewhat artificial one assumed in Schwarzschild's interior solution. Here we will return to the exterior solution and consider its dynamical and geometrical consequences. These turn out to have bearing on realistic tests that can be designed to test the validity of general relativity.

9.3 Motion of a test particle

Imagine a test particle moving in the spacetime of Schwarzschild's exterior solution. By a 'test' particle we mean a particle that is subjected to the gravitational influence of the central mass M, but which does not in turn contribute to any gravitational effects of its own mass. Thus we have introduced an artificiality into the picture, which can be justified only if the mass of our moving particle is negligibly small compared with M. In this case we use the result that our test particle follows a timelike geodesic.

Writing its equations in the standard form

$$\frac{d^2 x^i}{ds^2} + \Gamma^i_{kl} \frac{dx^k}{ds} \frac{dx^l}{ds} = 0, \tag{9.24}$$

with $x^0 = t, x^1 = r, x^2 = \theta, x^3 = \phi$, we get, for $i = 0$,

$$\frac{d^2 t}{ds^2} + \Gamma^0_{ij} \frac{dx^i}{ds} \frac{dx^j}{ds} = 0. \tag{9.25}$$

From our earlier computations on page 128, we have the only non-zero Γ^0_{ij} as $\Gamma^0_{01} = v'/2$. So we get for (9.25)

$$\frac{d^2 t}{ds^2} + \frac{dt}{ds} \frac{dr}{ds} v' = 0. \tag{9.26}$$

This easily integrates to

$$e^v \frac{dt}{ds} = \text{constant} = \gamma \text{ (say)}. \tag{9.27}$$

Likewise we get, for $i = 3$, the following first integral:

$$r^2 \sin \theta \, \frac{d\phi}{ds} = \text{constant} = h \, (\text{say}). \qquad (9.28)$$

We may identify γ as the energy per unit mass and h as the angular momentum per unit mass. The $i = 2$ (theta) equation reduces to

$$\frac{d^2\theta}{ds^2} + \frac{2}{r} \frac{d\theta}{ds} \frac{dr}{ds} - \sin \theta \, \cos \theta \left(\frac{d\phi}{ds}\right)^2 = 0. \qquad (9.29)$$

This somewhat complicated-looking equation has a simple solution:

$$\theta = \text{constant} = \frac{\pi}{2}. \qquad (9.30)$$

It may be easily verified that, provided we 'start' the particle moving in the $\theta = \pi/2$ plane, it will continue to move in that plane. In a spherically symmetric spacetime any plane through the centre may be chosen as a reference plane, without any loss of generality. Taking the $\theta = \pi/2$ plane as the chosen case, we see that the solution to the geodesic equations is

$$\frac{dt}{ds} = \gamma e^{-\nu}, \qquad r^2 \frac{d\phi}{ds} = \text{constant} = h \text{ and } \theta = \frac{\pi}{2}. \qquad (9.31)$$

Since the metric itself is a first integral of the geodesic equations, we get

$$1 = e^\nu \left(\frac{dt}{ds}\right)^2 - e^{-\nu} \left(\frac{dr}{ds}\right)^2 - r^2 \left(\frac{d\phi}{ds}\right)^2,$$

where $e^\nu = 1 - 2GM/r$. By substituting from (9.31), we can transform this equation to

$$\left(\frac{dr}{ds}\right)^2 = \gamma^2 - \left(1 - \frac{2GM}{r}\right)\left(1 + \frac{h^2}{r^2}\right) \equiv \gamma^2 - V^2(r), \qquad (9.32)$$

where

$$V^2(r) = \left(1 - \frac{2GM}{r}\right)\left(1 + \frac{h^2}{r^2}\right) \qquad (9.33)$$

is the 'effective potential'. For motion to be possible we need $V^2(r) \leq \gamma^2$.

(i) The Newtonian approximation
In this case,

$$|\gamma - 1| \ll 1, \qquad r \gg 2GM \text{ and } h \ll r. \qquad (9.34)$$

The first inequality tells us that the total energy γc^2 is not much different from the rest-mass energy, as is the case in a slow-motion approximation. The $r \gg 2GM$ inequality relates to Schwarzschild's solution, identifying a weak-field approximation. The last inequality tells us that the transverse velocity h/r is small compared with $c \, (=1)$.

In this case we have

$$V^2 \cong 1 + 2V_{\mathrm{N}}, \tag{9.35}$$

where

$$V_{\mathrm{N}} = \frac{h^2}{2r^2} - \frac{GM}{r} \tag{9.36}$$

is the effective Newtonian potential for radial motion:

$$\frac{1}{2}\left(\frac{dr}{dt}\right)^2 + V_{\mathrm{N}} = \frac{\gamma^2 - 1}{2} = E_{\mathrm{N}} \tag{9.37}$$

(say).

In Figure 9.1, V_{N} is plotted against r to illustrate the typical Newtonian situation. Notice that V_{N} drops from an infinite value to zero as

Fig. 9.1. The 'Newtonian' approximation of motion in the empty exterior Schwarzschild solution is described by the potential–distance plot shown. The dotted curve represents radial motion. See the text for interpretation of the curves $E_{\mathrm{N}} <, =, > 0$.

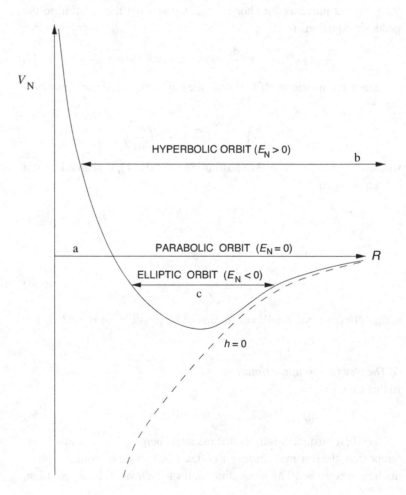

r increases from zero to a finite value $r_p = h^2/(2GM)$. Thereafter it continues to decrease as r increases. As r approaches $2r_p = h^2/(GM)$, V_N reaches a negative minimum. Beyond $2r_p$, as r increases to infinity, V_N increases but stays negative. It asymptotically approaches zero.

Against this background we have three lines a, b and c drawn horizontally, corresponding to three values of E_N, the 'total' energy of the particle. Line a corresponds to $E_N = 0$ and to motion in a parabolic trajectory. Line b has $E_N > 0$ and describes a hyperbolic trajectory. Line c with $E_N < 0$ describes the elliptical orbits typically followed by planets around the Sun. Notice that, because the kinetic energy has to be positive, the condition to be satisfied by the trajectory is $V_N \leq E_N$.

(ii) The relativistic orbits

We now turn to orbits that might not satisfy (9.34). In this case we have to use the full equation (9.32). To facilitate the algebra, define dimensionless parameters

$$x = r/(GM), \qquad \eta = h/(GM), \qquad \sigma = s/(GM). \qquad (9.38)$$

We then have (9.32) written as

$$\left(\frac{dx}{d\sigma}\right)^2 = \gamma^2 - V^2, \qquad (9.39)$$

where

$$V^2 = \left(1 - \frac{2}{x}\right)\left(1 + \frac{\eta^2}{x^2}\right). \qquad (9.40)$$

The function $V(x)$ has a maximum V_{max} at $x = x_{max}$ and a minimum V_{min} at $x = x_{min}$. Both x_{min} and x_{max} are given by the equation $\partial V^2/\partial x = 0$, i.e., by the quadratic equation

$$x^2 - \eta^2 x + 3\eta^2 = 0. \qquad (9.41)$$

We therefore have

$$x_{max} = \frac{1}{2}\left\{\eta^2 - \eta\sqrt{\eta^2 - 12}\right\}, \quad x_{min} = \frac{1}{2}\left\{\eta^2 + \eta\sqrt{\eta^2 - 12}\right\}. \qquad (9.42)$$

The maxima and minima coincide for $\eta^2 = 12$, i.e., for

$$h = 2\sqrt{3}GM, \qquad x_{min} = x_{max} = 6. \qquad (9.43)$$

From the considerations of stability of orbits we deduce that circular orbits at $x = x_{max}$ are unstable, whereas those at $x = x_{min}$ are stable. From (9.43) we see that stable circular orbits are possible for $r \geq 6GM$, the smallest such orbit having radius $6GM$. If $\eta \to \infty$, $x_{min} \to \infty$ but

Fig. 9.2. The square of the effective potential for relativistic motion is plotted against distance scaled to the size of half the Schwarzschild radius. See the text for a discussion of the various curves shown here.

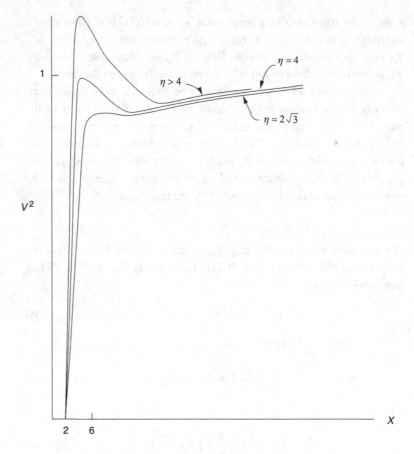

$x_{\max} \to 3$. Thus $r = 3GM$ is the lower limit to the size of circular orbits all of which are unstable. Figure 9.2 illustrates these features.

Another point of interest is the value of η for which $V_{\max} = 1$. This happens at $\eta = 4$, $x_{\max} = 4$.

In Figure 9.2, V^2 is plotted against x for various values of η. We note that, for $\eta < 2\sqrt{3}$, V^2 has no real turning points and it increases from $V^2 = 0$ at $x = 2$ to $V^2 = 1$ at $x \to \infty$. Thus an incoming particle with $\gamma > 1$ will fall in without a bounce. The same conclusion applies to an incoming particle with $\gamma > V$ to start with. For $2\sqrt{3} < \eta < 4$ there are bound orbits like the Newtonian ellipses, provided that $\gamma < V_{\max} < 1$. For $\gamma > V_{\max}$ the incoming particle falls in to be sucked into the object.

For $\eta > 4$ there are three types of orbits. Those with $\gamma > V_{\max}$ represent particles that, if coming in, fall into the object. Incoming particles with $1 < \gamma < V_{\max}$ bounce at the potential barrier (at some minimum r) and then move out again like the hyperbolic orbits. Similarly, for $\gamma < 1 < V_{\max}$ the orbits are bound as for Newtonian ellipses.

Example 9.3.1 *Problem.* Calculate the Schwarzschild time and the proper time elapsed when a particle moves once round a circular path of size $r = a$. What happens when $r \to 3GM$?

Solution. For travel along a circular path $dr/ds = 0$, $d^2r/ds^2 = 0$. From Equation (9.32) we have by differentiation

$$\frac{d^2r}{ds^2} = -\frac{1}{2}\frac{d}{dr}V^2(r).$$

Hence, at $r = a$ we require

$$\gamma^2 = V^2(a) \qquad \text{and} \qquad \frac{d}{dr}V^2(r)|_{r=a} = 0.$$

So we have two equations:

$$\gamma^2 = \left(1 - \frac{2GM}{a}\right)\left(1 + \frac{h^2}{a^2}\right) \qquad (A)$$

and

$$\frac{2GM}{a^2}\left(1 + \frac{h^2}{a^2}\right) - \frac{2h^2}{a^3}\left(1 - \frac{2GM}{a}\right) = 0. \qquad (B)$$

From (B) we get

$$h^2 = \frac{a^2 GM}{a - 3GM}, \qquad 1 + \frac{h^2}{a^2} = \frac{a - 2GM}{a - 3GM}.$$

Hence from (A) we get

$$\gamma^2 = e^{2\nu}\frac{a}{a - 3GM}.$$

Let T be the time taken by one revolution, as measured by the t time coordinate. Then

$$T\frac{d\phi}{dt} = 2\pi.$$

However, $r^2\,d\phi/ds = h$, i.e., for $r = a$ we have

$$h = a^2\frac{d\phi}{dt}\cdot\frac{dt}{ds}.$$

Since $e^\nu(dt/ds) = \gamma$, we have $dt/ds = \gamma e^{-\nu} = \sqrt{a/(a - 3GM)}$. We therefore have

$$\frac{d\phi}{dt} = \frac{h}{a^2}\sqrt{\frac{a - 3GM}{a}} = \frac{1}{a}\sqrt{\frac{GM}{a - 3GM}}\cdot\sqrt{\frac{a - 3GM}{a}}$$

$$= \frac{1}{a}\sqrt{\frac{GM}{a}}.$$

Therefore

$$T = \frac{2\pi a^{3/2}}{(GM)^{1/2}}.$$

Since

$$\frac{dt}{ds} = \sqrt{\frac{a}{a - 3GM}},$$

the time taken as measured by the proper time of the observer is

$$\tau = \sqrt{\frac{a - 3GM}{a}} \cdot \frac{2\pi a^{3/2}}{(GM)^{1/2}}$$

$$= 2\pi a \sqrt{\frac{a - 3GM}{GM}}.$$

As $a \to 3GM$ this shrinks to zero since the observer tends to have a null geodesic.

9.4 Trajectories of photons

The null geodesics describe trajectories of photons and, following our earlier work (*vide* Section 5.2) and analogously to Equation (9.31), we set up their equations as follows:

$$e^{\nu} \frac{dt}{d\lambda} = \text{constant} = \frac{1}{b}, \qquad r^2 \frac{d\theta}{d\lambda} = 1, \qquad \theta = \frac{\pi}{2}. \qquad (9.44)$$

We have taken λ to be an affine parameter and scaled it so that the second of the above equations has unity on the right-hand side. Likewise it is convenient to write the constant in the first equation as $1/b$, rather than γ.

The first integral of the geodesic equation then becomes

$$\left(\frac{dr}{d\lambda}\right)^2 + V^2(r) = \frac{1}{b^2}, \qquad V^2(r) = \frac{1}{r^2}\left(1 - \frac{2GM}{r}\right). \qquad (9.45)$$

Figure 9.3 shows a plot of V^2 against the dimensionless 'distance', $x = r/(GM)$. Starting at zero value for $x = 2$, V^2 rises to a maximum value of $(27G^2M^2)^{-1}$ at $x = x_{\max} = 3$ and then falls off to zero as $x \to \infty$. What does this potential behaviour imply for an incoming photon?

If the photon travels with $b < 3\sqrt{3}GM$, then it has $1/b^2 > (27G^2M^2)^{-1}$ and such a photon cannot bounce at a potential wall and go out again: it drops into the object. A photon with $b > 3\sqrt{3}GM$ will bounce and go out. What about $b = 3\sqrt{3}GM$? At this value the line touches the peak of the potential curve and the point of contact corresponds to a circular trajectory with radius $x = 3$ ($r = 3GM$). However, this trajectory is unstable and the photon, on slight disturbance, either falls in or spirals out to $r = \infty$.

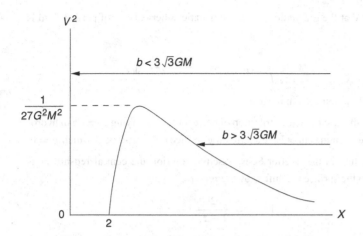

Fig. 9.3. The curve similar to Figure 9.2 is drawn here for describing the motion of photons in Schwarzschild's spacetime. See the text for a discussion of the various horizontal levels here.

These considerations make an important point: that gravitation affects the path of light. Isaac Newton had wondered about this effect, when he wrote

> Do not bodies act upon light at a distance? And by their action bend its Rays: and is not this action [caeteris paribus] strongest at the least distance?
>
> Optics: Query 1.

Einstein's general relativity returns an affirmative reply to the question 'Does gravitation affect the light track?'. Is this reply in conformity with reality? To find out, we move to the next chapter to discuss the experimental tests of general relativity.

Exercises

1. By considering the isometries of the spherically symmetric spacetime deduce that, with the notation used in the text,

$$T_2{}^2 = T_3{}^3$$

and that all components of $T_k{}^i$ with $i \neq k$, except $T_0{}^1$, are zero.

2. Calculate the non-zero components of R_{iklm} in the Schwarzschild spacetime. Verify that $R_{iklm} R^{iklm}$ is finite at the Schwarzschild radius.

3. Show that there exists a path of the light ray in the form of a terminating spiral given by

$$\frac{1}{r} = -\frac{1}{6GM} + \frac{1}{2GM} \tanh^2\left(\frac{\phi}{2}\right)$$

in the gravitational field of a spherical object of mass M.

4. Show that the gravitational mass of a static spherical star of perfect fluid is given by

$$M = \int_0^{R_0} (\rho + 3p)e^{(\nu+\lambda)/2} \cdot 4\pi r^2 \, dr,$$

with the notation used in the text.

5. Show that a spherical distribution of perfect fluid satisfying $p = k\rho$ in hydrostatic equilibrium cannot have a bounding surface $r = R_b < \infty$ at which $p = 0$.

6. Show that in the interior Schwarzschild solution the central redshift z_c is related to the surface redshift z_s by the relation

$$1 + z_c = \frac{2}{2 - z_s}(1 + z_s).$$

As $z_s \to 2$, $z_c \to \infty$. (For a definition of gravitational redshift see Chapter 10, Section 10.3.)

7. Write down the equation of geodesic deviation for test particles falling freely along radial trajectories onto the central gravitating point mass M. Interpret the two cases in which the initial deviation is in (i) the ϕ-coordinate and (ii) the T-coordinate. Compare your results with the Newtonian theory.

8. Show that a freely and radially falling tachyon (described by a *spacelike* geodesic) bounces at a finite Schwarzschild T-coordinate in the gravitational field of a central gravitating mass M.

9. Show that the test particles experience no gravitational forces inside a self-gravitating hollow spherical shell.

10. Give dynamical arguments to show that the orbit $r = 3GM$ is unstable, whereas $r = 6GM$ is stable.

Chapter 10
Experimental tests of general relativity

10.1 Introduction

The general theory of relativity, like any other physical theory, must submit itself for experimental verification. It started with a disadvantage in that it was competing with a well-established paradigm, viz. the Newtonian laws of motion and gravitation. Any test that could be designed for testing general relativity had at the same time to show ways of distinguishing its predictions from those of the Newtonian framework. Here the situation has been different from the case of special relativity. Many laboratory tests have been designed (see some in the opening chapter) to study the dynamics of fast-moving particles. For, in this case, the crucial factor γ, distinguishing relativity from Newtonian dynamics, is significantly different from unity. For really testing general relativity we need situations of strong gravitational fields that cannot be arranged in a terrestrial laboratory. The differences from Newtonian predictions can and do exist in relatively weak fields, however, provided that we look at astronomical situations. Therefore astronomical tests have figured prominently in establishing the general theory.

In the early days Einstein proposed three tests, which are known as the *classical tests* of general relativity. More tests emerged in later years, although their number is still small. In this chapter we will disuss both classes of tests. All except one require an astronomical setting.

To place matters in proper perspective, let us see how 'strong' or 'weak' the Earth's gravitational field is at its maximum, i.e., on the surface of the Earth. Putting in the numbers for M and R, the mass and the radius of the Earth, we get the dimensionless parameter

$$\eta \equiv \frac{2GM}{c^2 R} \cong 1.2 \times 10^{-9}. \tag{10.1}$$

This ratio is close to 4×10^{-6} when evaluated for the Sun on the solar surface and to 2×10^{-8} when evaluated at a typical point on the Earth's orbit around the Sun. The smallness of these numbers conveys the challenge facing an experimenter attempting to distinguish between the predictions of Newton and Einstein. For, to detect any effect characteristic of non-Euclidean geometry as expected in general relativity, these numbers should be closer to unity.

The physicist cannot help making another comparison. For testing the electromagnetic theory, both in its classical and in its quantum version, a large number of experiments could be performed. The more experimental checks there are the greater the confidence inspired by a physical theory. Because of its somewhat esoteric nature the general theory falls far short in such a comparison. This was reportedly one reason why Einstein was not given Nobel Prizes for the special and general theories of relativity: apparently the experts were not convinced that enough experimental proofs had been provided for these theories. This was somewhat surprising, at least in the case of special relativity; for the discoverers of other effects (such as the Compton effect) involving the dynamics of special relativity were awarded the Nobel Prize.

We mention these perspectives so that the reader will appreciate the few tests that there are!

10.2 The PPN parameters

Most of the present tests of general relativity are based on the Schwarzschild solution, and they seek to measure the fine differences between the predictions of Newtonian gravitation and those of general relativity. These form the main part of this chapter.

Before confronting the experimental situation, however, it is necessary to clarify how to attach meanings to measurements in a spacetime that is non-Euclidean. We have already seen that coordinates have no absolute status, hence blindly relying on them might lead to incorrect results. The Schwarzschild metric (*vide* Chapter 9) can be used to illustrate the concept of measurement as seen in the example below.

Example 10.2.1 Suppose that an observer is located at a point with $r = $ constant, $\theta = $ constant, $\phi = $ constant. How does he relate the time τ kept by his watch to the coordinate t? From the principle of equivalence we know that, since $d\tau = ds/c$ measures the observer's proper time in a locally inertial frame, being a scalar, it does so in all frames. For our observer, $dr = 0$, $d\theta = 0$, $d\phi = 0$; so from (9.11) we get

$$d\tau = \left(1 - \frac{2GM}{c^2 r}\right)^{1/2} dt.$$

This gives the required answer.

The experimental tests mostly revolve around the Schwarzschild line element as applied to objects in the Solar System. However, beyond comparing the relativistic predictions with the corresponding Newtonian ones, there has also been interest in *other* theories of gravitation. Some of these, such as the Brans–Dicke theory, will be discussed in Chapter 18. These theories use a spacetime metric as in relativity, but come up with line elements different from Schwarzschild's. All these can be simultaneously looked at in their weak-field limits and by comparing their predictions in the various experiments. A series of parameters can be used to specify the various components of the metric with reference to these tests. Since we are looking at a level of approximation one step beyond the Newtonian limit, the procedure is called *parametrized post-Newtonian* approximation or simply the PPN approximation. The parameters are denoted by γ, β, ξ, $\alpha_1, \alpha_2, \alpha_3$, and $\zeta_1, \zeta_2, \zeta_3, \zeta_4$. We will not discuss the details of how these parameters are derived in a particular theory (see Table 10.1 from Reference [18], a review by C. M. Will, which is given at the end of this chapter), except to identify the first two, which have values of unity in general relativity and occur explicitly in the classical tests of this theory. The rest have value zero in relativity.

To identify β and γ, we express the Schwarzschild line element in the isotropic form, in which the spatial part of the metric is the Euclidean one multiplied by a radial function:

$$ds^2 = e^\mu c^2\, dt^2 - e^\eta[dR^2 + R^2(d\theta^2 + \sin^2\theta\, d\phi^2)], \qquad (10.2)$$

where μ and η are functions of the new radial coordinate R (see Example 9.1.1). By expanding these in powers of (M/R) we get

$$e^\mu = 1 - 2\frac{GM}{C^2 R} + 2\beta\left(\frac{GM}{C^2 R}\right)^2, \qquad e^\eta = 1 + 2\gamma\frac{GM}{C^2 R}, \qquad (10.3)$$

where, as mentioned earlier for general relativity, both β and γ are unity. For some other theories they may have different values. We will later summarize the present status of the measured values of these parameters.

10.3 The gravitational redshift

Imagine two observers A and B moving in spacetime exchanging light signals with each other. In the general situation illustrated in Figure 10.1, we have B transmitting two successive wavefronts at instants B_1 and B_2 corresponding to successive peaks of intensity. The proper time elapsed between these instants as measured by B is, say, Δ_B. To B, these measurements bring the information that the wavelength of light just released by him is $\Delta_B \times c$, where c is the speed of light.

In the corresponding geometrical optics we may argue that these wavefronts reach A along null rays ζ_1 and ζ_2, reaching points A_1 and A_2 on the world line of A. To A the proper time gap between the receipts of these two wave peaks is Δ_A. To him therefore the wavelength received will appear to be $\Delta_A \times c$. Denoting the wavelengths emitted (by B) and received (by A) as λ_B and λ_A, respectively, we define the *spectral shift* by

$$z = \frac{[\lambda_A - \lambda_B]}{\lambda_B} = \frac{\lambda_A - \lambda_B}{\lambda_B}. \tag{10.4}$$

In the optical spectrum, the red colour is towards the long-wavelength end and the blue/violet colour is at the short-wavelength end. Hence, if in optical astronomy the observer finds that $z > 0$, the result is described as *redshift*. Likewise, if $z < 0$, the result is called *blueshift*.

We will encounter various applications of this result under different physical conditions. The application most physicists are familiar with is that due to pure motion, known as the *Doppler effect*. As described in

Fig. 10.1. Signal communication between two observers A and B, in a general spacetime with an inhomogeneous gravitational field.

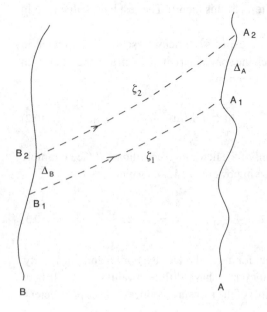

Chapter 1, this arises when there is relative motion between A and B. In a later chapter we will consider the cosmological redshift arising from an overall expansion of the Universe. Here we describe the spectral shift arising from passage of light between static inhomogeneous gravitational fields.

10.3.1 The gravitational spectral redshift

Consider any static line element – that is, one in which g_{ik} do not depend on $x^0 \equiv ct$. Suppose we have two observers A and B with world lines

$$x^\mu = \text{constant} = a^\mu, \quad b^\mu, \tag{10.5}$$

respectively. Let ζ_1 be a null geodesic from B to A, with parametric equations given by

$$x^i = x^i(\lambda), \tag{10.6}$$

with $x^\mu(0) = b^\mu, x^\mu(1) = a^\mu, x^0(0) = ct_B, x^0(1) = ct_A$. What does our geodesic correspond to in physical terms?

It describes a light ray leaving observer B at time t_B and reaching observer A at time t_A. Because of the static nature of the line element, we also have another null geodesic solution given by

$$x^\mu = x^\mu(\lambda), \quad \mu = 1, 2, 3, \tag{10.7}$$

$$x^0 = x^0(\lambda) + \Delta c, \quad \Delta = \text{constant}.$$

This describes a light ray ζ_2 leaving B at $t_B + \Delta$ and reaching A at $t_A + \Delta$. Figure 10.2 illustrates this result.

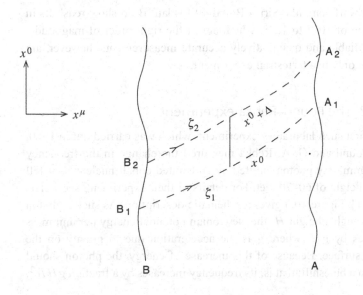

Fig. 10.2. Signal communication in a static spacetime. If B is in a stronger gravitational field than A, the signals from B to A will show a redshift.

Now, in the rest frame of B, the time interval Δ corresponds to a proper time interval (measured by B) of

$$\delta\tau_B = \Delta[g_{00}(b^\mu)]^{1/2}.$$

If n light waves have left B in this time interval, then the frequency of these waves as measured by B is

$$\nu_B = \frac{n}{\Delta}[g_{00}(b^\mu)]^{-1/2}.$$

Since the same *number* of waves is received by A in the corresponding proper time interval $\delta\tau_A$, we get the ratio of frequencies measured by B and A as

$$\frac{\nu_B}{\nu_A} = \left[\frac{g_{00}(a^\mu)}{g_{00}(b^\mu)}\right]^{1/2}. \tag{10.8}$$

This is also the ratio of the wavelengths $\lambda_A : \lambda_B$ measured by A and B, respectively.

If in the Schwarzschild solution B is an observer located on the surface of a star, at $r = r_s$, say, and A is a distant observer with $r \gg 2GM/c^2$, we get

$$\frac{\lambda_A}{\lambda_B} \cong \left(1 - \frac{2GM}{c^2 r_s}\right)^{-1/2}. \tag{10.9}$$

Thus spectral lines from a massive compact star should be *redshifted*. For $2GM/(c^2 r_s)$ small compared with unity, the redshift

$$z = \frac{\lambda_A - \lambda_B}{\lambda_B} \approx \frac{GM}{c^2 r_s}. \tag{10.10}$$

White dwarf stars like Sirius B and 40 Eridani B do show redshifts in the range of 10^{-4} to 10^{-5}, which are of the right order of magnitude. More reliable and quantitatively accurate measurements, however, are possible only in a terrestrial experiment.

10.3.2 The Pound–Rebka experiment

In the first such laboratory experiment, which was carried out in 1960, R. V. Pound and G. A. Rebka measured the change in the frequency of a gamma-ray photon emitted by an excited cobalt nucleus as it fell from a height of 60–70 feet. For details of their experiment, see Reference [11]. Figure 10.3 gives a schematic description. As such a photon falls through a height H, the Newtonian potential energy per unit mass increases by gH, where g is the acceleration due to gravity on the Earth's surface. Because of this increase of energy, the photon should undergo a blueshift; that is, its frequency increases by a fraction gH/c^2.

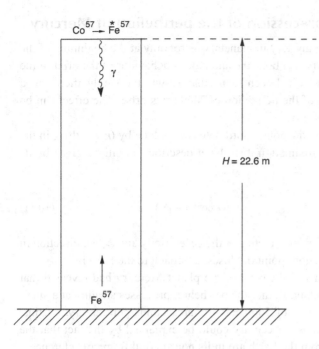

Fig. 10.3. A schematic diagram describing the Pound–Rebka experiment on the measurement of gravitational spectral shift. The upward arrow indicates the alignment of the iron nucleus to receive the gamma ray coming from above.

Although this fraction is as small as 10^{-15}, it can be measured by modern laboratory techniques.

The trick is to have an iron nucleus as the absorber at the bottom. By moving the nucleus at a suitable speed away from the approaching gamma-ray photon falling from above, one can effectively *reduce* its apparent relative frequency with respect to the iron nucleus. When the frequency matches the energy gap between the cobalt and iron nuclei absorption occurs. The speed of the iron nucleus then tells us what blueshift the gamma ray had.

This experiment had been thought of much earlier, but ensuring that there would be no recoil problems with the absorbing nucleus had proved to be difficult, until the discovery of the Mössbauer effect. The nucleus could then be held in a crystal. The recoil was largely borne by the holder as per the Mössbauer effect.

The Pound–Rebka experiment and later work have confirmed the gravitational redshift effect to a high level of accuracy.

Notice, however, that we used Newtonian gravity applied to the photon to derive the expected result. We could have also used the relativistic formula (10.8) with the approximation on the surface of the Earth, with the same result. Thus a defender of Newtonian gravity could argue that the formula (10.10) does not uniquely confirm general relativity. So we now turn to two other tests, which do precisely that.

10.4 The precession of the perihelion of Mercury

An outstanding mystery in planetary astronomy at the beginning of the twentieth century had been the anomalous behaviour of the orbit of the planet Mercury. It had been found that, slowly but surely, there was a secular motion of the perihelion of Mercury's orbit. The effect can be described as follows.

If we denote the polar coordinates on a plane by (r, ϕ), then, in the Newtonian approximation, the planet describes an ellipse given by its polar equation

$$\frac{l}{r} = 1 + e \cos(\phi - \phi_0). \tag{10.11}$$

Here l is the semi latus rectum, e the eccentricity and ϕ_0 the direction in which its perihelion (point of closest approach to the Sun) lies.

Observations of the orbit of the planet Mercury had revealed that ϕ_0 is not a constant. Rather the perihelion precesses steadily at a small but perceptible rate of 5600 ± 0.401 arcseconds per century. Of this, ~ 5025 arcseconds per century could be explained by the fact that the observations from the Earth are in its non-inertial frame of reference – the Earth spins about an axis and also goes round the Sun. Thus there was a remaining amount of about 575 arcseconds to explain. Of this, all but 43 arcseconds per century could be accounted for by the pertur- bation of Mercury's orbit by the Newtonian gravitational effect of other planets. How should one account for the remaining 43 arcseconds per century? Notice that the amount to be explained is minuscule: it works out that Mercury's perihelion is seen to advance by about a 35 000th part of a degree at the end of one orbit (see Figure 10.4). The fact that

Fig. 10.4. A grossly exaggerated picture of the advance of the perihelion of Mercury. The actual effect as described in the text is minute but significant.

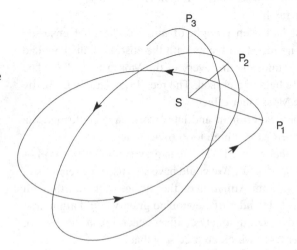

this discrepancy was a matter for worry shows the high expectations scientists (astronomers and physicists) had of the Newtonian paradigm.

Indeed in the 1860s, when the discrepancy became a matter of concern, U. J. J. Leverrier in Paris had offered a solution to the problem on the basis of his experience two decades earlier when a discrepancy had been noticed in the orbit of Uranus. At that time Leverrier had suggested that the orbit of Uranus was being perturbed by a new planet in the vicinity. (A similar prediction had been made by John Couch Adams.) The explanation had worked and a new planet, later called *Neptune*, was identified as the cause of the discrepancy. This time, Leverrier suggested a similar explanation: look for an intramercurial planet as the perturber of Mercury's orbit. He even named the new planet *Vulcan*. Alas, no such planet was found despite exhaustive searches. So the discrepancy remained a possible demonstration of a failure of Newton's laws of gravitation and motion.

This was taken as a challenge by proponents of the general theory of relativity. Taking the Sun as the source mass M in the Schwarzschild solution and Mercury as a test particle, one can work out the orbit of Mercury. Following the treatment of the problem given in Section 9.3, we arrive at the following equation for a planetary orbit:

$$\frac{1}{r^4}\left(\frac{dr}{d\phi}\right)^2 = \frac{\gamma^2}{h^2} - \left(1 - \frac{2GM}{r}\right)\left(\frac{1}{h^2} + \frac{1}{r^2}\right). \qquad (10.12)$$

We simplify this by writing $u = 1/r$:

$$\left(\frac{du}{d\phi}\right)^2 = \frac{\gamma^2 - 1}{h^2} - u^2 + \frac{2GMu}{h^2} + 2GMu^3. \qquad (10.13)$$

Differentiate this relation with respect to ϕ. After taking out the common factor $du/d\phi$, we get the equation as

$$\frac{d^2u}{d\phi^2} + u = \frac{GM}{h^2} + 3GMu^2. \qquad (10.14)$$

A comparison with Newtonian orbital dynamics will reveal that here we have an extra term on the right-hand side. In the Newtonian framework such a term would have arisen from an *extra force* obeying an inverse fourth-power (r^{-4}) law. The extra force is small compared with the Newtonian force. Thus its effect on the Newtonian orbit would be small. We will try to estimate it with the problem of Mercury in view.

Let us write the solution of the purely Newtonian equation (i.e., without the last term) as

$$u_0 = \frac{1}{l}[1 + e\cos(\phi - \phi_0)], \quad l = \frac{h^2}{GM}. \qquad (10.15)$$

Next we try a perturbative approach to solve Equation (10.14) by substituting for u the above solution for the (small) second term on the right-hand side of the equation. Thus we write

$$\frac{d^2 u}{d\phi^2} + u = \frac{GM}{h^2} + \frac{3GM}{l^2}[1 + 2e\cos(\phi - \phi_0) + e^2\cos^2(\phi - \phi_0)]. \quad (10.16)$$

Try a solution of this equation in the following form:

$$u = u_0 + u_1 + u_2, \quad (10.17)$$

where

$$\frac{d^2 u_1}{d\phi^2} + u_1 = \frac{6GM}{l^2} e\cos(\phi - \phi_0),$$

$$\frac{d^2 u_2}{d\phi^2} + u_2 = \frac{3GM}{l^2}[1 + e^2\cos^2(\phi - \phi_0)].$$

The complementary functions for these equations are $\cos\phi$ and $\sin\phi$. The right-hand side of the second of the above two equations does not contain any of these functions, so we will get a bounded oscillatory solution from this equation. Such a solution will not explain a secular behaviour like that shown by the perihelion of Mercury.

The other equation, namely the first of the above two equations, does have these functions on the right-hand side, so we do expect a secular solution here. Indeed, a particular integral of the differential equation is

$$u_1 = \frac{3GMe}{l^2}\phi\sin(\phi - \phi_0). \quad (10.18)$$

On adding this to the Newtonian solution, we get the approximate secular solution as

$$u \cong u_0 + u_1 = \frac{1}{l}\left[1 + e\left(\cos(\phi - \phi_0) + \frac{3GM}{l}\phi\sin(\phi - \phi_0)\right)\right]$$

$$\cong \frac{1}{l}[1 + e\cos(\phi - \phi_0 - \epsilon)],$$

where we have assumed that

$$\epsilon = \frac{3GM}{l} \times \phi \quad (10.19)$$

is a small quantity so that $\cos\epsilon$ is taken as unity and $\sin\epsilon$ is approximated by ϵ. Thus we see that the direction to the perihelion is not constant at the value ϕ_0, but changes its magnitude at a steady rate as illustrated. This precession of perihelion at a rate computed over a period T of one orbit (when ϕ changes by 2π) around the Sun ($M = M_\odot$) is given by

$$n = \frac{6\pi GM_\odot}{lT c^2}. \quad (10.20)$$

On putting in the values $l = 5.53 \times 10^{12}$ cm and $GM_\odot = 1.475 \times 10^5$ cm and one century $= 415T$, we get n as 43.03 arcseconds (per century). This was an excellent resolution of a long-standing anomaly and it went a long way towards establishing the credibility of general relativity in the minds of physicists and astronomers. The value of n is largest for Mercury, which of all the planets has the orbit which is most eccentric and closest to the Sun. Given the same major axis, the latus rectum l in the denominator of formula (10.20) is small for an orbit of high eccentricity.

10.4.1 The binary pulsar

In the late 1970s, a more dramatic example of such a precession was observed for the periastron of the binary star system that houses the pulsar PSR 1913+16. Here the gravitational effects are stronger than in the Sun–Mercury system, and the precession rate is as high as 4.23 degrees per year – about 3.6×10^4 times the value for Mercury. (See Reference [12].) We should caution the reader, however, that, unlike in the Sun–Mercury case where, because of the large disparity of their masses, the Sun could be considered at rest and Mercury moving around it as a 'test' particle, in the binary pulsar case the two stars have comparable masses and hence the Schwarzschild solution is not strictly applicable. Ideally one should solve the relativistic two-body problem. This has not been possible so far, so only an approximate extrapolation of the Sun–Mercury problem is generally used for the above theoretical comparison.

10.5 The bending of light

The perihelion precession of Mercury was the best of the three tests of general relativity in establishing a clear stamp of the theory. However, in terms of a popular impact, the test to be described now played a key role in establishing the superiority of general relativity over Newtonian gravity and making Einstein a celebrity.

Does gravity affect light by bending its direction? When Newton described his law of gravitation as universal, he meant it to be applicable to all forms of matter, large and small. But what about light? Did the universality extend to light rays also? Newton was not sure. We have mentioned his query on this issue in Section 9.4.

Nevertheless, if we take some liberties with the Newtonian concept, we can get an affirmative reply to this query. Imagine light made of particles, e.g., photons. A photon of frequency ν would have an energy $h\nu$ and hence a mass equivalent of $h\nu/c^2$. Let such a particle approach

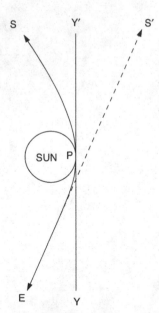

Fig. 10.5. The direction from the Earth (E) to the star (S) is changed because of the 'bending' of light by gravity. The star image accordingly is shifted (to S'). This shift, in reality, is quite small and is shown in an exaggerated form in this figure.

a mass M from infinity such that its asymptotic direction of motion passes it at a distance b. Assume that the particle had velocity c at infinity and it travelled like a typical particle in Newtonian dynamics. What will be the final direction of the particle as it recedes from M at infinity? The bending angle works out as $2GM/(c^2R)$, where R is the distance of closest approach of the particle. In what we will refer to as the 'Newtonian prediction' we imply this value.

While developing the general theory of relativity, Einstein had an earlier version that gave exactly this answer. In 1911 he put it up for testing should a suitable opportunity arise. The way to test the result is explained in Figure 10.5. The source of light, say a star S, is viewed from the Earth, E, on two occasions, once when the Sun is grazing its line of sight as shown in the figure and once when the Sun is nowhere near the line ES. As shown in Figure 10.5, the light ray from the star approaches the Sun at a closest point P on its surface. Although the ray is said to be bent by gravity, as we know, it is really following a null geodesic in the curved spacetime produced by the Sun's gravity. Clearly, under normal circumstances it is not possible to sight the star with the Sun in the foreground. The exceptional situation when we can see the star is when the Sun is totally eclipsed. If one measures the direction of the star at this stage and compares it with the direction under normal circumstances (when the Sun is nowhere near) one should find a small difference, corresponding to the bending angle predicted in Figure 10.5.

There was an eclipse due in 1914 that would be visible from Russia and a team of scientists from Germany went to observe it and to perform this experiment. But World War I broke out and the team members were interned as aliens from an enemy country. This turned out to be fortunate for Einstein, in a way; for in 1915, when he arrived at the final form of relativity, which we are studying here, he found that under it the correct answer was *twice* what he had got earlier. In short, his prediction of the bending angle had changed. It would have been embarrassing for him had the 1914 expedition gone ahead and found a result that disagreed with his then prediction. Meanwhile, amongst the few astronomers who understood what relativity was all about was Arthur Stanley Eddington at Cambridge. Eddington proposed an eclipse expedition that would test Einstein's claim after the war had ended. A total solar eclipse in 1919, visible from a band in the southern hemisphere, was of sufficiently long duration to attempt the observations. Fortunately the war ended in 1918 and a financial grant of one thousand pounds from the British Government enabled Eddington to execute his plans.

The trials and tribulations of the experiment and its report to the joint meeting of the Royal Society and the Royal Astronomical Society on 6 November 1919 have been described in a very absorbing account by Peter Coles [13]. We will next carry out a brief calculation of the expected result.

Just as timelike geodesics determine the tracks of planets, we can calculate the track of a light ray by solving the equations of its null geodesic. These equations were written down in general and solved in the case of the Schwarzschild exterior spacetime geometry in the last chapter: *vide* Equations (9.44). These can be combined to form a single equation given by

$$\frac{1}{r^4}\left(\frac{dr}{d\phi}\right)^2 + \frac{1}{r^2} - \frac{2GM}{r^3} = \frac{1}{b^2}. \tag{10.21}$$

In terms of $u = 1/r$ this equation takes the form

$$\left(\frac{du}{d\phi}\right)^2 + u^2 = \frac{1}{b^2} + 2GMu^3. \tag{10.22}$$

Differentiation with respect to ϕ gives

$$\frac{d^2u}{d\phi^2} + u = 3GMu^2. \tag{10.23}$$

We will solve this non-linear equation in a perturbative fashion as we did in the case of Mercury's orbit. Looking at Figure 10.5, we first represent the tangential straight line YPY' as the zeroth-order (no bending) approximation:

$$u_0 = \frac{1}{R_0}\cos\phi. \tag{10.24}$$

Here P is the point of closest approach to the Sun and it is given by the maximum value of u_0, i.e., by $\phi = 0$. R_0, the closest distance, is, ideally, the Sun's radius. On substituting this solution into the right-hand side of Equation (10.23) we get the next approximation satisfying

$$\frac{d^2u}{d\phi^2} + u = \frac{3GM}{R_0^2}\cos^2\phi, \tag{10.25}$$

which has the solution

$$u = \frac{1}{R_0}\cos\phi + \frac{GM}{R_0^2}(2 - \cos^2\phi). \tag{10.26}$$

Now, looking at Figure 10.5, we see that the above equation describes the curve EPS, the asymptotes of which are denoted by $u = 0$, $\phi = \pm\phi_0$, where

$$\phi_0 = \left\{\frac{\pi}{2} + \frac{2GM}{R_0}\right\} \tag{10.27}$$

since we expect $2GM/R_0$ to be small compared with unity. The net deflection of the light ray is therefore given by

$$\Delta\phi = \frac{4GM}{R_0 c^2} = 1.75 \text{ arcseconds,} \qquad (10.28)$$

when we substitute the values for the Sun. To enable the numerical computation we have restored the speed of light c to its proper place.

In his report to the scientific meeting in November 1919, Eddington compared his values of $\Delta\phi$ with the theoretical predictions of 1.75 arcseconds (Einstein) and 0.875 arcsecond (Newton). He had measurements from two places, from Sobral in Brazil and Principe in Guinea. His conclusion was in favour of Einstein. This observational confirmation, besides its high impact on the media, went a long way towards establishing the credibility of general relativity in the eyes of the physicists.

Yet, looking back at those results in a dispassionate way, one could point to several uncertainties that might have weakened that conclusion! The experiment itself was not performed under the best of conditions and the equipment used had room for improvement. The positions of the stars when the Sun was not in the vicinity could not be measured from the same location, thus introducing a possible source of error. Also, the optical refraction effects in the upper layers of the solar atmosphere were not adequately estimated. Thus there was a strong reason for repeating the experiment whenever the eclipse opportunity presented itself.

10.5.1 Measurement at longer wavelengths

Subsequent attempts by optical astronomers yielded somewhat inconclusive results, largely because of the limited sensitivity of the measuring equipment and the uncertain nature of systematic errors. In the 1970s, however, measurements with microwaves confirmed the above bending angle much more precisely with only about 5% experimental error. Pioneering measurements were made by Counselman and others in 1974 (see Reference [14]) and by E. B. Fomalont and R. A. Sramek [15] in 1975. The technique used was to observe the quasar 3C-279, whose line of sight intersects the Sun every October. Since the Sun is not a powerful radiator of energy at these wavelengths, there is no need to wait for a solar eclipse in order to make the measurements. The direction to another source, 3C-273, was used as a reference point for measuring the shift in angle.

This technology has subsequently been improved to reduce the error bars further, as can be seen in Table 10.1 at the end of this chapter.

10.6 Radar echo delay

Just as the direction of a light ray is altered by the Sun's gravity, so is its apparent travel time. This effect, which was first highlighted by L. I. Schiff, can also be calculated in a straightforward manner. We first derive the result and then discuss how it is put to test. We again refer to the null-geodesic equations (9.44) in the Schwarzschild spacetime. From these we get, by eliminating the independent variable λ, the following equation:

$$\left(\frac{dr}{dt}\right)^2 = \left(1 - \frac{2GM}{r}\right)^2 - \frac{b^2}{r^2}\left(1 - \frac{2GM}{r}\right)^3. \tag{10.29}$$

Let KPE denote a null-geodesic track from a planet K to Earth E, grazing the Sun's surface at P as shown in Figure 10.6. The radial Schwarzschild coordinate r is measured from the Sun's centre. At P, the conditions are $r = R_0$, the Sun's radius, while the 'shortest-distance' requirement means $dr/dt = 0$ at P. Let $r = R_1$ at K and $r = R_2$ at E. What is the time taken by light to travel the path KPE?

We break the answer into two bits, the time T taken by light to go from K to P and the time taken from P to E:

$$T = f(R_1, R_0) + f(R_0, R_2), \tag{10.30}$$

where the f functions are formally similar and defined by the integrals

$$f(R_1, R_0) = \int_{R_0}^{R_1} \frac{dr}{(1 - 2GM/r)}\left\{1 - \frac{b^2}{r^2}\left(1 - \frac{2GM}{r}\right)\right\}^{-1/2} \tag{10.31}$$

and

$$f(R_0, R_2) = \int_{R_0}^{R_2} \frac{dr}{(1 - 2GM/r)}\left\{1 - \frac{b^2}{r^2}\left(1 - \frac{2GM}{r}\right)\right\}^{-1/2}. \tag{10.32}$$

Consider the first integral and let $D(r)$ denote the denominator in it. Requiring that $dr/dt = 0$ at P means that the function $D(r)$ vanishes at $r = R_0$. This determines b and we have

$$D(r) = \left(1 - \frac{2GM}{r}\right)\left\{1 - \frac{R_0^2}{r^2}\left(1 - \frac{2GM}{r}\right)\right\}^{1/2}. \tag{10.33}$$

We now implement the weak-field approximation by using the fact that $r > R_0 \gg 2GM$. Some straightforward but tedious algebra then leads us to

$$D(r) \cong \left(1 - \frac{R_0^2}{r^2}\right)^{1/2}\left\{1 - \frac{GMR_0}{r(r + R_0)} - \frac{2GM}{r}\right\}. \tag{10.34}$$

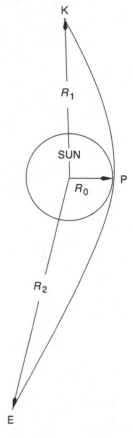

Fig. 10.6. The echo of a radar signal bounced off a Solar-System body K may arrive late if the signal path goes close to the Sun. This, as explained in the text, happens because of the geometry of curved spacetime near the Sun.

Therefore, we have

$$f(R_1, R_0) \cong \int_{R_0}^{R_1} \left(1 - \frac{R_0^2}{r^2}\right)^{-1/2} \left\{1 + \frac{GMR_0}{r(r + R_0)} + \frac{2GM}{r}\right\} dr. \quad (10.35)$$

In the absence of any local gravitational field we would have got the simple expression of Euclidean geometry for $f(R_1, R_0)$:

$$f(R_1, R_0) \cong \int_{R_0}^{R_1} \left(1 - \frac{R_0^2}{r^2}\right)^{-1/2} dr. \quad (10.36)$$

We thus see that the time taken by light to travel is *increased* in the presence of local gravitational sources. In principle we could measure this effect by bouncing a radar signal from a planet when it is in superior conjunction with respect to the Sun and the Earth; that is, when the radar signal to and from the planet grazes the Sun. By comparing the to and fro time for the signal with the radar time taken when the Sun is nowhere near the signal path one can test the above prediction. The estimated effect for Mercury is the highest and close to 200 μs. However, in practice there are several sources of errors in this procedure. The distance of the planet must be known to an accuracy of 1.5 km or so to ensure that the error of the travel time does not exceed 10 μs. Also the bounce region on the surface of the planet should be relatively small to minimize the spread in the arrival of the return signal.

In the 1970s, the first serious measurements were made by bouncing radar signals emitted from the spacecraft Mariner 6 and 7 off the surface of the Earth as the signals grazed the solar limb. The expected delays of the order of 200 μs were observed within 3% error bars [16]. This test has also been made more accurate with time and Table 10.1 gives the updated information. However, the experiment performed by the Cassini spacecraft on 10 October 2003 during its mission to Saturn improved the accuracy of the experiment so as to reduce the error bars to 0.002% [17]. Here signals were bounced between the spacecraft and the Earth as they grazed the Sun in between.

10.7 The equality of inertial and gravitational mass

An important consequence of the principle of equivalence is the equality of inertial and gravitational mass. A little thought will convince us that Galileo's experiment from the Leaning Tower of Pisa, which demonstrated that all bodies fall freely with equal rapidity, is an essential part of Einstein's thought experiment involving the freely falling lift. Both experiments are possible because the same quantity enters the law of motion as inertial mass and the law of gravitation as gravitational mass.

Experiments with lunar laser ranging have been successful in measuring the distance of the Moon from the Earth to within a few centimetres. Such experiments also demonstrate that the Moon moves around the Earth as predicted by the equations of general relativity. In particular, these experiments ruled out certain alternative theories of gravitation, like the Brans–Dicke theory, that allow for the variation of the inertial mass of a moving object as a function of its distance from another mass.

Laboratory experiments of the torsion-balance type have been conducted very accurately with various materials to establish this equality with high sensitivity. Such experiments place stringent upper limits on the possible presence of a 'fifth force' operating at a range of a few metres. For a review of the measured accuracy of the principle of equivalence, see the article by C. M. Will [18].

10.8 Precession of a gyroscope

Although the Schwarzschild solution describes the gravitational effects of the Sun or the Earth with great accuracy, there is scope for further improvement. For instance, a rotating mass would introduce a $d\phi\, dt$ term into the metric. Although the effects of such terms are very small for the Earth or the Sun, modern technology should be able to measure them.

A proposed experiment that can measure the effect of a rotating mass makes use of gyroscopes. The axis of a gyroscope sent on an equatorial orbit around the Earth will slowly precess. An estimated rate of precession of ~ 7 arcseconds per year can be detected with present technology, and such an experiment has been on the drawing board for four decades, but not yet performed. The Gravity Probe B mission at present in space has promised a result by the year of writing of this account (2009).

Table 10.1 gives the measured values of the PPN parameters or, rather, the limits set on their deviation from the predictions of general relativity. Although the experiments described go beyond what we have outlined above, it is clear from the entries of Table 10.1 that the theory of relativity comes out with flying colours.

Exercises

1. A photon of energy 1 MeV travels from the Earth to the Moon. By looking up physical data on the Earth and the Moon, calculate its energy upon arrival at the Moon.

2. Use formula (10.20) to calculate the perihelion precession of Pluto, for $e \cong 0.25$, perihelion distance 29.7 AU, aphelion distance 49.1 AU and orbital period 248 years.

Table 10.1. *Limits on the measured values of the PPN parameters (based on data reviewed by C. M. Will)*

Parameter	Effect	Limit	Remarks
$\gamma - 1$	Time delay	2×10^{-3}	Viking ranging
	Light deflection	3×10^{-4}	VLBI
$\beta - 1$	Perihelion shift	3×10^{-3}	$J_2 = 10^{-7}$ from helioseismology
	Nordtvedt effect	6×10^{-4}	$\eta = 4\beta - \gamma - 3$ assumed
ξ	Earth tides	10^{-3}	Gravimeter data
α_1	Orbital polarization	4×10^{-4}	Lunar laser ranging
		2×10^{-4}	PSR J2317 + 1439
α_2	Spin precession	4×10^{-7}	Solar alignment with ecliptic
α_3	Pulsar acceleration	2×10^{-20}	Pulsar $\dot{P}$ statistics
η	Nordtvedt effect[1]	10^{-3}	Lunar laser ranging
ζ_1	–	2×10^{-2}	Combined PPN bounds
ζ_2	Binary acceleration	4×10^{-5}	$\ddot{P}_p$ for PSR 1913+16
ζ_3	Newton's third law	10^{-8}	Lunar acceleration
ζ_4	–	–	Not independent

[1] Here $\eta = 4\beta - \gamma - 3 - 10\xi/3 - \alpha_1 - 2\alpha_2/3 - 2\zeta_1/3 - \zeta_2/3$.

N.B. The general theory of relativity predicts that all entries in this table are zero.

3. The Newtonian escape velocity of a massive star is v. Show that it bends light by an angle of $2v^2/c^2$.

4. An equilateral triangle is described by the space-tracks of light rays grazing a spherical object of mass M and Schwarzschild coordinate radius R_0. Show that the sum of the three angles of this triangle exceeds π by an amount

$$\frac{21\sqrt{3}}{4} \frac{GM}{R_0 c^2}$$

in the approximation $2GM \ll R_0 c^2$.

5. A source of light moves in a circular orbit of Schwarzschild coordinate radius $2GM/\zeta$ about a spherical mass M in an otherwise empty space. It emits light in the forward direction of its motion, of frequency ν_0 in its rest frame. This is received by a remote observer located at rest at a Schwarzschild coordinate radius $R \gg 2GM$, with a frequency ν in *its* rest frame. Show that, for ζ approaching $\frac{2}{3}$ from below,

$$\frac{\nu}{\nu_0} = \frac{(1 - \frac{3}{2}\zeta)(2 - 2\zeta)^{1/2}}{(2 - 2\zeta)^{1/2} - \zeta^{1/2}}.$$

6. The Newtonian potential for an oblate sun has the form

$$\phi = \frac{GM_\odot}{R} \left\{ 1 - J \left(\frac{R_\odot}{R} \right)^2 P_2(\cos\theta) \right\}$$

where J is the quadrupole-moment parameter. Show that this produces a per-ihelion precession (of purely Newtonian origin) of a planet with orbital latus rectum $2l$ by an angle

$$\frac{3\pi R_\odot^2}{l^2} J$$

per orbit ($R_\odot$ is the radius of the Sun). Estimate this effect for Mercury in terms of J. (The present estimate of J is $J \lesssim 2.5 \times 10^{-5}$.)

7. Assume Newtonian physics and imagine a particle sent towards a spherical mass M with velocity c along a direction such that the perpendicular distance of the centre of the mass from that line of motion is R. If $GM \ll c^2 R$, find the distance R_0 of closest approach of the particle to M and show that the particle will eventually be moving away from the gravitating mass in a direction asymptotically making an angle $2GM/(c^2 R_0)$ with the original direction of motion. (Note that this is *half* the value predicted by general relativity.)

Chapter 11
Gravitational radiation

11.1 Introduction

Do Einstein's equations permit the existence of gravitational waves? As in the case of electromagnetism, where Maxwell's field equations led to the important deduction of electromagnetic waves carrying energy and momentum with the speed of light, one expects the relativistic equations to imply the existence of gravitational waves that do the same. However, several issues intervene to make the answer to our question non-trivial.

The first problem is posed by the non-linearity of the Einstein field equations. In the wave motion discussed in electromagnetic theory, acoustics, elastic media, etc. the basic equations are linear and a superposition principle holds. There is no corresponding situation in general relativity. Secondly, there is no corresponding vector or tensor in relativity that plays the role of the Poynting vector in the transport of electromagnetic energy.

A third difficulty arises from the general covariance of the field equations. With the facility available to use *any* coordinate system as per convenience, it is not clear whether a particular 'wavelike' solution is a real physical effect or a pure coordinate effect. Thus one has to be on guard against solutions that describe coordinate waves that may travel 'with the speed of thought'.

Even during Einstein's lifetime, the above question did not receive an unequivocal answer. His long-time coworker Leopold Infeld has narrated one incident when Einstein thought that he had a proof that disproved the existence of gravitational waves, only to find at the last moment before he was to give a seminar on this finding that his proof broke down in a tautology.

In the post-Einstein years, however, relativists like Hermann Bondi, Ivor Robinson, Felix Pirani and Roger Penrose were able to sort out real physics from the coordinate effects and generate confidence that gravitational waves exist. Still, the general non-linear description of a wave is too complicated for this elementary treatise and we will study only the linearized version in this chapter. It is this version that is relevant to the practical issue of *detecting* gravitational waves passing through a man-made receiver.

11.2 Linearized approximation

We return to the 'weak-field' approximation discussed in Chapter 8; but this time we drop the restriction of 'slow motion' and retain all time derivatives. Then we have

$$R_{iklm} \cong \frac{1}{2}[h_{im,kl} + h_{kl,im} - h_{km,il} - h_{il,km}]. \tag{11.1}$$

There is, however, a freedom of coordinate transformation still available. This allows us to choose certain auxiliary conditions as follows. Define

$$\psi_i^k = h_i^k - \frac{1}{2}h\delta_k^i, \quad h = h^k{}_k \tag{11.2}$$

and choose coordinates such that

$$\psi_{i,k}^k = 0. \tag{11.3}$$

In analogy to a similar transformation of the electromagnetic potentials A_k, which ensures that $A^k{}_{,k} = 0$, this transformation is called a *gauge transformation* and the ψ_i^k are called gravitational potentials. The conditions (11.3) are called the gauge equations. We still have a freedom of coordinates available. Thus, if we try a coordinate transformation given by

$$x'^i = x^i + \xi^i; \quad \Box\xi^i = 0, \tag{11.4}$$

we still satisfy (11.3) for the primed coordinates. We shall have occasion to use this facility later.

With (11.3) holding, we get for the various relevant tensors

$$R_{ik} \cong \frac{1}{2}\Box h_{ik}, \quad R \cong \frac{1}{2}\Box h,$$

$$R_{ik} - \frac{1}{2}\eta_{ik}R \cong \frac{1}{2}\Box\psi_{ik}.$$

The wave operator of course relates to the Minkowski spacetime of special relativity. From Einstein's field equations, the above relation leads to

$$\Box \psi_{ik} = -16\pi G T_{ik}. \tag{11.5}$$

We can write a formal solution to this equation, assuming that the sources (T_{ik}) are confined to a bounded compact 3-volume V:

$$\psi_{ik}(t, \mathbf{r}) = 4G \int_V \frac{T_{ik}(t - |\mathbf{r} - \mathbf{R}|)}{|\mathbf{r} - \mathbf{R}|} \, d^3\mathbf{R}. \tag{11.6}$$

What type of sources does one talk about? We shall return to this important issue shortly.

11.2.1 Plane waves

A different type of solution to that given by a compact source is one of plane waves. (Compare this case with that of a plane electromagnetic wave.)

Take a coordinate system as $x^0 = t$, $x^1 = x$, $x^2 = y$, $x^3 = z$, with, as usual in this discussion, $c = 1$. For the present problem we assume that apart from the wave the space is empty. So the equations (11.5) yield

$$\Box h_{ik} = 0. \tag{11.7}$$

For a plane wave travelling in the x direction all h_{ik} are functions of $(t - x)$ only. Hence the gauge conditions become

$$0 = \frac{\partial \psi_i^1}{\partial x} + \frac{\partial \psi_i^0}{\partial t} = -\frac{\partial \psi_i^1}{\partial t} + \frac{\partial \psi_i^0}{\partial t}. \tag{11.8}$$

so that $[\psi_i^0 - \psi_i^1]$ is a function of x only. However, if we are admitting only wave-type physical functions, then this may be set equal to zero. Thus we have $\psi_i^0 = \psi_i^1$.

Next, using the freedom provided by Equation (11.4), we can use the arbitrary ξ^i as a function of $(t - x)$ to make the following quantities vanish:

$$\psi_1^0, \ \psi_2^0, \ \psi_3^0, \ \psi_2^2 + \psi_3^3.$$

Since $\psi_i^0 = \psi_i^1$, we also have zero values for ψ_1^1, ψ_2^1 and ψ_3^1. Further, $\psi_0^0 = \psi_0^1 = -\psi_1^0 = 0$. Hence $\psi_i^i = 0$ and we have $h_{ik} = \psi_{ik}$. In short, the only quantities that cannot be rendered zero are ψ_2^3 and $\psi_2^2 - \psi_3^3$. In terms of the h_{ik}, we can say that the plane wave is characterized by two functions,

$$h_{22} = -h_{33} \quad \text{and} \quad h_{23}.$$

This is the most elementary plane-wave solution. It may be compared with the plane electromagnetic wave travelling in the x direction having an electric field E_2 in the y direction and a magnetic field B_3 in the z direction, with $|E_2| = |B_3|$.

What is the energy flux in such a wave, analogous to the Poynting vector

$$\Phi = \frac{c}{4\pi}(\mathbf{E} \times \mathbf{B})$$

for the electromagnetic case? In Newtonian gravitation one can define an energy density of gravitational field in terms of the potential ϕ as follows:

$$\epsilon = -\frac{(\nabla\phi)^2}{8\pi G}.$$

The notion of gravitational energy in general relativity has been a subject of a lot of discussion and controversy, especially since relativity discards the notion of force in the context of gravity. Nevertheless, a limited meaning can be attached to the concept of gravitational energy density and flux of that energy in a wave. Since a detailed description of the topic would lead us into technical details away from the main theme we are describing, we will simply quote the result, leaving the reader to look up advanced texts such as References [19, 20]. In the case of the plane wave described above, the *energy flux of a plane gravitational wave* is given by

$$\mathcal{F} = \frac{c^2}{16\pi G} \times \left[\dot{h}_{23}^2 + \frac{1}{4}(\dot{h}_{22} - \dot{h}_{33})^2 \right].$$
(11.9)

The over dot denotes differentiation with respect to t. Thus we have a well-defined expression for the energy carried by a gravitational wave as it travels with the speed of light in vacuum. We will next look at *sources* that can produce gravitational waves and the way they can be detected.

11.3 Radiation of gravitational waves

Again we refer the reader to advanced texts (e.g., References [20, 21]) for the somewhat intricate manipulation needed to derive an apparently simple result, which we just state below. Assuming that we have a compact time-dependent source confined to a 3-volume V, the gravitational wave emerging from it will appear to a remote observer as a plane wave passing by him. Using local coordinates in which the x-axis is along the direction of propagation of the wave, we can use the results of the previous section. Calculations give, for a source located at distance R_0,

$$h_{23} = \frac{2G}{3R_0}\dddot{D}_{23}, \qquad h_{22} - h_{33} = \frac{2G}{3R_0}(\dddot{D}_{22} - \dddot{D}_{33}), \qquad (11.10)$$

where we define the quadrupole-moment tensor of the source by the standard formula:

$$D_{\alpha\beta} = \int \rho(3x^\alpha x^\beta - \delta_{\alpha\beta}r^2)dV. \qquad (11.11)$$

Here the coordinates x^α are Cartesian and r denotes the Euclidean distance from the origin of a point with these coordinates.

As in the case of the oscillating electric dipole radiating electromagnetic waves, we can work out the net loss of energy by the source above through gravitational radiation. The answer comes out to be

$$\mathcal{P} = \frac{G}{45c^5}\dddot{D}_{\alpha\beta}^2. \qquad (11.12)$$

Thus $\mathcal{P}$ is the power radiated by the source leading to its energy reservoir being depleted.

We have deliberately restored the velocity of light to its rightful place in the above formula. Its high power (5) serves to tell us that, unless the third time derivative of the quadrupole moment is enormously high, the power radiated is insignificant.

Let us estimate the energy radiated by a laboratory source that is in the form of two masses of M kg each, going round each other with a period of a millisecond, the overall length scale of the apparatus being L metres. A crude estimate of the quadrupole moment of this system is

$$D \cong M \times L^2 \times 10^7 \, \text{g cm}^2,$$

while the triple time derivative of it would be as high as $ML^2 \times 10^{16}$ c.g.s. units. Now multiply by the coefficient $G/(45c^5)$, which is approximately 5×10^{-62}. Thus we get a minuscule power of $\sim 5 \times 10^{-24}$ erg per second for $M = 10$ and $L = 10$, say. Clearly it is very unlikely that a terrestrial technology can in the near future produce a laboratory-based source of gravity waves of any practical significance.

11.4 Cosmic sources of gravitational waves

Given the above calculation, it is clear that we need to look to the cosmos for sources strong enough to be noticed. We outline some that are likely to play a significant role in gravitational-radiation astronomy.

11.4.1 Coalescing binaries

Consider a binary star system of two stars with masses m_1 and m_2 moving round one another. As is well known, the Newtonian equations of motion are given by

$$m_1\ddot{\mathbf{r}}_1 = Gm_1m_2\frac{(\mathbf{r}_2 - \mathbf{r}_1)}{|\mathbf{r}_2 - \mathbf{r}_1|^3},$$

$$m_2\ddot{\mathbf{r}}_2 = Gm_1m_2\frac{(\mathbf{r}_1 - \mathbf{r}_2)}{|\mathbf{r}_2 - \mathbf{r}_1|^3},$$

These are solved by going to the barycentric frame of reference. The centre of mass has coordinates

$$\mathbf{R} = \frac{m_1\mathbf{r}_1 + m_2\mathbf{r}_2}{m_1 + m_2}, \tag{11.13}$$

and it satisfies the condition $\dot{\mathbf{R}} = $ constant. Since gravitational radiation involves the third time derivative of the quadrupole moment (*vide* Equation (11.12)), without loss of generality we may set $\mathbf{R}$ as constant. We also have, from (11.13),

$$\mathbf{r}_1 = \mathbf{R} + \left(\frac{m_2}{m_1 + m_2}\right)\mathbf{r},$$

$$\mathbf{r}_2 = \mathbf{R} - \left(\frac{m_1}{m_1 + m_2}\right)\mathbf{r}.$$

As shown in Figure 11.1, $\mathbf{r}$ is the vector $(\mathbf{r}_1 - \mathbf{r}_2)$. A general solution for this vector is the same as for a particle of reduced mass $m_1m_2/(m_1 + m_2)$ moving under the gravitational field of a single mass $(m_1 + m_2)$. Such a particle can describe a bound (elliptical) or unbound (parabolic or hyperbolic) orbit. For a binary star the former type of orbit is relevant. We take the special case of a circular orbit. Assuming an angular speed of ω and orbital radius r, we get the rectangular coordinates of the two

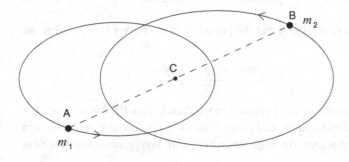

Fig. 11.1. This figure shows stars A and B following their elliptical orbits while their centre of mass C remains stationary. Such a system emits gravitational waves. With loss of energy, the orbits shrink. A and B come closer and closer and ultimately coalesce. BA is the vector **r**.

stars respectively as

$$x_1 = r\cos(\omega t), \qquad y_1 = r\sin(\omega t),$$

$$x_2 = -r\cos(\omega t), \qquad y_2 = -r\sin(\omega t).$$

Using these coordinates and taking note of the above transformations we get for the two masses

$$|\dddot{D}_{11}| = 24\frac{m_1 m_2}{m_1 + m_2}r^2\omega^3|\sin(2\omega t)|;$$

$$|\dddot{D}_{22}| = 24\frac{m_1 m_2}{m_1 + m_2}r^2\omega^3|\sin(2\omega t)| \tag{11.14}$$

so that, averaged over a period, the value of $\dddot{D}_{\alpha\beta}^2$ is

$$288\left(\frac{m_1 m_2}{m_1 + m_2}\right)^2 r^4\omega^6.$$

Thus the radiation rate is

$$\mathcal{P} = \frac{32G}{5c^5}\left(\frac{m_1 m_2}{m_1 + m_2}\right)^2 r^4\omega^6. \tag{11.15}$$

Now the energy of the binary is given by

$$\mathcal{E} = -\frac{Gm_1 m_2}{2r}. \tag{11.16}$$

We also have the third Kepler law from orbital dynamics:

$$r^3\omega^2 = G(m_1 + m_2). \tag{11.17}$$

So, with the loss of energy through gravitational radiation, the magnitude of $\mathcal{E}$ increases and this means that r decreases. From the relations derived above we have

$$-\frac{d\mathcal{E}}{dt} = \mathcal{P}$$

and a little manipulation leads to the rate of shrinkage of the orbit as

$$\dot{r} = -\frac{64G^3 m_1 m_2(m_1 + m_2)}{5c^5 r^3} \tag{11.18}$$

and this in turn tells us that the period P of the binary is reduced at the rate

$$\dot{P} = -\frac{192\pi G^{2.5}(m_1 + m_2)^{1/2}m_1 m_2}{5c^5 r^{2.5}}. \tag{11.19}$$

This phenomenon becomes more dramatic as the binary stars get closer and move faster and faster. The radiation rate also increases and becomes dramatically large when the stars finally coalesce. Needless to say, those looking for evidence for gravitational radiation opt for coalescing binaries as their most favoured option.

Although an example of this kind is still to come, the above result was strikingly but indirectly verified when observations were made of the binary pulsar PSR 1913+16. This is a pulsar that forms a pair with another neutron star. The above result was applied to this binary system. Studies by J. H. Taylor and J. M. Weisberg in 1984 led to the conclusion that the period of this binary system was *decreasing* at the rate of 2.4 picoseconds per second [22]. Such a tiny measurement was possible because pulsars are good timekeepers. The observations of the relative positioning of the two members in the binary system give further information on the shrinking of the orbit. It was shown that the result was consistent with the general relativistic theory of gravitational radiation and did not give such a good fit to some of the alternative theories of gravitation.

11.4.2 Explosive sources

More dramatic than binaries are sources like supernovae, active galactic nuclei, mini-creation events (predicted by an alternative cosmology as mentioned in Chapter 18), etc. in which a large mass undergoes a rapid redistribution over a substantial volume. Recalling the formula (11.12), we carry out a crude order-of-magnitude calculation as follows.

Let M be the mass involved and R its characteristic linear size. Then the quadrupole moment of the distribution will be of the order of $\eta M R^2$. Here η is a dimensionless number, which also includes information on the anisotropy of the matter distribution. For example, a spherically symmetric matter distribution will have $\eta = 0$. Further, let T denote the characteristic time scale for change of the system. Then the third time derivative of the quadrupole moment is given by

$$\dddot{D} \cong \frac{\eta M R^2}{T^3}$$

and formula (11.12) gives the radiation rate as

$$\mathcal{P} \cong \frac{G}{45c^5} \times \frac{(\eta M R^2)^2}{T^6}. \qquad (11.20)$$

Let us take a supernova and set $M = 10 M_\odot$, $R \cong 10^{12}$ cm and $T \cong 10^4$ s. With $\eta \cong 10^{-1}$ the above formula gives

$$\mathcal{P} \cong 2.4 \times 10^{29} \text{ erg s}^{-1}.$$

Compared with the Sun's luminosity, this is a fraction as small as 10^{-4}. It may of course be possible that in a particular frequency band over a short time scale the emission may be much higher than the average calculated above. This now brings up the question of detectors: how are we to detect the signals, which may be very weak astronomically?

11.5 Experimental detection of gravitational waves

How do we detect gravitational waves passing through space in our neighbourhood? As in the case of any wave motion, we need an interacting substance to 'respond' to the passing wave in such a way that we can spot and measure the response. In this case the clue is given by the equation of geodesic deviation derived in Chapter 5. Equation (5.24) may be rewritten in a slightly different notation, as

$$\frac{\delta^2 V^i}{\delta u^2} = R^i_{klm} U^k V^l U^m. \tag{11.21}$$

To recapitulate, as shown in Figure 11.2, two neighbouring geodesics Γ_A and Γ_B, describing two free test particles A and B separated by a small separation vector V^l, come closer or move apart depending on the nature of the ambient spacetime geometry. The above equation shows the variation of V^l with respect to the affine parameter, u, as the particles move along their geodesics. The passage of a wave will change the geometry and hence the value of the driving term $R^i{}_{klm}$. So the mechanical movement of A and B will indicate the presence of the wave.

The practical problem which a detector has to handle is how to translate the acceleration produced by (11.21) so that it can be measured, given that it is a very small effect amidst other environmental and hence noisy effects of larger magnitudes. Since the early 1960s mechanical detectors have been designed to capture the small and elusive gravitational waves. We describe three major attempts.

11.5.1 Bar detectors

Joseph Weber played a pioneering role in designing a detector and using it for measurements. He used cylindrical bars, each with length 153 cm, diameter 66 cm and weight 1.4 tons. Each bar was suspended by a wire in vacuum and mechanically decoupled from its surroundings. Ideally the bar should be completely isolated from its surroundings. The bar has a fundamental frequency of 1660 Hz for lengthwise oscillation. Figure 11.3 shows a bar detector.

The bar has piezoelectric strain transducers. When a gravitational wave passes through the cylinder, different parts of it feel the acceleration and the tendency to be displaced from their normal positions causes them to be strained. The strain transducers respond to these strains and produce

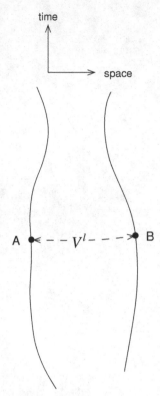

Fig. 11.2. The world lines of two particles A and B are shown. If a gravitational wave passes by, the separation vector V^l of these world lines would change.

Fig. 11.3. A bar detector of gravitational waves located in Frascati near Rome.

corresponding electric fields. Measurements of these fields lead to the estimate of the intensity of the gravitational wave. Because of resonance at 1660 Hz, the detector is most sensitive to waves of frequencies near this value.

Weber had sited a detector in the University of Maryland and another in the Argonne National Laboraotory near Chicago. If the source of gravitational radiation were a distant cosmic one, it would affect both detectors in the same way. Thus Weber took as significant only those results which had the two detectors responding simultaneously. The rest could be dismissed as part of local noise.

Even so, judging by the level of effects he got as significant, their sources had to be much more powerful than those we have looked at here. In short, the gravitational-wave community doubted, despite the care taken with the measurements, that real gravitational waves had been detected. Of course, had there been another detector with a different technology also reporting positively, the results would have been accepted. The controversies relating to the reality of Weber's finding remained unsettled throughout his lifetime, despite the high regard he was held in by his peers.

Nevertheless, although bar detectors exist in at least six laboratories now, it was generally felt that new technology was needed to improve the sensitivity and to reduce background noise. We next describe the interferometer technology that has been employed in several present-generation detectors.

11.5.2 Laser interferometers

We encountered Michelson's interferometer in Chapter 1 while discussing the Michelson–Morley experiment. It creates two paths into which a beam of light splits. After following the paths, the two beams recombine. At that stage there will be interference between the waves and how they combine will depend on their phases. If there is a slight difference in the lengths of the two paths followed by the waves, that difference will determine the outcome of interference.

Since an interferometer is very sensitive to small changes in pathlengths, it is ideal for measurements of gravitational waves. The idea is that, as a gravitational wave crosses the interferometer, it causes changes in pathlengths because of the changes in geometry (*vide* Equation (11.21)). If the effect can be measured, the intensity of the original wave can be estimated.

The layout of the most ambitious of these detectors, the Laser Interferometric Gravitational Observatory (LIGO) is shown in Figure 11.4. The two paths, shown there at 90°, form the letter L. A partially reflecting/transmitting mirror serves as a 'beam splitter' at the vertex of L. The light beam is in the form of a laser. The arms of the L are each of length 4 km and the beams are allowed to make several rounds along the two paths. To keep the laser beams focussed, and not dispersed, it is desirable to make them travel through vacuum. The higher the level of vacuum, the less the dispersion of the laser waves and the sharper their tracks.

If the two split beams meet with the same phase, they are allowed to continue making the round of the interferometer. If there is a path

Fig. 11.4. The LIGO detector (photograph by courtesy of the LIGO team).

difference, a part is diverted to a photomultiplier and measured. The chances are that what is collected there arises from the passage of gravitational waves. In a way, the interferometer measures the gravitational signals through their conversion to electrical ones, just as a microphone converts sound waves into electrical disturbances. How large are the expected signals?

We may express the answer in the form of the magnitude of h_{ik} generated by a typical source. For typical 'strong' sources the magnitudes are about 10^{-20}. Thus, for a pathlength of 10 km, the length fluctuations to be measured are about 10^{-14} cm. A typical optical wavelength is 500 nm, i.e., 5×10^{-7} cm. This puts into perspective the precision needed in these measurements.

At present LIGO is functioning and taking routine observations. It has two identical detectors, one in Washington state and the other in New Orleans; thus they form a long baseline in the SE–NW direction of the mainland USA. It is hoped that simultaneous detection of signals by both the instruments would lend credibility to the finding. It is believed that LIGO's sensitivity is just below the threshold for detection and upgrading of its capabilities is going on. There are other similar but smaller detectors in three other places, two in Europe and one in Japan.

In short, there is considerable interest in the campaign to detect gravitational waves. The question is whether current technology is capable of doing so. Perhaps this thought has prompted scientists to be even more ambitious and think of space as the place of detection. We will look at this possibility next.

11.5.3 LISA from space

A major problem with terrestrial detectors is that of 'noise', which includes seismic disturbances and disturbances produced by terrestrial sources. Because the signal to be expected is very small, these noises have to be minimized and/or accounted for by sophisticated techniques of data analysis. Indeed, data-analysis techniques have played a major 'software' role in the case of terrestrial detectors.

Nevertheless, to minimize the noise, it is now proposed to have a space-based mission called the Laser Interferometric Space Antenna (LISA). As shown in Figure 11.5, it is made up of a trio of spacecraft forming an equilateral triangle, which follows at a distance of 20° behind the Earth on the same orbit. Its plane will make an angle of 60° with the Earth's orbit and the triangle will face the Sun. Like the Earth, it will take one year to orbit the Sun.

The vertices of the triangle will carry two mirror reflectors each for reflecting laser rays so that they describe the equilateral triangle. The

Fig. 11.5. A schematic picture of the LISA project (photograph by courtesy of NASA and the ESA).

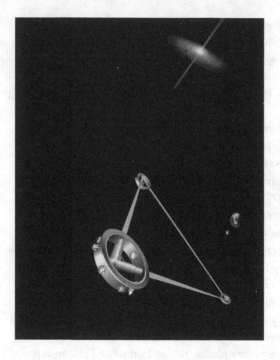

arrival of a gravitational wave will modify the geometry in the vicinity of the triangle, which will be detected through the laser interferometry.

This ambitious project is currently scheduled for completion in 2016, if funding is made available. It has been initiated by NASA and the ESA, although other space organizations are expected to join to make it a 'world project'.

11.6 Concluding remarks

Gravitational waves pose a challenge to human intellect and technical achievements. If general relativity is right, one can find support for it in the detection of gravitational waves emitted by cosmic sources. Yet, the task is not an easy one, as we saw. The minute effects to be found and measured require the present technology to be pushed beyond its cutting edge.

Exercises

1. Using formula (11.19), estimate the rate of decrease of the period of a close binary star system with component stars of masses $M_\odot$ and $2M_\odot$, moving in a circular orbit with separation 10^{15} cm.

2. A supernova core starts to collapse at $t = 0$ while retaining its ellipsoidal shape with its axes maintaining the same ratio. The starting values of the principal

quadrupole moments of the core were $(5, 4, 3)$ in units of $M_\odot R_\odot^2$. During collapse all dimensions of the core homologously shrink $\propto (t_0 - t)^{1/3}$ with $t_0 = 100$ s. Calculate the gravitational radiation emitted by the core per second when its linear size is half its starting value.

3. Assuming that the rate of radiation of gravitational waves by a source of quadrupole moment $D_{\alpha\beta}$ is proportional to $|\dddot{D}_{\alpha\beta}|^2$, deduce formula (11.12) except for its numerical coefficient, using dimensional analysis.

4. A cylinder of length L, cross-sectional radius R and uniform density ρ is made to spin at rate ω about an axis perpendicular to the length of the cylinder and passing through its midpoint. Show that it radiates gravity waves at the rate $2GM^2L^4\omega^6/(45c^5)$, where M is the mass of the cylinder. Estimate this rate for $R = 1$ m, $L = 20$ m, $\rho = 7.8$ g cm^{-3} and $\omega = 28$ radians s^{-1}. (For an iron cylinder a spin rate of $\omega = 28$ rad s^{-1} is the maximum spin that can be borne by its tensile strength.) (This problem has been taken from *Gravitation* by C. W. Misner, K. S. Thorne and J. A. Wheeler, Freeman (1970).)

5. From (11.18) estimate the time within which the binary stars will coalesce.

Chapter 12
Relativistic astrophysics

12.1 Strong gravitational fields

So far we have been concerned with gravitational effects that are weak, even when we were talking of effects requiring post-Newtonian approximation. To give the example of Schwarzschild's solution, in the term

$$e^\lambda = \left(1 - \frac{2GM}{c^2 R}\right)$$

we assume that the departure from unity is small. Even for the compact white dwarf stars the difference $|e^\lambda - 1|$ is less than 10^{-3}. Thus we have been able to 'get away with' linearizing Einstein's equations. While this has served our purpose in the limited applications of the theory, we have not been confronted with its inherent non-linearity. One reason why physicists and mathematicians studying relativity did not get into such confrontations for many years after the inception of the theory was because nature did not present a scenario where the full non-linear impact of the theory could be felt. However, one may look upon the year 1963 as a watershed when nature did oblige the relativist with such examples.

In 1963, thanks to the cooperation between optical and radio astronomers, the so-called quasi-stellar objects (QSOs or *quasars* in brief) were discovered. These are compact-looking sources emitting optical and radio radiation. The first two quasars to be discovered, 3C273 and 3C48, were at first mistaken for stars in our Galaxy.[1] Later studies

[1] 3C273 is the 273rd source in the third Cambridge catalogue of radio sources.

revealed features that did not fit in with this interpretation and led to the conclusion that they are extragalactic, located far away like some of the most distant galaxies. Here we emphasize their role as radiators of huge amounts of energy from within a compact region. What 'energy machine' led a quasar to generate so much radiation from within a compact volume?

Prior to this discovery, theoreticians such as Fred Hoyle and William A. Fowler had conjectured about the existence of very massive stars. *Supermassive objects*, as they came to be known, had masses in the range 10^6–10^{10} solar masses. $M_\odot$ is the symbol for a solar mass and it serves as a mass unit in astrophysics: $M_\odot \cong 2 \times 10^{33}$ g. These authors had realized that such massive stars, if they exist, cannot sustain a luminous phase for long, since their nuclear resources would be unable to generate enough pressure to withstand the self-gravity of the 'star'. For a Sunlike star the equilibrium can be sustained for a long time because the pressures generated by the nuclear reservoir can effectively counter the gravitational contraction. However, if we increase the mass in the calculation, the nuclear reservoir grows at a rate proportional to star-mass M, whereas the gravitational energy grows as M^2. Clearly such supermassive stars would evolve fast and consume their nuclear energy in a matter of a few thousand years. Being unable to maintain a steady volume, such a supermassive star would shrink and shrink, until its gravitational environment became very powerful. Hoyle and Fowler suggested that the contraction of these objects would be very rapid and the energy so generated (which was gravitational in origin) would be radiated by the star [23].

In the early months after the discovery of quasars, astrophysicists realized that a follow-up of the Hoyle–Fowler idea would lead them into territory they had never trod before, viz., an environment of a strong gravitational field that demanded the full application of general relativity. Thus there was convened an international conference on 'Relativistic Astrophysics', to which general relativists as well as theoretical and observational astronomers were invited. This meeting, held in Dallas, Texas, was to be the first of a biennial series of meetings known as 'Texas Symposia'. The name recalls the early association of Texas with these meetings, which have since been held in different parts of the world. Quasars such as 3C273 (see Figure 12.1) dominated the discussion of the first Texas meeting, although the scope of relativistic astrophysics has since expanded and widened.

We now look at some aspects of supermassive objects to demonstrate how relativity makes a significant difference to the evolution of a massive system, compared with the Newtonian theory.

Fig. 12.1. A photograph of the optical image of the quasar 3C273. A sign of the unusual nature of the source is the jet seen emerging from the source. (Photograph by courtesy of NASA and the ESA.)

12.2 Equilibrium of massive spherical objects

To see how relativity makes a difference, we first look at the Newtonian equations describing the equilibrium of a spherical star of mass M and radius R_b.

We choose radial coordinate r indicating distance from the centre, and denote by $m(r)$ the mass contained within a concentric sphere of radius r. We denote by $p(r)$ the pressure at distance r and by $\rho(r)$ the density at that distance from the centre. We expect these quantities to decrease monotonically from the centre outwards, with pressure vanishing at the boundary given by $r = R_b$.

We then have the mass–density relation for $r < R_b$, which is purely geometrical:

$$\frac{dm(r)}{dr} = 4\pi r^2 \rho(r), \qquad (12.1)$$

The second equation is of *hydrostatic equilibrium*. At any interior point with radial coordinate r, the mass within, $m(r)$, pulls (gravitationally) any material at r inwards, whereas the pressure gradient at r seeks to prevent this. The balance is described by the differential equation

$$\frac{dp}{dr} = -\rho(r)\frac{Gm(r)}{r^2}.$$ (12.2)

These equations have to be supplemented by an equation of state relating pressure p to density ρ. For example, if the star can be approximated by a 'polytrope' of index n, the relation is

$$p \propto \rho^{1+1/n}.$$ (12.3)

Usually, for an ordinary star, $n = 3$ is a good approximation. For details of stellar structure see, for example, References [24, 25]. For the relativistic discussion we will follow Fowler [26].

In the relativistic case, we start with the Schwarzschild coordinates with the line element written as

$$ds^2 = e^\nu\, dt^2 - e^\lambda\, dr^2 - r^2(d\theta^2 + \sin^2\theta\, d\phi^2),$$ (12.4)

where λ and ν are functions of r only in a static situation. Given the energy tensor of a perfect fluid with bulk velocity u^i, pressure p and density ρ,

$$T^{ik} = (p + \rho)u^i u^k - pg^{ik},$$ (12.5)

we get the following solution of the Einstein equations for λ in the interior of the supermassive star as defined by $0 < r < R_b$:

$$e^{-\lambda(r)} = 1 - \frac{2Gm(r)}{r},$$ (12.6)

where, for any R satisfying $0 < R \le R_b$,

$$m(R) = \int_0^R 4\pi r^2 \rho\, dr.$$ (12.7)

Note that the integrand above includes the term $T^0{}_0$, which in this case is ρ. Thus our equation (12.7) above finds an echo in the Newtonian equation (12.1). We also have $M = m(R_b)$ as the gravitational mass of the star. The Newtonian equation of hydrostatic equilibrium has a more complicated relativistic counterpart, however. For writing the energy-conservation equations $T^{ik}{}_{;k} = 0$ for $i = 1$ gives the following relationship between the pressure gradient and mass:

$$\frac{dp}{dr} = -\frac{4\pi\, Gr(p + \rho)}{1 - \dfrac{2Gm(r)}{r}} \times \left(p + \frac{m(r)}{4\pi r^3}\right).$$ (12.8)

This equation is exact and in the 'Newtonian limit' of $p \ll \rho$ and $Gm(r) \ll r$ it does reduce to Equation (12.2). Of course, as in the Newtonian case, we still need an equation of state if we are to be able to solve these equations.

The Schwarzschild interior solution (see Chapter 9) may be seen as an extreme example of this approach since it supposes that $\rho =$ constant. Sometimes this is considered an example of the *hard-core nucleon density*. The core is incompressible and therefore has a constant density. We will take it as a limiting case in what follows.

12.3 Gravitational binding energy

The notion of binding energy, as it exists in the Newtonian framework, can be defined in relativity also. This leads to the definitions of two masses: the nucleonic mass M_n and the gravitational mass M. We have already come across M as the value $m(R_b)$

To understand the relationship of these notions, imagine the super-massive star to be broken into its basic constituents, the nucleons (protons and neutrons), which are then transported to 'infinity'. To this end, work has to be done against the gravitational force of the star. The mass of the star in this infinitely dispersed state includes the energy equivalent of this work and will therefore be *greater* than the mass in the compact bound state. The mass at infinity is simply the sum of the masses of all nucleons in it. (Strictly speaking, we should include electrons also in this count; but their contribution is negligible (about 5×10^{-4}).) This is defined to be the quantity M_n mentioned before.

The quantity

$$B = M_n - M \tag{12.9}$$

is the *binding energy* of the star, and this is the work done in distributing it to infinity. Thus, for a bound object, $B > 0$ and it tends to zero in the infinitely dispersed state of the object. We may define M_n more concretely by the integral

$$M_n = 4\pi \int_0^{R_b} \rho_0 R^2 e^{\lambda/2} \, dr, \tag{12.10}$$

where ρ_0 is the rest-mass density (see Chapter 7): that is, in a unit proper volume, count the number of nucleons and multiply it by their rest masses before adding up. The volume element multiplying ρ_0 in the integral above is the proper volume at constant t of a spherical shell sandwiched between coordinates r and $r + dr$. Assuming that there is no change in the number of nucleons as the object expands or contracts, we take M_n to be constant.

12.3.1 The Schwarzschild interior solution revisited

Let us apply these concepts to the Schwarzschild interior solution derived in Chapter 9. Using the notation of Chapter 9, we have the inequality

(because $\rho_0 \leq \rho$)

$$M_n \leq \int_0^{R_b} 4\pi \rho r^2 e^{\lambda/2} \, dr$$

$$= \frac{2\pi \rho}{\alpha^{3/2}} \left[\sin^{-1} \left(\alpha R_0^2 \right) - \sqrt{\alpha R_0^2} \sqrt{1 - \alpha R_0^2} \right]. \quad (12.11)$$

For the limiting case of infinite pressure at the centre given by (9.23) we have the maximum of the right-hand side and hence, for that case,

$$M_n \leq \frac{2\pi \rho}{\alpha^{3/2}} \left(\sin^{-1} \sqrt{\frac{8}{9}} - \frac{8}{9} \right). \quad (12.12)$$

Taking the case of the hard-core nucleon potential, we set the constant density ρ equal to the nuclear density $\sim 10^{15}$ g cm^{-3}. Then the above inequality will give

$$M_0 \leq 3 M_\odot. \quad (12.13)$$

This means that the maximum mass that can be supported in this way is no more than three solar masses. Hence we get a perspective on how difficult it is for supermassive stars with masses millions of times the solar mass to remain in equilibrium after their nuclear resources have been exhausted. We consider such stars next.

12.3.2 Supermassive stars

We follow the discussion given by Fowler [26] and write the difference between actual and rest-mass densities as the internal (thermal) energy density

$$u = \rho - \rho_0. \quad (12.14)$$

Then the binding energy is given by

$$B = \int_0^{R_b} 4\pi r^2 e^{\lambda/2} (\rho - u) dr - \int_0^{R_b} 4\pi r^2 \rho \, dr$$

$$= \int_0^{R_b} 4\pi r^2 \rho \left[\left(\sqrt{1 - \frac{2Gm(r)}{r}} \right)^{-1} - 1 \right] dr$$

$$- \int_0^{R_b} 4\pi r^2 u \left(\sqrt{1 - \frac{2Gm(r)}{r}} \right)^{-1} dr. \quad (12.15)$$

These relations apply for the static case when the object is in equilibrium. Let us consider the Newtonian situation first. In the Newtonian approximation of the above expressions we assume that $u \ll \rho$ and neglect the term $Gm(r)/r$ in the second integral but not the first:

$$B = \int_0^{R_b} \frac{Gm(r)}{r} \rho \cdot 4\pi r^2 dr - \int_0^{R_b} 4\pi r^2 u \, dr. \qquad (12.16)$$

The first integral is the gravitational binding energy, which for a polytrope of index n is given by (see Reference [25])

$$\Omega = \frac{3GM^2}{(5-n)R_b}. \qquad (12.17)$$

On writing $u = 3\epsilon p$, the second of the above integrals becomes (for a constant ϵ)

$$-\int_0^{R_b} 4\pi r^2 u \, dr = -\int_0^{R_b} 4\pi r^2 \times 3\epsilon p \, dr = \int_0^{R_b} 4\pi \epsilon r^3 \frac{dp}{dr} dr.$$

Using (12.2), the equation of hydrostatic equilibrium, we get

$$-\int_0^{R_b} 4\pi r^2 u \, dr = -\int_0^{R_b} 4\pi \frac{Gm(r)}{r^2} \cdot r^3 \rho \epsilon \, dr$$

$$= -\int_0^{R_b} \epsilon \frac{Gm(r)}{r} \rho \cdot 4\pi r^2 \, dr = -\epsilon \Omega.$$

When the coefficient ϵ varies within the star, we may replace the constant ϵ by an average value $\langle \epsilon \rangle$. Then, from Equation (12.16) and the above relation, we get the binding energy as

$$B = -(\langle \epsilon \rangle - 1)\Omega. \qquad (12.18)$$

If $B > 0$, the star has negative total energy, relative to the case of infinite separation. If $B < 0$, the star has positive total energy, which normally should come from the star's nuclear reactions. If the star cannot supply this energy, it cannot be in equilibrium: it will contract.

12.3.3 The post-Newtonian approximation

Let us now look at the same problem from the next level of approximation, viz. the *post-Newtonian* approximation. In this case Equation (12.8) becomes

$$\left(1 + \frac{Gm(r)}{r}\right) \frac{dp}{dr} \cong -\rho(r) \frac{Gm(r)}{r^2} \left(1 + \frac{p}{\rho} + \frac{4\pi pr^3}{m(r)} + \frac{3Gm(r)}{r}\right). \qquad (12.19)$$

Using this relation together with (12.1), which remains unchanged in the relativistic case, we get back to the two integrals of Equation (12.16). In analogy to Equation (12.18), we get

$$B = -(\langle \epsilon \rangle - 1)\Omega - 8\pi G \langle \epsilon \rangle \int_0^{R_b} pm(r)r\, dr$$

$$- 12\pi G^2 \left(\langle \epsilon \rangle - \frac{1}{2} \right) \int_0^{R_b} [m(r)]^2\, dr.$$

Here the average $\langle \epsilon \rangle$ is with weights different from those in the Newtonian case discussed earlier.

For supermassive stars, i.e., for stars of a million times the solar mass (or even more), we use the convective-polytrope approximation. This implies that the energy transport from the centre to outer layers of the star is through convection and the index is $n = 3$. In such a case $\epsilon - 1$ is constant in the star and small compared with unity. Writing it as $\beta/2$, after some manipulation we get the above equation in the following form:

$$\frac{B}{M} = \frac{3}{4}\beta \frac{GM}{R_b} - 5.1 \frac{G^2 M^2}{R_b^2}. \qquad (12.20)$$

Suppose we express the right-hand side of this equation in terms of the Schwarzschild radius $R_s = 2GM$ introduced in Chapter 9:

$$\frac{B}{M} = \frac{3}{8}\beta \frac{R_s}{R_b} - 1.3 \left(\frac{R_s}{R_b} \right)^2. \qquad (12.21)$$

How high is the average density of the object? A short calculation leads to the result

$$\langle \rho \rangle \cong 1.8 \times 10^{16} \left(\frac{M_\odot}{M} \right)^2 \left(\frac{R_s}{R_b} \right)^3. \qquad (12.22)$$

Thus, for a star with mass $10^8 M_\odot$, we have a density comparable to that of water even for $R_b \sim R_s$. Also, from Equation (12.21) we see that the post-Newtonian term is comparable to the Newtonian one in magnitude when

$$\frac{3}{8}\beta \frac{R_s}{R_b} = 1.3 \left(\frac{R_s}{R_b} \right)^2,$$

that is, when $R_b \cong 0.35\beta^{-1} R_s$. The parameter β is estimated to be about 10^{-3} for stars as massive as those with $M = 10^8 M_\odot$. Then our calculation above tells us that, for $R_b \cong 350 R_s$, the general relativistic contribution becomes significant. In short, one does not have to wait until R_b becomes comparable to R_s for general relativity to become relevant.

It is convenient to use the central temperature T_c instead of the stellar radius and we refer to [24] for the polytropic-model formula:

$$R_b = \frac{5.83 \times 10^{18}}{T_c} \left(\frac{M}{M_\odot} \right)^{1/2} \text{ cm.} \qquad (12.23)$$

For the $n = 3$ polytrope considered here, the formula (12.21) for the binding energy becomes

$$\frac{B}{M} \cong 1.6 \times 10^{-13} T_c - 3.3 \times 10^{-27} \left(\frac{M}{M_\odot} \right) T_c^2. \qquad (12.24)$$

Note that the binding energy has a maximum at a temperature

$$T_c \cong 2.5 \times 10^{13} \left(\frac{M_\odot}{M} \right) \text{ K.} \qquad (12.25)$$

Figure 12.2 shows the variation of binding energy with T_c. The binding energy becomes zero at $T_c^0 \cong 5 \times 10^{13} (M_\odot/M)$ K and then turns negative as the temperature is further increased. This means that stars centrally hotter than this value need to have an energy source to keep an overall positive energy.

Normally about $10^{-3} M$ energy is available for nuclear fusion, provided that the central temperature is adequate to trigger these reactions. Assuming that the second term of (12.24) dominates at high enough T_c,

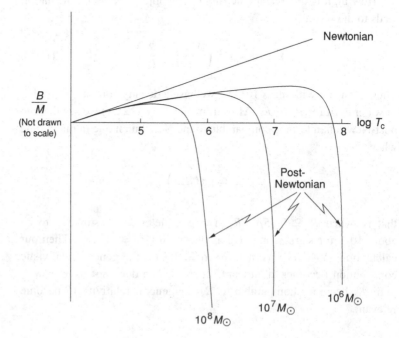

Fig. 12.2. Variation of binding energy per unit mass of supermassive stars, with their central temperatures.

the energy requirement is

$$\frac{B}{M} \sim 3.3 \times 10^{-27} \left(\frac{M}{M_\odot} \right) T_c^2 \leq 10^{-3}. \qquad (12.26)$$

The central temperature is as high as 8×10^7 K at the densities involved in the central region. On setting this value in the above relation, we get the upper limit on the mass:

$$M \leq \sim 10^8 M_\odot. \qquad (12.27)$$

What does this mean? For supermassive objects with masses greater than this value, nuclear reactions are not able to provide sufficient energy for hydrostatic support. Even for stars of masses below the above limit, when the nuclear reactions have exhausted all nuclear fuel, they too can no longer sustain equilibrium.

The equilibrium configurations need to be analysed from the stability point of view before they can be considered credible. S. Chandrasekhar carried out the small-oscillations analysis near the equilibrium states. He found that the models become unstable long before R_b becomes close to R_s (*vide* Reference [28]).

Figure 12.2 demonstrates why it is more difficult for a supermassive star to remain in equilibrium in the post-Newtonian relativistic regime. If there is an inadequate supply of energy, the star cannot maintain equilibrium and will start shrinking. As the central temperature rises with shrinking of the star because of its gravity, its binding energy sinks further and this makes it more difficult for the star to supply the required energy. So the star shrinks even faster. This leads to what is commonly called *gravitational collapse*.

Thus we see that supermassive stars take us to regions of strong gravity, where general relativity will apply in full. As seen here, we first encounter the post-Newtonian phase wherein additional terms are taken from relativity to supplement the Newtonian discussion. If we follow the path of gravitational collapse we should encounter even stronger gravitational fields and would need not a post-Newtonian approximation but the full use of general relativity. We will consider this phase in the following chapter.

As part of relativistic astrophysics we must also consider the effect of gravity on light. That gravity affects light was demonstrated by the eclipse experiment of Chapter 10. How does the interaction manifest itself in nature?

12.4 The first gravitational lens

The bending of light rays due to the gravity of a massive object gives rise to a variety of phenomena now known as *gravitational lensing*. The lensing is caused by gravity rather than by the refraction of light passing through an inhomogeneous medium. We will discuss this topic next, since it has played a major role in astronomy since 1979.

A brief historical description of the path that led in 1979 to the discovery of the first gravitational lens involving quasars may be in order.

The first paper [29] on the subject, entitled 'Nebulae as gravitational lenses', was published by Fritz Zwicky in 1937. He clearly stressed the role of galaxies as light-deflecting objects that could produce *multiple* images of background sources. He pointed out the possibility of ring-shaped images, of flux amplification and of the use of this phenomenon for understanding the large-scale structure of the Universe [29]. Zwicky was ahead of his time in assuming that the nebulae (i.e., galaxies) would be several hundred billion times as massive as the Sun and, in another paper, he also estimated the probability of lensing occurring in extra-galactic astronomy [30].

In the 1960s and 1970s, S. Refsdal, J. M. Barnothy, R. K. Sachs, R. Kantowski, C. C. Dyer, R. C. Roeder, N. Sanitt and several others published papers highlighting various aspects of gravitational lensing, ranging from purely theoretical investigations in general relativity to observational predictions in astronomy. We refer the reader to the book by Schneider, Ehlers and Falco [31] for further details.

In a different context, Chitre and Narlikar [32] invoked the gravitational bending of radio waves from the VLBI components of a quasar by an intervening galaxy to explain the apparent superluminal separation of these components. If the galaxy is suitably located (i.e., close to the critical point of the lensing system) the apparent magnification of the separation between two components due to the lensing can be enormous and can convert a real subluminal speed into a superluminal one (see Figure 12.3). (However, the generally accepted interpretation of superluminal motion involves relativistic beaming.)

The quasars, galaxies etc. entering our discussion in the rest of this chapter are far-away objects located well beyond our Galaxy. Such objects, as will be discussed in detail in Chapters 14–17, participate in the expansion of the Universe. One important consequence of this is that their spectra show redshift that increases in proportion to their distance from us. Thus, in what follows, we will have occasion to refer to such *cosmological redshifts* as indicators of distance. The reader unfamiliar with cosmology may wish to familiarize himself with cosmological redshifts by taking a quick look at Chapters 14 and 15.

The real stimulus to the work on gravitational lensing came from the discovery of the first lens involving the quasars 0957+561 A and B by Walsh, Carswell and Weymann [33]. The quasars A and B showed

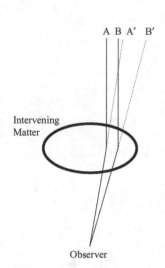

Fig. 12.3. The magnification produced by gravitational lensing is shown in the figure as the increase in the linear separation of a radio source from AB to A' B'. Very-long-baseline interferometry (VLBI) of quasars shows A' B' increasing at speeds many times the speed of light. In reality the source AB may not increase at superluminal rate, however.

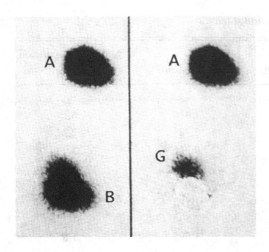

Fig. 12.4. A photograph of the twin quasar images 0957+561 A and B in which the image of one component is 'subtracted' from the other. The balance reveals the image of a galaxy, G, that is believed to have acted as a lens producing the two images of a single source. (From Stockton A. 1980, *Ap. J.*, Part 2, Letters to the Editor, **242**, pp. L141–L144.)

very similar features and spectra at a redshift of ∼1.4. Their angular separation was ∼6 arcsec. Although the existence of two quasars with very similar features at close separation cannot be ruled out, the circumstantial evidence indicated a gravitational lens doubly imaging one source. The discovery of a lensing galaxy at a redshift of ∼0.36 later lent further credibility to this scenario. The quasars and lensing galaxy are shown in Figure 12.4, while a ray diagram of the bending of light by the lens is shown in Figure 12.5.

The basic features of a gravitational-lens system are described in the following section. By now there are several known lens systems and probable candidates, as listed in Table 12.1. The original expectations of Zwicky have been fully borne out.

12.5 The basic features of a gravitational lens

Figure 12.6 is a schematic diagram of a lens system wherein S is the source, a spherical mass M provides the deflector lens d and O is the observer. (We deplore the notation of denoting the lens by d as it is

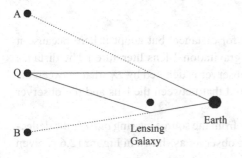

Fig. 12.5. The ray diagram showing how gravitational bending by lensing a galaxy can produce two images A and B of a single source Q.

Table 12.1. *A partial list of interesting cases of gravitational lenses*

System	No. of images	Lens redshift	Image redshift	Maximum separation (arcsec)[a,b]
Quasar images				
0957+561	2	0.36	1.41	6.1
0142−100	2	0.49	2.72	2.2
0023+171	3	?	0.946	5.9
2016+112	3	1.01	3.27	3.8
0414+053	4	0.468[c]	2.63	3.0
1115+080	4	0.29	1.722	2.3
1413+117	4	1.4[c]	2.55	1.1
2237+0305	4	0.0394	1.695	1.8
Arcs				
Abell 370		0.374	0.725	
Abell 545		0.154	?	
Abell 963		0.206	0.77	
Abell 2390		0.231	0.913	
Abell 2218		0.171	0.702	
Cl0024+16		0.391	0.9	
Cl0302+17		0.42	0.9	
Cl0500−24		0.316	0.913	
Cl2244−02		0.331	2.237	
Rings				
MG1131+0456		?	?	2.2
0218+357		?	?	0.3
MG1549+3047		0.11	?	1.8
MG1654+1346		0.25	1.75	2.1
1830−211		?	?	1.0

[a] For arcs the maximum separation is the diameter of the corresponding Einstein ring.
[b] For rings this corresponds to the diameter of the ring.
[c] Assumed or still to be confirmed.

a symbol normally reserved for distance, but adopt it here because it has become common in the gravitational-lens literature.) The distance between the source and the observer is denoted by D_s, that between the source and the lens by D_{ds} and that between the lens and the observer by D_d.

The condition that the ray from the source passing *outside* the deflector at a distance ξ reaches the observer as shown in Figure 12.6 is given

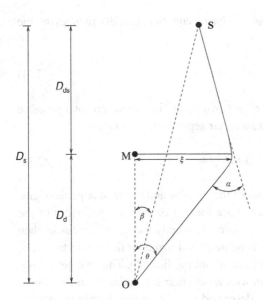

Fig. 12.6. A schematic diagram of a gravitational lens. For details, refer to the text.

by the rules of projection:

$$\beta D_{\mathrm{s}} = \frac{D_{\mathrm{s}}}{D_{\mathrm{d}}}\xi - \frac{2r_{\mathrm{S}}}{\xi}D_{\mathrm{ds}}. \qquad (12.28)$$

Here $r_{\mathrm{S}} = 2GM/c^2$ is the Schwarzschild radius of the deflector mass. We have tacitly assumed that the gravitational bending of light is small and so the angles β and α (the bending angle of the original ray) are both small compared with unity. Also, when applying this relation over cosmological distances, we have to take due note of the non-Euclidean measures of redshift-related distance. Thus in general $D_{\mathrm{ds}} \neq D_{\mathrm{s}} - D_{\mathrm{d}}$. The deflected ray in Figure 12.6 makes an angle θ with the line OM. Hence, with our small-angle approximation, $\theta D_{\mathrm{d}} = \xi$. Therefore the above equation becomes

$$\beta = \theta - \frac{2r_{\mathrm{S}}D_{\mathrm{ds}}}{D_{\mathrm{d}}D_{\mathrm{s}}}\theta^{-1}. \qquad (12.29)$$

This equation can be generalized to a full three-dimensional case in which the vectors β and θ lie in different planes. We will continue with the two-dimensional simplification.

It is convenient to define an angle α_0 and a length ξ_0 by

$$\alpha_0 = \sqrt{\frac{2r_{\mathrm{S}}D_{\mathrm{ds}}}{D_{\mathrm{d}}D_{\mathrm{s}}}}, \qquad \xi_0 = \alpha_0 D_{\mathrm{d}}. \qquad (12.30)$$

With these definitions the basic equation (12.29) reduces to the quadratic

$$\theta^2 - \beta\theta - \alpha_0^2 = 0. \tag{12.31}$$

This tells us that there are two roots, i.e., there are two possible locations of images, whose angular separation is given by

$$\Delta\theta = \sqrt{4\alpha_0^2 + \beta^2}. \tag{12.32}$$

Note that the two values of the roots θ_1 and θ_2 are of opposite signs, implying that the two images are located on the opposite sides of the source. Note also that, if the source, lens and observer are collinear, then the angle $\beta = 0$ and there is no preferred plane for the rays to take. The geometry is then axisymmetric about the line SO. Thus we get a ring-like distribution of images with an angular separation from the source of $\theta = \alpha_0$. What we have described is, of course, a highly symmetric situation involving a symmetric matter distribution, a point source and a special alignment. In practice these conditions are not fully satisfied, but we may still get approximately ring-shaped images of extended sources, which are called *Einstein rings* in the literature.

A general lens is more difficult to quantify. However, we may make a few statements that can be proved using detailed mathematics, which the reader may look up in books or monographs specializing in gravitational lensing, e.g., Reference [31]. A general theorem proves that *any transparent matter distribution with a finite total mass and weak gravitational field produces an odd number of images*. Sometimes a few images are too faint to be seen, and we see an even number of images. In our simple case described above, we get two images. In this case, if the gravitating object is a point mass (black hole), the possibility exists that the incident rays would have small impact parameters, thus violating the weak-field condition. If the lens were a transparent sphere of matter, the theorem would still apply.

12.6 The magnification and amplification of images

Consider a simple example in Euclidean geometry of a spherical source of radius a and luminosity L located at a distance D. The flux of radiation received from the source is given by

$$f = \frac{L}{4\pi D^2}, \tag{12.33}$$

while the solid angle subtended by the source at the observer is

$$\Omega = \frac{\pi a^2}{D^2}. \qquad (12.34)$$

The surface brightness of the source is given by

$$\sigma = \frac{L}{4\pi a^2}. \qquad (12.35)$$

It is easy to verify from these results that

$$f = \Omega\sigma/\pi. \qquad (12.36)$$

In other words, the surface brightness is proportional to the ratio of the flux to the solid angle subtended. Since gravitational bending does not introduce any additional spectral shift, we may assume this result to be valid for lensed sources. Since the surface brightness of a small source is not changed by lensing, the ratio of the flux of an image to that of the source (in the absence of lensing) would simply equal the ratio of their solid angles:

$$\mu = \Omega/\Omega_0, \qquad (12.37)$$

the zero suffix standing for the unlensed situation. This is valid for cases in which the images and sources are not extended so that one may use a constant surface brightness. For extended sources one integrates over the source with suitable weighting of the surface brightness at the point.

Returning to the small source, if $\vec{\beta}$, the source position, is related to $\vec{\theta}$ by a generalization of Equation (12.29) to the full three-dimensional case, we have the angular magnification given by the Jacobian

$$\frac{\Omega_0}{\Omega} = J[\beta;\theta] \equiv \det\left\|\frac{\partial\beta}{\partial\theta}\right\|. \qquad (12.38)$$

Thus the amplification of the flux is given by the reciprocal of the above Jacobian.

The Jacobian has great significance in the lensing calculations. The parity of the image is decided by the sign of the Jacobian for that image. If it is positive, the sense of its curves (i.e., clockwise or anticlockwise) is preserved with respect to the source. For negative parity it is reversed. Thus regions of opposite parity are separated in the lens plane. The *critical* curves separating them are those on which the factor μ diverges. This is, however, an idealization since the sources are in general extended and infinite amplification of image brightness does not take place. These critical curves in the source plane are called *caustics*. It can be shown that *the number of images changes by two if and only if the source position is changed in such a way that it crosses a caustic.* If the locations of caustics are known, then, for a given source position, the number of images can be determined using this property. Another result that is useful in this respect is that for any transparent distribution of matter with finite mass the number of images of a point source sufficiently misaligned with the lens is *unity*.

Even though we do not in practice have enormously bright images (μ tending to infinity), another theorem guarantees that, *for a transparent lens, there is one image with positive parity having an amplification factor not less than unity (i.e., the image is at least as bright as the source)*. Thus lensing may mislead the observer into thinking that he or she has found a very bright source.

Exercises

1. A quasar shows time variation on the scale T. Assuming that special relativistic causality limits the size of its diameter, show that the maximum mass the quasar can have is $c^3 T/(4G)$. Express the answer in units of solar masses when T is expressed in hours.

2. Assuming an interior Schwarzschild solution to apply for the hard-core nucleon potential, estimate the maximum redshift from the surface of such an object.

3. Explain why nuclear energy generation cannot effectively support supermassive stars beyond a limiting mass. Compare this limit with the way the Chandrasekhar limit operates for white dwarfs.

4. Using the following definitions given by H. Bondi, derive the equations of Section 12.2 in the form given below:
Definitions:

$$u = \frac{m(r)}{r}, \qquad w = 4\pi r^2 p(r).$$

Results to prove:

$$r\frac{dv}{dr} = \frac{2(u+w)}{1-2u}, \qquad \frac{dp}{dr} = -\frac{1}{2}(p+\rho)\frac{dv}{dr},$$

$$r\frac{du}{dr} = \frac{H}{(1-2u)\left(\dfrac{dw}{du} - \alpha\right)}, \qquad \rho = \frac{u}{4\pi r^2}\left(\frac{dw}{du} - \beta\right)\left(\frac{dw}{du} - \alpha\right)^{-1},$$

where

$$H = 2w - (u^2 + 6uw + w^2), \qquad \alpha = -\frac{u+w}{1-2u}, \qquad \beta = -\frac{w}{u}\frac{z - 5u - w}{1-2u}.$$

5. Show that in the interior Schwarzschild solution the central redshift z_c is related to the surface redshift z_s by the relation

$$1 + z_c = \frac{2}{2 - z_s}(1 + z_s).$$

As $z_s \to 2$, $z_c \to \infty$.

6. Show that the interior Schwarzschild solution is conformally flat.

7. Show that a star located exactly along the line of sight from the Earth to the Sun can under certain circumstances be seen as an Einstein ring of radius $4GM_\odot/(c^2 R_\odot)$.

Chapter 13
Black holes

13.1 Introduction

We found in the previous chapter that, if a massive star runs out of nuclear fuel, it would lose its equilibrium and begin to shrink. Even when nuclear fuel is available to the star, it may be insufficient to meet the demands for the star's equilibrium. In the early 1930s the young astrophysicist Subrahmanyan Chandrasekhar had encountered a somewhat similar situation when discussing the state of stars like the Sun, after they run out of their nuclear fuel. He found that the star can still sustain equilibrium if its internal matter can attain the degenerate state. Degeneracy can arise if the density of matter is so high that all available energy levels of atoms are filled up, up to some low energy. In such a situation further compression of matter is not possible and gravity is held at bay. This is an excellent example of a macroscopic effect of quantum mechanics: a star as massive as the Sun feels an effect whose origin is in quantum mechanics. We cannot describe it in detail since that would take us farther away from our present interest.

The early work on degenerate matter by R. H. Fowler had shown that *every star* on sufficient compression attains degeneracy, thereby ensuring that the star would rest in peace in a state of very high density and small radius. It was felt that *white dwarf stars* are precisely the stars which are in this state. They are faint and very compact stars with radius typically 1% of the solar radius.

Chandrasekhar, however, introduced a modification into the Fowler calculation. He noticed that, for large-mass stars, the filled levels are so high in energy that the electrons occupying them *would be relativistic* and this would alter the degeneracy criteria. With his modification Chandrasekhar [27] found *an upper limit* to the mass of a star existing as a white dwarf. This limit, known as the *Chandrasekhar limit*, is $1.44 M_\odot$. This result means that stars more massive than this limit would have central temperatures so high that the electrons there will not have become degenerate. Without the degeneracy pressure the star cannot remain in a state of equilibrium.

This conclusion implies that white dwarfs should not be found with mass greater than the Chandrasekhar limit. So far this result has held firm. The concept of relativistic degeneracy has become accepted and astrophysicists have extended it to the neutron stars also. There neutrons are tightly packed in a small volume and become degenerate, again provided that the mass of the star does not exceed a limit close to $2M_\odot$.

Nevertheless, in the early days Chandrasekhar faced considerable opposition to his result. His main opponent was no less a person than Arthur Stanley Eddington, who had played a pioneering role in stellar astrophysics. Eddington castigated Chandrasekhar's use of relativistic degeneracy in the following words:

> [If Chandrasekhar is right, then . . .] the star has to go on radiating and radiating until, I suppose, it gets down to a few km. radius, when gravity becomes strong enough to hold in the radiation, and the star can at last find peace [. . .] I think there should be a law of Nature to prevent a star from behaving in this absurd way . . .

Eddington was visualizing a situation in which a star finds itself without any counterforce to its own gravity, which makes it contract and continue to contract with ever increasing force. For, with its 'inverse-square behaviour', gravity grows stronger as the object shrinks, with the result that the contraction enters a run-away mode. This phenomenon is called *gravitational collapse*. It is ironic that the *reductio ad absurdum*-type argument used by Eddington against the continued contraction of a massive star can be turned round to predict the existence of a new genre of objects. An object of this type develops a gravitational force so strong that it pulls back even the light originating in the object, thus rendering it invisible to external observers.

Such an object is called a *black hole* today. We will spend this chapter summarizing the properties of black holes within the framework of general relativity. We begin with a discussion of gravitational collapse, the phenomenon that is supposed to lead to the formation of a black hole.

13.2 Gravitational collapse

Before taking up the general relativistic problem, we briefly outline its Newtonian counterpart. For we will find it of interest to compare and contrast the descriptions of the phenomenon by these two leading theories of gravity.

13.2.1 The Newtonian problem

Let us consider a ball of matter of mass M and radius R having a spherical symmetry in its physical parameters such as density ρ and pressure p. Suppose that it is undergoing a gravitational collapse, i.e., continued and ever-increasingly fast contraction. We ignore the effect of pressure as an opposing agency to gravitation and write the equation of motion as

$$\ddot{R} = -\frac{GM}{R^2}. \tag{13.1}$$

This equation represents the acceleration of a test particle on the surface of the ball, and it can be easily solved with the initial conditions $R = R_0$, $\dot{R} = 0$ at $t = 0$. We find that the time taken for R to reach zero is

$$t_0 = \frac{\pi}{2}\sqrt{\frac{R_0^3}{2GM}}. \tag{13.2}$$

For a Sun-like star this works out at 29 minutes! The short time scale indicates how powerful the collapse phenomenon can be. The above time scale can be written in terms of the starting density as follows:

$$t_0 = \frac{\pi}{2\sqrt{\alpha}}, \quad \alpha = \frac{8\pi G\rho_0}{3}. \tag{13.3}$$

We will now look at the relativistic problem, which leads to a surprisingly similar answer.

Example 13.2.1 *Problem.* In the above example let $m(r)$ denote the mass of the ball within radius r. Imagine the star as made of layers of different densities and find the condition that the layers do not cross as the ball collapses.

Solution. From (13.2) we see that the collapse time for $m(r)$ is

$$t = \frac{\pi}{2}\sqrt{\frac{r_0^3}{2Gm(r)}},$$

where we assume that initially the value of r was r_0. For 'no crossing', $m(r) = m(r_0)$ and we expect t to be an increasing function of r_0. (Thus the outer layers collapse later.) So the condition is that $r^3/m(r)$ should increase with r. On defining by $\bar{\rho}(r)$ the average density of $m(r)$, we have

$$\frac{r^3}{m(r)} = \frac{3}{4\pi}\bar{\rho}(r)^{-1}.$$

Hence our requirement for no crossing is that $\bar{\rho}(r)$ should *decrease* with r; i.e., there is no density inversion in the ball. The density should steadily decrease outwards.

13.2.2 The general relativistic problem

We begin with a discussion of spherically symmetric collapse since this is the only case that has been dealt with exactly. The line element for the spacetime is given by

$$ds^2 = e^\nu \, dt^2 - e^\omega \, dr^2 - e^\mu (d\theta^2 + \sin^2\theta \, d\phi^2). \qquad (13.4)$$

Here ν, ω and μ are functions of r and t. The energy-momentum tensor for the ball made of perfect fluid will be as given in (7.26):

$$T^{ik} = (p + \rho)u^i u^k - pg^{ik}. \qquad (13.5)$$

The energy-conservation relations $T^{ik}{}_{;k} = 0$ then lead to two equations:

$$\dot{\omega} + 2\dot{\mu} = -\frac{2}{p + \rho}\dot{\rho} \qquad (13.6)$$

and

$$\frac{\partial \nu}{\partial r} = -\frac{2}{p + \rho}\frac{\partial p}{\partial r}. \qquad (13.7)$$

We next simplify the problem by assuming that pressures are unimportant during collapse. Thus, ignoring the pressure gradient in Equation (13.7), we get ν independent of r and therefore a function of t only. A time/time transformation can then be used to set $\nu = 0$. We have used this trick before. Ignoring pressure reduces the problem to that of 'dust' and allows the coordinates to be given a 'comoving' interpretation. Thus we assume that a comoving observer falling in with the collapsing ball has constant coordinate values for r, θ and ϕ. Thus such an observer has t as his proper time.

We next consider the field equations and look at the R_{14} component. It reduces to the equation

$$2\dot{\mu}' + \dot{\mu}\mu' - \dot{\omega}\mu' = 0. \qquad (13.8)$$

This integrates to

$$e^{\omega} = \frac{e^{\mu} \mu'^2}{4(1+g)},\tag{13.9}$$

where g is an arbitrary function of r only. Next, the $(1, 1)$ component of the field equations becomes

$$\frac{1}{4}\mu'^2 e^{-\omega} - \left(\ddot{\mu} + \frac{3}{4}\dot{\mu}^2\right) - e^{-\mu} = 0.\tag{13.10}$$

Using (13.9) this can be integrated to

$$\dot{\mu}^2 = 4g(r)e^{-\mu} + 4F(r)e^{-3\mu/2}.\tag{13.11}$$

Here $F(r)$ is a function of r only. Finally, from Equations (13.9) and the $(0, 0)$ component of the field equations, we find that

$$F'(r) = \frac{4\pi}{3}G\rho e^{3\mu/2}\mu'.\tag{13.12}$$

Now suppose that the dust ball was of uniform density ρ_0 at $t = 0$ and that it was at rest then. Thus we assume that at that initial moment the time derivatives of ω and μ were zero. We can still choose the r coordinate and do so at that initial moment by requiring that a sphere of constant r has the surface area $4\pi r^2$. This requirement leads to

$$e^{\mu(0,r)} = r^2.\tag{13.13}$$

We will specify the extent of the collapsing mass by requiring that it is limited by $r \le r_{\rm b}$. For $r > r_{\rm b}$ we may assume that the space is empty and describable by the Schwarzschild solution.

By applying (13.12) to the situation at $t = 0$, we get

$$F(r) = \frac{8\pi G\rho_0}{3}r^3 = \alpha r^3,\tag{13.14}$$

say. There is an arbitrary constant of integration that corresponds to a point mass at $r = 0$, which we set equal to zero. Also, from Equation (13.11), at $t = 0$, $\dot{\mu} = 0$ we get

$$g(r) = -\frac{1}{r}F(r) = -\alpha r^2.\tag{13.15}$$

For $t > 0$ we get a solution for e^{μ} by writing

$$e^{\mu/2} = rS(t)\tag{13.16}$$

and using (13.11), (13.14) and (13.15), we get

$$\dot{S}^2 = \alpha\left(\frac{1-S}{S}\right).\tag{13.17}$$

The initial conditions are $S(0) = 1$, $\dot{S}(0) = 0$. A comparison with the Newtonian problem that we briefly looked at earlier shows that,

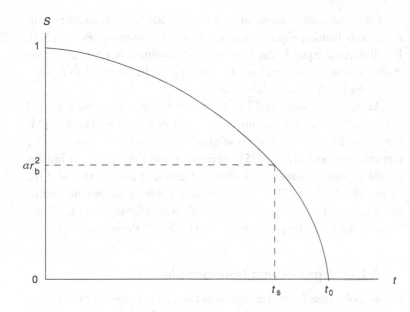

Fig. 13.1. A plot of the scale factor $S(t)$ of a supermassive dust ball undergoing gravitational collapse. The time t_s when the outer surface of the ball crosses the Schwarzschild radius is shown by a dotted line.

had we written there $R = S(t)R_0$, we would have got exactly the same equation for $S(t)$ as (13.17). The solution also matches the Newtonian one when we get the time of collapse to $S = 0$ as

$$t_0 = \frac{\pi}{2\sqrt{\alpha}}. \tag{13.18}$$

Equation (13.9) determines the function e^ω in terms of the above quantities as

$$e^\omega = \frac{S^2(t)}{1 - \alpha r^2}. \tag{13.19}$$

The line element inside the dust ball thus takes the form

$$ds^2 = dt^2 - S^2(t)\left[\frac{dr^2}{1 - \alpha r^2} + r^2(d\theta^2 + \sin^2\theta\, d\phi^2)\right]. \tag{13.20}$$

We have encountered this line element in Chapter 6 as an example of a maximally symmetric space of three spatial dimensions. We will encounter it again in Chapter 14 as a cosmological spacetime metric.

Figure 13.1 shows the function $S(t)$ plotted between $t = 0$ and $t = t_0$. We can match it to an exterior Schwarzschild solution for a mass

$$M = 4\pi r_b^3 \rho_0/3. \tag{13.21}$$

For, at $t = 0$, the proper radius of the ball is r_b and its density is ρ_0. As the ball contracts by a factor $S(t)$, its density goes up by a factor $S(t)^{-3}$, compensating for the reduction of its proper volume. Thus its mass remains the same during collapse.

For an external observer at a constant radial Schwarzschild coordinate, the collapsing object would have an effective radius $R_b = r_b S(t)$. It will become equal to the Schwarzschild radius when $S = \alpha r_b^2$. This radius is crossed in a finite time as measured on the t-scale. *This is when the object has become a black hole.*

At this stage, we would like to emphasize that the technique used here to solve the collapse problem is the one used by B. Datt [34] from Kolkata in 1938. The same problem was solved a year later by Oppenheimer and Snyder [35]. Because it was published in a higher-profile journal the latter work received greater publicity than did the earlier one by Datt. Thus the collapse problem is commonly called the 'Oppenheimer–Snyder problem'. We will refer to it as the 'Datt–Oppenheimer–Snyder problem' or simply the 'DOS problem'.

13.2.3 Collapse viewed from outside

Let us look at the DOS problem from the vantage point of the external Schwarzschild observer. The line element outside the object is, of course,

$$ds^2 = \left(1 - \frac{2GM}{R}\right) dT^2 - \frac{dR^2}{\left(1 - \frac{2GM}{R}\right)} - R^2(d\theta^2 + \sin^2\theta \, d\phi^2).$$

(13.22)

Note that we have departed from our earlier notation by changing the coordinates (t, r) to (T, R) since we want to relate this line element to the one considered for the collapsing dust ball. Thus the line element (13.22) has to be matched to the line element (13.20) at the boundary $r = r_b$ of the collapsing object. So we need at the boundary the condition

$$R = r_b S(t). \tag{13.23}$$

Next consider a test particle at the boundary. It is falling freely and so follows a timelike geodesic. From our general solution of the geodesics in Schwarzschild's spacetime we use Equation (9.27) to deduce that for radial free fall

$$\frac{dT}{ds}\left(1 - \frac{2GM}{R}\right) = \text{constant}. \tag{13.24}$$

However, from the same geodesic equation for spacetime given by the line element (13.20) we have

$$\frac{dt}{ds} = 1. \tag{13.25}$$

Hence on the surface we must have

$$\frac{\mathrm{d}T}{\mathrm{d}t}\left(1 - \frac{2GM}{R}\right) = \text{constant} = \gamma \tag{13.26}$$

(say). Since the line element is a first integral of the geodesic equations, we use Equation (13.22) together with the above to get

$$\left(\frac{\mathrm{d}R}{\mathrm{d}t}\right)^2 = \gamma^2 - 1 + \frac{2GM}{R}. \tag{13.27}$$

On the boundary, $R = r_{\mathrm{b}} S(t)$, this equation becomes

$$\left(\frac{\mathrm{d}S}{\mathrm{d}t}\right)^2 = \frac{\gamma^2 - 1}{r_{\mathrm{b}}^2} + \frac{2GM}{r_{\mathrm{b}}^3 S(t)}. \tag{13.28}$$

A comparison with Equations (13.17) and (13.21) gives

$$2GM = \alpha r_{\mathrm{b}}^3, \qquad \gamma^2 = 1 - \alpha r_{\mathrm{b}}^2. \tag{13.29}$$

We had arrived at this value of M from earlier discussions of gravitational collapse.

Consider now a radially outward light signal from an observer B on the boundary of the body at $R = R_1$ leaving him at time T_1 and reaching another Schwarzschild observer A located at fixed $R = R_2$ at time T_2. Notice that R_1 decreases as the object collapses.

The radial null geodesic equation then yields

$$T_2 - T_1 = \int_{R_1}^{R_2} (1 - 2GMR)^{-1} \, \mathrm{d}R \tag{13.30}$$

This integral diverges as $R_1 \rightarrow 2GM \, (= R_{\mathrm{s}})$. Let, at $t = t_{\mathrm{s}}$, $R_{\mathrm{s}} = r_{\mathrm{b}} S(t_{\mathrm{s}})$. Then, as $t \rightarrow t_{\mathrm{s}}$, $T_1 \rightarrow \infty$. In short, the observer A at R_2 has to wait for ever for the signal sent out by B at t_{s}.

Figure 13.2 shows the signal propagation from B to A on the contracting object. If A has a means of measuring wavelength, he will find that light waves from B are increasingly redshifted. The shift is gravitational as well as Doppler.

We may liken the signal exchanges between A and B to a correspondence between an Applicant and a Bureaucrat. The applicant may think that the bureaucrat is being very dilatory... but this is really the trick played by curved spacetime! The time flows at different rates for the two protagonists.

It is clear that as B approaches R_{s} his signals begin to be more and more difficult to receive, both because of their redshift and owing to their faintness.

Fig. 13.2. A schematic diagram showing signal exchanges between observers A and B. For details, refer to the text.

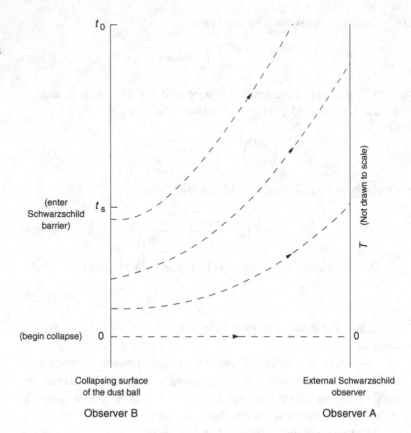

(enter Schwarzschild barrier) t_S

(begin collapse)   0

Collapsing surface of the dust ball

Observer B

External Schwarzschild observer

Observer A

t_0

T (Not drawn to scale)

Example 13.2.2 *Problem.* Show that the time measured by the comoving observer riding on the surface of the gravitationally collapsing dust ball registers a finite value as he crosses the Schwarzschild barrier.

Solution. Using the calculations in the text, we find that the time measured by B is given by t and Equation (13.17) describes the rate of collapse. The Schwarzschild barrier is crossed when $S = \alpha r_b^2$.

To solve (13.17) write $S = \cos^2\theta$. Then Equation (13.17) becomes

$$2\cos^2\theta \;\dot{\theta} = \sqrt{\alpha}.$$

This integrates to

$$\theta + \sin\theta\cos\theta = \sqrt{\alpha}t$$

so that, at $t = 0$, $\theta = 0$ and $S = 1$. The state $S = 0$ is reached when $\theta = \pi/2$. At the Schwarzschild barrier the observer B will have $\cos^2\theta = \alpha r_b^2$ so that the corresponding time is

$$t_S = \frac{1}{\sqrt{\alpha}}\left(\cos^{-1}\sqrt{\alpha r_b^2} + \sqrt{\alpha r_b^2(1 - \alpha r_b^2)}\right).$$

This is finite and less than $\pi/\left(2\sqrt{\alpha}\right)$.

Problem. Estimate α and r_b for a homogeneous dust ball with the Sun's mass and radius as the starting values for the DOS problem.

Solution. For the Sun $M_{\odot} = 2 \times 10^{33}$ g and $R_{\odot} = 7 \times 10^{10}$ cm are the mass and radius. So $r_b = R_{\odot} = 7 \times 10^{10}$ cm.

The starting density is $\rho_0 = 3M_{\odot}/(4\pi R_{\odot}^3) \cong 1.4$ g cm^{-3}. The formula for α then gives

$$\alpha = \frac{8\pi\rho_0 G}{3} \cong 7.7 \times 10^{-7} \text{ s}^{-2}.$$

The barrier at R_s is thus a one-way membrane. Having crossed it, B will continue to receive signals from A, but his own messages will not be able to cross the barrier, let alone reach A. This *Schwarzschild barrier* marks the boundary of what is called a *black hole*. It is also usual to refer to the boundary $R = R_s$ as the 'event horizon'. We will refer to this aspect more specifically later in this chapter.

The external observer does not see what happened to B after B has crossed this barrier. B's fate is not very pleasant. Besides any tidal effects that may tear him apart, at $t = t_0$ he hits the state described by $S = 0$. This is a state wherein the entire space 'shrinks' to zero volume with the density going to infinity. Parameters describing spacetime geometry diverge and there is no way of describing what is happening mathematically or physically. This extreme state of space, time and matter is called *singularity*. We will refer back to this strange aspect of spacetime geometry in Chapter 18.

13.3 The Schwarzschild solution in other coordinate systems

The line element (13.22) is somewhat inconvenient for discussing regions containing the Schwarzschild barrier. For example, we find that the metric components g_{00} and g_{11} become respectively zero and infinity at $R = 2GM$. Also, inside the barrier the coordinates T, R interchange their time/spacelike character. It is sometimes preferable to use other coordinates, which behave normally in this region. We have already seen that the comoving coordinates used for discussing collapse do not throw up any problem at $R = R_s$. However, they are not so convenient for relating to an external observer. We describe some other coordinate systems that have been found useful for connecting across the barrier at $R = R_s$.

13.3.1 Eddington coordinates

In these coordinates, the Schwarzschild coordinate T is replaced by the 'null' coordinate

$$V = T + R + 2GM \ln \left| \frac{R}{2GM} - 1 \right|. \tag{13.31}$$

It can be seen via a simple manipulation that $V = V_0$ (constant) describes a null geodesic corresponding to a radial null ray going in from outside. Also, the line element (13.22) is transformed to

$$ds^2 = \left(1 - \frac{2GM}{R} \right) dV^2 - 2\, dV\, dR - R^2(d\theta^2 + \sin^2\theta\, d\phi^2). \tag{13.32}$$

This coordinate was used by Eddington [36] to connect observers A and B, A outside the barrier and B inside, with light rays coming from A to B.

13.3.2 Kruskal–Szekeres coordinates

These coordinates were independently discovered in 1960 by M. D. Kruskal and G. Szekeres [37, 38]. We give the transformations from the Schwarzschild coordinates below. It will be clear that they carry the Eddington coordinates a step further in using null tracks.

The coordinate system involves a changeover from T, R coordinates to u, v coordinates while leaving the other two coordinates θ, ϕ unchanged. The transformations relate to four different but connected regions of spacetime, which we will denote by I, II, III and IV. Briefly, we have the following.

Region I: $R \geq 2GM, u \geq 0$:

$$u = \left(\frac{R}{2GM} - 1 \right)^{1/2} \exp\left(\frac{R}{4GM} \right) \cosh\left(\frac{T}{4GM} \right),$$

$$v = \left(\frac{R}{2GM} - 1 \right)^{1/2} \exp\left(\frac{R}{4GM} \right) \sinh\left(\frac{T}{4GM} \right).$$

Region II: $R \leq 2GM, v \geq 0$:

$$u = \left(1 - \frac{R}{2GM} \right)^{1/2} \exp\left(\frac{R}{4GM} \right) \sinh\left(\frac{T}{4GM} \right),$$

$$v = \left(1 - \frac{R}{2GM} \right)^{1/2} \exp\left(\frac{R}{4GM} \right) \cosh\left(\frac{T}{4GM} \right).$$

Region III: $R \geq 2GM, u \leq 0$:

$$u = -\left(\frac{R}{2GM} - 1\right)^{1/2} \exp\left(\frac{R}{4GM}\right) \cosh\left(\frac{T}{4GM}\right),$$

$$v = -\left(\frac{R}{2GM} - 1\right)^{1/2} \exp\left(\frac{R}{4GM}\right) \sinh\left(\frac{T}{4GM}\right).$$

Region IV: $R \leq 2GM, v \leq 0$:

$$u = -\left(1 - \frac{R}{2GM}\right)^{1/2} \exp\left(\frac{R}{4GM}\right) \sinh\left(\frac{T}{4GM}\right),$$

$$v = -\left(1 - \frac{R}{2GM}\right)^{1/2} \exp\left(\frac{R}{4GM}\right) \cosh\left(\frac{T}{4GM}\right).$$

For all four regions, the line element is, however, the same:

$$ds^2 = \frac{32G^3 M^3}{R} \exp\left(-\frac{R}{2GM}\right) \cdot (du^2 - dv^2) - R^2(d\theta^2 + \sin^2\theta \, d\phi^2).$$

$$(13.33)$$

The coordinates are thus u, v, θ, ϕ and we may look upon R in the above line element as a function of (u, v) given implicitly by

$$\left(\frac{R}{2GM} - 1\right) \exp\left(\frac{R}{2GM}\right) = u^2 - v^2. \qquad (13.34)$$

As can be seen from Equation (13.33), the line element is well behaved at $R = R_s$. Figure 13.3 is the so-called Kruskal–Szekeres diagram showing a radial constant θ, ϕ section with some important lines marked. The lines $R = $ constant are rectangular hyperbolae in the u–v plane. The lines $R = 2GM$ form a cross with one arm (SW to NE) having $T = \infty$ while the other arm (SE to NW) has $T = -\infty$. The four regions I, II, III and IV are within the sectors defined by the radial lines $u^2 - v^2 = 0$.

The collapsing object has the outer boundary shown in Figure 13.3 by a dotted line. The collapse starts in Region I and ends in Region II. The boundary point hits the singularity shown by the hyperbola $R = 0$. The same trajectory can be continued as in the figure from Region I to Region IV. What does it represent? It represents the time-reversed version of collapse: an eruption out of the singularity in Region IV, which is sometimes called a white hole. We will describe a *white hole* later.

The Kruskal–Szekeres diagram demonstrates the incompleteness of the Schwarzschild coordinate system, besides explaining the difference between a black hole and a white hole.

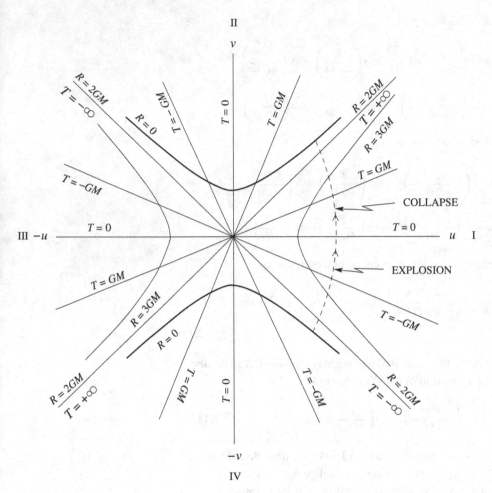

Fig. 13.3. The Kruskal–Szekeres diagram relates the null coordinates (u, v) to the Schwarzschild coordinates (R, T) and shows the incompleteness of the latter.

13.4 Non-spherical gravitational collapse

The only problem of gravitational collapse we could discuss in some detail was one of spherical symmetry. One could argue that stars are by and large spherical and will start their collapse in that state. Although they would have pressures, these may be neglected in the 'run-away' state of collapse, as mentioned earlier. The formalism developed by Datt can deal with any inhomogeneous (but spherically symmetric) initial state.

Nature, however, might not be so obliging always in giving us spherical symmetry as an initial condition. The departure from spherical symmetry complicates the problem enormously. Attempts are being made to solve the partial differential equations of collapse numerically on a large computer.

Nevertheless, if one sticks to 'small' departures from spherical symmetry, one can make progress, as R. H. Price's work has shown. The spirit of this work (not the details) is conveyed in the following paragraphs of this section.

Suppose the collapsing body generates an external physical field Φ, which can be characterized by an integral spin s and zero rest mass (so as to be of long range). Thus an electromagnetic disturbance is characterized by $s = 1$, whereas a gravitational one has $s = 2$. These disturbances may be treated as small, first-order, perturbations on an external Schwarzschild solution, which is assumed to be left unchanged at 'zeroth order'.

Suppose the external field is written as a power-series expansion over spherical harmonics:

$$\Phi = \sum_{n,s} \frac{1}{R} A_n^{(s)} S_n(\theta, \phi). \tag{13.35}$$

The coefficients $A_n^{(s)}$ are functions of R and T. When the time development of these coefficients is carried out, most of them die away as the object collapses. What does remain at the end?

Price found that all harmonics of order $n \geq s$ are radiated away and only those with $n < s$ remain. Thus nothing survives from a scalar field $(s = 0)$, only the electric charge survives as a source in the $s = 1$ electromagnetic case, while mass $(n = 0)$ and angular momentum $(n = 1)$ are left as sources in the $s = 2$ gravitational case.

This analysis is limited to small departures from spherical symmetry. However, if we stretch our belief to larger departures away from spherical symmetry, it tells us that if we are limited to gravitational and electromagnetic interactions only (these are the only long-range basic interactions known today) then the end point of gravitational collapse for an external observer is a black hole with mass (M), electric charge (Q) and angular momentum (H). It is therefore of interest to know whether such black holes exist and how they are described.

13.5 The Reissner–Nordström solution

H. Reissner in 1916 and G. Nordström in 1918 independently arrived at a solution for the metric exterior to a spherically symmetric distribution of charged matter with total mass M and electric charge Q. (See References [39, 40].) We will briefly show how the problem is solved and discuss the nature of the solution.

We start with a spherically symmetric line element as in the Schwarzschild case:

$$ds^2 = e^{\nu} \, dT^2 - e^{\lambda} \, dR^2 - R^2(d\theta^2 + \sin^2\theta \, d\phi^2). \tag{13.36}$$

The only difference is that we have a Coulomb field F_{ik}, generated by the charge Q at the origin, that has an energy-momentum tensor given by

$$T_{ik} = -\frac{1}{4\pi}\left[F_i{}^m F_{km} - \frac{1}{4}F_{lm}F^{lm}g_{ik}\right]. \tag{13.37}$$

Recall from Section 7.5.2 that this is the general expression for the energy-momentum tensor of the electromagnetic field.

From the condition that the source is static, we assume that the solution of the electromagnetic field as well as the spacetime geometry will be static too. Thus we try the solution for the 4-potential as $A_i \equiv [\psi(R), 0, 0, 0]$. Then, for the only non-zero field components $F_{01} = -F_{10}$, we have

$$F_{01} = -\psi', \qquad F^{01}\sqrt{-g} = e^{-\frac{1}{2}(\lambda+\nu)}R^2\sin\theta \times \psi'.$$

The condition for a point charge at the origin is that the covariant divergence of F^{ik} should vanish everywhere except at $R = 0$. From the above relation

$$\frac{\partial F^{01}\sqrt{-g}}{\partial R} = 0 \Rightarrow \psi' = \frac{E}{R^2}e^{\frac{1}{2}(\lambda+\nu)},$$

where E is a constant of integration. On substituting into the expression (13.37) for T^{ik} we get the non-zero components as

$$T_0^0 = T_1^1 = -T_2^2 = T_3^3 = \frac{1}{4\pi}e^{-\frac{1}{2}(\lambda+\nu)}\psi'^2. \tag{13.38}$$

We now substitute $-8\pi G T_k^i$ on the right-hand side of the Einstein equations written out for the above metric. Referring back to the Schwarzschild solution of Chapter 9, we again see that the equations with $i = k = 0$ and $i = k = 1$ taken together give as before $\lambda' + \nu' = 0$. As on that occasion, we can again simplify the solution by having $\lambda = -\nu$. So we are left with only one independent equation:

$$e^\nu(1 + R\nu') - 1 = -\frac{GE^2}{R^2}. \tag{13.39}$$

This simple differential equation can be solved and we get the solution as

$$e^\nu = 1 - \frac{B}{R} + \frac{GE^2}{R^2}.$$

Here B is a constant of integration. If we look at the asymptotic form of the line element at large R and demand that it look like that

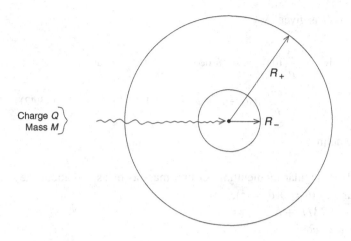

Fig. 13.4. A schematic view of the Reissner–Nordström solution. It has two spheres on which horizon-like properties are found.

for mass M at the origin, then we get $B = 2GM$. Likewise, if we ascribe a magnitude Q to the charge at the origin, then its asymptotic Coulomb field in Gaussian units would be Q/R^2. A comparison with our solution enables us to set $E = Q$. Thus we have the following Reissner–Nordström line element to describe a spacetime around a spherical mass M and charge Q:

$$ds^2 = \left(1 - \frac{2GM}{R} + \frac{GQ^2}{R^2}\right)dT^2 - \left(1 - \frac{2GM}{R} + \frac{GQ^2}{R^2}\right)^{-1} dR^2$$
$$- R^2(d\theta^2 + \sin^2\theta \, d\phi^2). \tag{13.40}$$

It is easy to see that there is an apparent problem where e^ν vanishes, i.e., at two values of R at

$$R_\pm = GM \pm \sqrt{G^2M^2 - GQ^2}. \tag{13.41}$$

We have not one but two surfaces where e^ν vanishes. The outer one ($R = R_+$) plays effectively the role of an event horizon of the black hole just as the $R = R_s$ surface does for a charge-free body. See Figure 13.4 for a schematic description of the Reissner–Nordström black hole.

13.6 The Kerr solution

In 1963 Roy Kerr obtained what may arguably be the most important exact solution of Einstein's field equations since the Schwarzschild solution. It describes the spacetime outside a spinning mass. Specifically, the line element for the empty spacetime outside a mass M having an angular momentum H is given by

$$ds^2 = \frac{\Delta}{\rho^2}(dT - h\sin^2\theta \, d\phi)^2 - \frac{\rho^2}{\Delta}\, dR^2 - \rho^2 \, d\theta^2$$

$$- \frac{\sin^2\theta}{\rho^2}[(R^2 + h^2)d\phi - h\, dT]^2, \qquad (13.42)$$

where we define

$h \equiv H/M =$ angular momentum per unit mass as measured about the
  polar axis $(\theta = 0; \theta = \pi)$,
$\Delta \equiv R^2 - 2GMR + h^2$,
$\rho^2 \equiv R^2 + h^2 \cos^2\theta$.

For details on how this line element was derived, see Kerr's original paper [41]. We will look at some important properties of the Kerr solution.

13.6.1 The static limit

As we go closer and closer to the Kerr black hole, we notice the effect of rotation in various ways. As we notice from the Earth, because of its spin, we see stars rise in the East and set in the West. If we desire to see the stars stationary as they really are, in the frame of reference in which the distant stellar background is at rest, we need to counteract the Earth's spin by travelling in a fast aircraft or in a space station from East to West.

The same would happen for the Kerr solution, but up to a limit. The coordinates (R, θ, ϕ) are constant for distant stars and, if an observer wishes to stay at rest in such a frame, he will encounter greater and greater difficulty as he approaches the object. He will be required to exert stronger and stronger force to stay in the same place relative to distant stars. Thus the line element shows that a world line having a constant (R, θ, ϕ) will be timelike provided that

$$R > R(\theta) = GM + \sqrt{G^2M^2 - h^2\cos^2\theta}. \qquad (13.43)$$

For $R \leq R(\theta)$ the observer will be dragged along past the constant R, θ, ϕ framework in the direction in which the mass is spinning. Even if the observer employs rocket power to counter this drag, it will be to no avail. This surface $R = R(\theta)$ is called the 'static limit'.

13.6.2 The ergosphere

Just as we have horizons for the Schwarzschild and the Reissner–Nordström black holes, here too we have an event horizon, provided, of course, that the object itself is not larger than it. The Kerr horizon is located at $\Delta = 0$, i.e., at

$$R = R_+ = GM + \sqrt{G^2 M^2 - h^2}. \tag{13.44}$$

It can be easily verified that, as in the Schwarzschild case, light rays may enter the horizon from outside ($R > R_+$), but they cannot emerge outwards from inside the horizon.

The surface signifying the static limit is larger than the above horizon. As shown in Figures 13.5(a) and (b), the two surfaces touch each other at the poles ($\theta = 0$, $\theta = \pi$); otherwise the static limit is strictly outside the horizon. The volume between the two surfaces, shown filled by dots, has been named 'the ergosphere'. The reason for the name is that the compulsive spin imposed on any piece of matter entering the ergosphere enbles us to 'extract' energy from the spinning black hole. The black hole thereby loses some of its rotational energy. The rapidly rotating piece will carry that energy away, if it is enabled to emerge from the ergosphere.

13.6.3 The Kerr–Newman black hole

Ted Newman and his colleagues at Pittsburgh University combined the Reissner–Nordström solution with the Kerr solution and generated the spacetime geometry for a charged spinning mass. Thus the line element for a mass M with angular momentum H and electric charge Q is given by Equation (13.42), but with the quantity Δ redefined as

$$\Delta = R^2 - 2GMR + h^2 + GQ^2. \tag{13.45}$$

Like the Kerr solution, this solution also exhibits the properties of a horizon, a static limit and the ergosphere. Further, if we set $h = 0$, we come back to the Reissner–Nordström black hole. This solution has one special significance.

As we saw in Section 13.4, Price's theorem indicates (but does not prove in the most general collapse case) that a black hole forming by gravitational collapse under the classical long-range forces of electromagnetism and gravitation will at most exhibit mass, electric charge and angular momentum. This state is precisely described by the Kerr–Newman black hole with its three parameters M, H and Q.

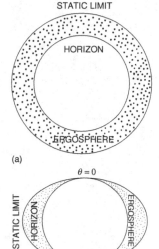

Fig. 13.5. In (a) we see a 'constant-latitude' section of the Kerr black hole. The portion between the horizon and the static limit belongs to the ergosphere. In (b) the section along a longitude great circle shows the location of the 'poles' at $\theta = 0, \pi$ and the section of the ergosphere (marked with dots).

We will now look at the general physical properties of black holes, drawing analogies with the laws of thermodynamics. While illustrating them quantitatively, we will use the Kerr–Newman black hole or the simpler Kerr black hole as the typical example of a black hole.

13.7 Black-hole physics

It has been noticed that the rules describing the dynamical properties of black holes bear a striking resemblence to the laws of thermodynamics.

Following the analogy with the laws of thermodynamics, we begin by stating the first law of black-hole physics:

In any process involving a black hole and other objects, the total energy, momentum, angular momentum and electric charge are conserved.

This simply means, for example, that, if a Schwarzschild black hole gobbles up a mass having total energy E, its mass will grow from the earlier value M to $M + E$.

This raises the question, can the process be reversed? That is, can we extract energy from the black hole and reduce its mass? The answer is given by the second law of black-hole physics:

In any physical interaction, the surface area of a black hole can never decrease.

The wording has a distinct similarity to the second law of thermo-dynamics, with the 'surface area' playing the role of entropy. Let us examine this law and also seek an answer to the question raised above.

The surface area of a black hole is the area of its horizon surface. For the Schwarzschild black hole, the horizon is given by $R = R_s = 2GM$, so the surface area of the black hole is simply

$$A = 4\pi R_s^2 = 16\pi G^2 M^2. \tag{13.46}$$

With this definition, A cannot decrease and so M cannot decrease. This in turn implies that we cannot extract energy from this black hole since such a process would tend to *reduce* rather than increase M. However, not all is lost! For, if we have a spinning black hole, we note from Equation (13.41) that the horizon is at $R = R_+$. From the line element we see that at $T = $ constant and $R = $ constant any surface described by $[\theta, \theta + d\theta] \times [\phi, \phi + d\phi]$ has area

$$\rho \, d\theta \sqrt{R_+^2 + h^2} \times \frac{\sin\theta}{\rho} \, d\phi = \sqrt{R_+^2 + h^2} \times \sin\theta \, d\theta \, d\phi.$$

By integrating over θ, ϕ, we get the surface area of the Kerr–Newman black hole as

$$A = 4\pi (R_+^2 + h^2) = 4\pi [2G^2 M^2 - GQ^2 + 2GM \sqrt{G^2 M^2 - GQ^2 - h^2}].$$

(13.47)

We now take the differential of the above expression, using the fact that the variables to change are A, M, Q and H. Since H has directionality like a vector, we should write it as a three-dimensional spatial vector $\mathbf{H}$. The same applies to $\mathbf{h}$. Thus, after some manipulation, we get

$$\frac{\delta A}{8\pi G} = \frac{1}{\sqrt{G^2 M^2 - GQ^2 - h^2}} \times [(R_+^2 + h^2)\delta M - R_+ Q \, \delta Q - \mathbf{h} \cdot \delta \mathbf{H}].$$

(13.48)

As in thermodynamics, we assume that the most efficient way of running a process is to ensure that it is running reversibly, insofar as the area is concerned. In short, we set δA as zero and simplify the above expression to

$$\delta M = \frac{R_+ Q \, \delta Q}{R_+^2 + h^2} + \frac{\mathbf{h} \cdot \delta \mathbf{H}}{R_+^2 + h^2}.$$

(13.49)

Is it possible to *reduce* the mass of a black hole? For that is the only way we can extract energy from it. The above equation tells us that if we reduce the electric charge or the angular momentum of the black hole we can achieve this feat. The Penrose mechanism described below is a way to do this.

13.7.1 The Penrose process

The process proposed by Penrose is in fact a thought experiment designed to extract energy from the rotating Kerr black hole by using the properties of the ergosphere discussed in the text. As shown in Figure 13.6, the process involves dropping a mass into the ergosphere, arranging for it to split into two bits, with one bit falling inside the horizon and the other escaping outside the ergosphere.

What happens here is that the mass entering the ergosphere is made to rotate along with the black hole, as discussed in the text. It therefore acquires energy as well as angular momentum from the black hole. When it splits and part of it falls into the black hole, it loses a fraction of the acquired energy and angular momentum back to the black hole.

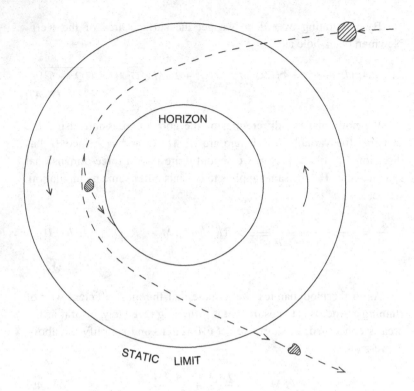

Fig. 13.6. A schematic illustration of the Penrose process. See the text for details.

HORIZON

STATIC   LIMIT

The escaping portion *can*, however, emerge with so much energy that it exceeds the energy of the total original mass.

13.7.2 Surface gravity

The analogy with thermodynamics can be pushed further by comparing the standard relation in thermodynamics,

$$dE = T\,dS - P\,dV, \tag{13.50}$$

with Equation (13.48) rewritten as

$$\delta M = \kappa \frac{\delta A}{8\pi G} + \frac{\mathbf{h} \cdot \delta \mathbf{H}}{h^2 + R_+^2} + \frac{R_+ Q\,\delta Q}{h^2 + R_+^2}, \tag{13.51}$$

where the function κ is defined by

$$\kappa = \frac{\sqrt{G^2 M^2 - G Q^2 - h^2}}{h^2 + R_+^2}. \tag{13.52}$$

What is this function supposed to represent? It is known as the *surface gravity* of the black hole. If we do a naive Newtonian calculation for a Schwarzschild black hole of mass M, its radius at the horizon is

$2GM$. The Newtonian acceleration due to gravity at this distance is

$$\kappa = \frac{G \times M}{(2GM)^2} = 1/(4GM).$$

The expression above reduces to this value for the Schwarzschild case. Thus κ in Equation (13.51) measures surface gravity and in our analogy with thermodynamics it plays the role of temperature. Just as in equilibrium temperature is constant, we have the corresponding 'zeroth law' of black-hole physics telling us of the existence of surface gravity, which happens to be constant over the horizon. This law was proposed by Bardeen, Carter and Hawking [42], whereas the second law was proposed by Hawking [43].

The analogy with thermodynamics was carried a step further by Bardeen *et al.* when they argued that it is impossible to reduce κ to zero by any finite set of operations. This statement matches the third law of thermodynamics.

Consider the Kerr black hole. What is the maximum angular momentum it can have for a given mass M? Since R_+ must be real, we need to have $h \leq GM$, i.e., $H \leq GM^2$. The state in which $H = GM^2$ describes what is called an *extreme Kerr black hole*. Notice that it has zero surface gravity and, by virtue of the theorem of Bardeen, Carter and Hawking, such a state cannot be attained in nature by any finite sequence of operations. It corresponds to the state of absolute zero temperature of thermodynamics.

With the help of the laws of black-hole physics we can understand the limitations on the Penrose process. Note that, as we reduce the mass and the angular momentum of the black hole, we can at best keep its area A constant. In general we see that if we travel along a constant-area curve we finally end with the Schwarzschild black hole of zero angular momentum. If the *third* law of black-hole physics holds, then we can at best start this process when the Kerr black hole is in what is known as the *extreme* state (when its surface gravity is zero). Here its angular momentum is so large that $h = GM$. Let us denote by M_0 the starting mass of the black hole in this extreme state. Then its surface area is (by the formula given in the text)

$$A_0 = 2\pi R_+^2 = 8\pi \frac{G^2 M_0^2}{c^4}.$$

In the final state its mass is M_i, say. The area of a Schwarzschild black hole of this mass is

$$A_i = 16\pi \frac{G^2 M_i^2}{c^4}.$$

By the second law of black-hole physics A_i cannot be less than A_0. At best we may equate A_0 to A_i, giving

$$M_i = \frac{M_0}{\sqrt{2}}.$$

Thus the Penrose process can at most extract $(M_0 - M_0/\sqrt{2})c^2$ of the original mass energy $M_0 c^2$. The available fraction of energy is therefore $(\sqrt{2} - 1)/\sqrt{2}$, i.e., nearly 30% of the total energy. (In contrast, the hydrogen-fusion process yields only $\sim 0.7\%$ of the total energy.)

The mass M_i is known as the *irreducible mass* of the Kerr black hole.

So, in principle, we can extract energy from a Kerr–Newman black hole until it reaches the state of no charge ($Q = 0$) and no angular momentum ($H = 0$). This state is that of the Schwarzschild black hole and it is characterized by its mass only. We call it *irreducible* mass since henceforth the second law prohibits any energy extraction. Since, in reaching this state, we have not changed the surface area (*vide* the condition $\delta A = 0$), we can relate the final irreducible mass to the area of the black hole. Thus we have the irreducible mass M_i as given by

$$4G^2 M_i^2 = h^2 + R_+^2,$$

i.e.,

$$M^2 = \left(M_{ir} + \frac{Q^2}{4G M_i} \right)^2 + \frac{H^2}{4G^2 M_i^2}. \tag{13.53}$$

Evidently this is the 'most efficient scenario'! If there is any irreversibility in the energy-extraction process, the irreducible mass would increase.

We next consider the problem of interest to astronomers, viz. how to detect black holes.

13.8 Detection of black holes

Given the fact that a black hole cannot be seen by detecting any form of light, how does one know that a black hole is located in some specified region? The answer is indicated by the following thought experiment. Suppose the Sun becomes a black hole. It will no longer be visible from the Earth. Nevertheless, we on the Earth would be able to deduce from the orbit of the Earth that there exists a source of attraction at the location of the Sun, with the mass of the Sun. This is because the black hole continues to exert gravitational force even if not seen.

Following this example, an ideal scenario for the detection of a black hole is if it has a companion that is easily visible. For example, if the black hole is a member of a binary star system, then by watching its companion move we can deduce the presence of an invisible mass. If from the dynamics of the system we are able to place a lower limit on the mass of the invisible object, and it exceeds, say, $3M_\odot$, then we can assert that the object is a black hole. (*Vide* the limits placed on stellar masses existing as white dwarfs or neutron stars, in Chapter 12.)

The first strong case for a black hole in a binary system was that of Cygnus X-1, an X-ray source. (See Reference [44].) Figure 13.7 illustrates the typical binary X-ray-source scenario. Here we have a supergiant star and a black hole going round their common barycentre. The black hole exerts an attractive force strong enough to pull the loosely attached outer layers of plasma from the companion. The plasma goes round and round the black hole as it spirals in and ultimately falls in across the horizon. The infalling material, because of its viscosity, gets heated and radiates through the process known as *bremsstrahlung*. At the high temperature of about a million degrees, this radiation is in the form of X-rays. Cygnus X-1, which was first found in the mid 1970s, proved to be typical of several similar examples of X-ray binaries in which the companion was invisible. However, a large fraction of these turned out to be neutron stars rather than black holes, since their masses did not exceed $2M_\odot$. The invisible star in Cygnus X-1, in contrast, has mass more than $8M_\odot$.

During the mid 1980s, observers of galaxies began reporting massive black holes (of $(10^8 - 10^9)M_\odot$) in the centres of galaxies

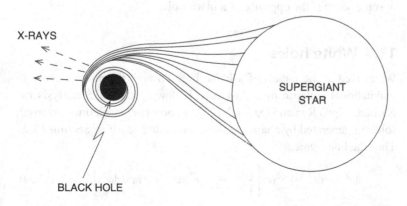

X-RAYS

SUPERGIANT
STAR

BLACK HOLE

Fig. 13.7. The binary X-ray scenario which in the case of the X-ray source Cygnus X-1 provides indirect evidence for its invisible component being a black hole (shown in the figure as a dark sphere).

Fig. 13.8. M87 photographed with its nucleus that is believed to house a black hole. An indication of violent activity in the nucleus is the emergence of the jet seen. (Photograph by courtesy of NASA.)

like M87 (see Figure 13.8). The presence of such massive black holes is inferred from the large dynamical activity of nearby stars as indicated by their large spectral shifts, or by the abnormal rise in luminosity of the region. The latter effect arises because of the concentration of stars near the black hole, which is brought about by its powerful attraction. The latter fact may appear paradoxical in the sense that the black hole itself is rendered invisible because of its strong attraction!

We end this chapter with a brief description of a *white hole*, which in some sense is the opposite of a black hole.

13.9 White holes

We arrived at the notion of a black hole through the phenomenon of gravitational collapse of a dust ball. Following the 1975 analysis by Narlikar, Appa Rao and Dadhich [45], we now consider a time-reversed solution generated by changing the coordinate t to $-t$ in Section 13.2. Thus the line element is

$$ds^2 = dt^2 - S^2(t)\left[\frac{dr^2}{1 - \alpha r^2} + r^2(d\theta^2 + \sin^2\theta \, d\phi^2)\right] \qquad (13.54)$$

with the function $S(t)$ satisfying the differential equation

$$\dot{S}^2 = \alpha \left(\frac{1-S}{S} \right) \tag{13.55}$$

and the boundary conditions $S = 0$ at $t = -t_0$ and $S = 1, \dot{S} = 0$ at $t = 0$. Instead of gravitationally collapsing, the dust ball erupts as an explosive event at $t = -t_0$. While the behaviours of the collapsing ball and expanding ball can be related through time symmetry in the t-coordinate, the two solutions look asymmetrically different from the vantage point of a distant Schwarzschild observer. The collapsing ball is seen to disappear slowly into the event horizon of the black hole. Let us now see how the exploding ball looks from outside.

We consider a radial signal leaving the surface of the expanding ball at the Schwarzschild time T_1 and reaching a distant observer at constant Schwarzschild $R = R_2$ coordinate at time T_2. The relation between these quantities is

$$\int_{R_1}^{R_2} \left(1 - \frac{2GM}{R} \right)^{-1} dR = T_2 - T_1. \tag{13.56}$$

R_1 is (as before) the changing value of the Schwarzschild coordinate of the white-hole surface.

Next consider a signal sent out a short time ΔT_1 later, arriving at the observer at $T_2 + \Delta T_2$. During this period R_2 has not changed, but R_1 has. Since $R_1 = r_b S(t)$, we can write

$$\Delta T_2 - \Delta T_1 = -\frac{r_b \dot{S}(t_1) \Delta t_1}{1 - \dfrac{2GM}{R_1}}. \tag{13.57}$$

From Equations (13.26) and (13.29) we can deduce that

$$\Delta T_1 = \frac{\sqrt{1 - \alpha r_b^2}}{1 - \dfrac{2GM}{R_1}} \Delta t_1. \tag{13.58}$$

Therefore, we have from the above two relations the result

$$\frac{\Delta T_2}{\Delta t_1} = \frac{\sqrt{1 - \alpha r_b^2} - r_b \dot{S}(t_1)}{1 - \dfrac{2GM}{R_1}}. \tag{13.59}$$

At $R_2 \gg 2GM$ we may treat T_2 as the proper time of the observer at R_2. Hence a light wave sent from the surface of the expanding ball undergoes a spectral shift z given by

$$(1+z)^{-1} = \frac{1 - \dfrac{2GM}{R_1}}{\sqrt{1 - \alpha r_b^2} - r_b \dot{S}(t_1)}. \tag{13.60}$$

This expression is well behaved outside the Schwarzschild horizon. However, it comes as a surprise to find that it is well behaved on the horizon and inside too! On expressing $\dot{S}(t_1)$ in terms of S, we get, after a limiting process,

$$(1+z)^{-1} = \sqrt{1 - \alpha r_b^2} + \sqrt{\alpha r_b^2 \left(\frac{1}{S} - 1\right)}. \tag{13.61}$$

The value of this ratio at the event horizon is $2\sqrt{1 - \alpha r_b^2}$.

On looking at the Kruskal–Szekeres diagram (Figure 13.3), we see that such rays are coming from Region IV into Region I. The T-coordinate behaves strangely, but the t and u, v coordinates are well behaved. The result in the above equation is finite. Since we have used 'bad' coordinates (R, T) we have arrived at two infinite integrals whose difference is finite. If we had used the (u, v) coordinates we could have got the result without subtraction of infinities.

From Equation (13.61) we see that signals not only come out from within the $R = 2GM$ surface but also can be blueshifted for

$$S(t_1) < \frac{1}{2} \left(1 + \sqrt{1 - \alpha r_b^2}\right). \tag{13.62}$$

Thus our dust ball, at least during the early stages of expansion, resembles a very shiny object with high-energy radiation coming out. Hence such objects are called *white holes*.

Compared with black holes, the white holes enjoy certain advantages and also suffer from disadvantages. The advantages are that they are readily visible, are exceptionally bright and may have an appearance that varies rapidly with time. The disadvantage is that a white hole as described here originates in a singularity and physicists are in general not happy with systems whose origin they cannot understand. (A major exception to this statement is the Universe as a whole, whose most popular model, the big-bang model, also originates in a singularity, *vide* Chapter 15.) One could advance a white hole as a source behind transient explosive phenomena like the gamma-ray bursts. A signature of white holes is that the frequency of their radiation declines with time: the factor $1 + z$ increases. The softening of radiation from a gamma-ray burst indicates just that effect.

Exercises

1. By considering a test particle on the surface of the collapsing dust ball discussed in the text as falling freely in the external Schwarzschild spacetime, deduce the relation of Equation (13.17).

2. Consider a more general solution of Equation (13.11). Define $e^{\mu/2} = R(r, t)$. Then this equation becomes

$$\dot{R}^2 = \frac{F(r)}{r}\left(\frac{r}{R} - 1\right).$$

Solve this by writing $R = r \cos^2 \Psi$. Show that the general solution includes cases wherein the radial displacements diverge while the transverse ones shrink. In all cases show that the typical proper volume element converges to zero.

3. A collapsing dust ball emits radiation radially outwards from its surface. Show that as its surface approaches the Schwarzschild barrier the redshift z of the radiation received by a distant Schwarzschild observer using the T-coordinate increases as

$$1 + z \propto \exp[T/(4GM)],$$

where M is the mass of the collapsing dust ball.

4. In the *Eddington* coordinates, the Schwarzschild T-coordinate is replaced by

$$V = T + R + 2GM \ln\left|\frac{R}{2GM} - 1\right|.$$

Show that $V =$ constant describes an 'ingoing' radial null geodesic and that the Schwarzschild line element is transformed to

$$ds^2 = \left(1 - \frac{2GM}{R}\right) dV^2 - 2\, dV\, dR - R^2(d\theta^2 + \sin^2\theta\, d\theta^2).$$

Construct the 'outgoing' Eddington coordinates along the same lines.

5. Show that the following metric describes Schwarzschild's spacetime:

$$ds^2 = dt^2 - \frac{4}{9}\left[\frac{9GM}{2(r-t)}\right]^{2/3} dr^2$$

$$+ \left[\frac{9GM}{2}(r-t)^2\right]^{2/3} (d\theta^2 + \sin^2\theta\, d\phi^2).$$

This metric arises on solving the exterior solution for a dust ball collapsing from a state of rest at infinite dispersion.

6. Show that, in the Kerr–Newman black hole, for $R < R_0(\theta)$ no physical observer can have constant R, θ, ϕ. If an observer has a constant R and θ

then he must have an angular velocity

$$\frac{d\phi}{dT} > \frac{h \sin\theta - \sqrt{\Delta}}{(R^2 + h^2)\sin\theta - \sqrt{\Delta}h \sin^2\theta}.$$

7. An 'extreme' Kerr–Newman black hole is defined by the relation $G^2 M^2 = G Q^2 + h^2$. If $G^2 M^2 < G Q^2 + h^2$, there is no horizon (since $\Delta = 0$ has no real roots). Show that, if a black hole is initially in the extreme form, it cannot evolve into a state of no horizon under the usual laws of black-hole physics.

Chapter 14
The expanding Universe

14.1 Historical background

In 1915 Einstein put the finishing touches to the general theory of relativity. The Schwarzschild solution described in Chapter 9 was the first physically significant solution of the field equations of general relativity. It showed how spacetime is curved around a spherically symmetric distribution of matter. The problem solved by Schwarzschild was basically a local problem, in the sense that the deviations of spacetime geometry from the Minkowski geometry of special relativity gradually diminish to zero as we move further and further away from the gravitating sphere. This result can be easily verified from the Schwarzschild line element by letting the radial coordinate go to infinity. In technical jargon a spacetime satisfying this property is called *asymptotically flat*. In general any spacetime geometry generated by a local distribution of matter is expected to have this property. Even from Newtonian gravity we expect an analogous result: that the gravitational field of a local distribution of matter will die away at a large distance from the distribution. Can the Universe be approximated by a local distribution of matter?

Einstein rightly felt that the answer to the above question would be in the negative. Rather, he expected the Universe to be filled with matter, howsoever far we are able to probe it. A Schwarzschild-type solution cannot therefore provide the correct spacetime geometry of such a distribution of matter. Since we can never get away from gravitating matter, the concept of asymptotic flatness must break down. A new type of solution was therefore needed to describe a Universe filled everywhere with matter. Einstein published such a solution (cf. Ref. [46]) in 1917. It was to launch the subject of *theoretical cosmology*, that is, a subject dealing with theoretical modelling of the Universe in the large.

14.1.1 The Einstein universe

It is evident from the field equations of general relativity that their solution in the most general form – the solution of an interlinked set of non-linear partial differential equations – is beyond the present range of techniques available to applied mathematics. It is necessary to impose simplifying symmetry assumptions in order to make any progress towards a solution. Just as Schwarzschild assumed spherical symmetry in his local solution, Einstein assumed homogeneity and isotropy in his cosmological problem. He further assumed, like Schwarzschild, that spacetime is static. This enabled him to choose a time coordinate t such that the line element of spacetime could be described by

$$ds^2 = c^2\, dt^2 - \alpha_{\mu\nu}\, dx^\mu\, dx^\nu, \tag{14.1}$$

where $\alpha_{\mu\nu}$ are functions of space coordinates x^μ ($\mu, \nu = 1, 2, 3$) only.

Note that the constraint of homogeneity implies that the coefficient of dt^2 can only be a constant, which we have normalized to c^2. We may further assume as hitherto that $c = 1$. Similarly, the condition of isotropy tells us that there should be no terms of the form $dt\, dx^\mu$ in the line element. This can be seen easily in the following way. If we had terms like $g_{0\mu}\, dt\, dx^\mu$ in the line element, then spatial displacements dx^μ and $-dx^\mu$ would contribute oppositely to ds^2 over a small time interval dt, and such directional variation would be observable and would be inconsistent with isotropy. Can we say anything more about $\alpha_{\mu\nu}$?

We go back to Chapter 6, where we discussed spacetime symmetries in general. Referring to the maximally symmetric spaces of three dimensions, to which the homogeneous and isotropic model of Einstein belonged, we can write down the most general line element for such spaces, *vide* Equation (6.31). However, Einstein felt that the matter-filled Universe will have a positive curvature, which will make it close onto itself. Thus, of the three alternatives for the curvature parameter k, he opted for $k = 1$, and so the line element of his spacetime became

$$ds^2 = c^2\, dt^2 - S^2\left[\frac{dr^2}{1 - r^2} + r^2(d\theta^2 + \sin^2\theta\, d\phi^2)\right]. \tag{14.2}$$

Given this line element, it is now straightforward to compute Christoffel symbols and the Ricci tensor. The calculation leads to the following non-zero components of the Einstein tensor:

$$R_0^0 - \frac{1}{2}R = -\frac{3}{S^2},$$ (14.3)

$$R_1^1 - \frac{1}{2}R = R_2^2 - \frac{1}{2}R = R_3^3 - \frac{1}{2}R = -\frac{1}{S^2}.$$ (14.4)

To complete the field equations, Einstein used the energy tensor for dust derived in (7.20). For dust at rest in the above frame of reference u^i has only one component, the time component, non-zero. We therefore get

$$T_0^0 = \rho_0 c^2,$$

$$T_1^1 = T_2^2 = T_3^3 = 0.$$ (14.5)

Thus the two equations (14.3) and (14.4) lead to two independent equations:

$$-\frac{3}{S^2} = -\frac{8\pi G}{c^2}\rho_0, \qquad -\frac{1}{S^2} = 0.$$ (14.6)

Clearly no sensible solution is possible from these equations, thus suggesting that no static homogeneous isotropic and dense model of the Universe is possible under the regime of Einstein equations as stated in (8.3).

This was a setback, for it indicated that either Einstein's assumptions about the Universe (homogeneous, isotropic and static) were wrong or that his set of basic equations was incomplete. The option to get out of the conundrum that Einstein adopted was the latter. He modified his field equations, by introducing the so-called 'λ-term', to the form

$$R_{ik} - \frac{1}{2}g_{ik}R + \lambda g_i k = \kappa T_{ik}.$$ (14.7)

We briefly discussed this modification in Chapter 8. The constant λ needed for cosmology turns out to be very small, of the order of 10^{-56} cm^{-2}. It therefore does not affect the observational checks on the theory from the Solar-System data. It makes a difference in cosmology, however, as we will find in the following chapter.

The additional term in the field equations now led Einstein to the following modified equations for his static model:

$$\lambda - \frac{3}{S^2} = -\frac{8\pi G}{c^2}\rho_0$$ (14.8)

and

$$\lambda - \frac{1}{S^2} = 0.$$ (14.9)

He now did have a sensible solution. He got

$$S = \sqrt{\frac{1}{\lambda}} = \frac{c}{2\sqrt{\pi G\rho_0}}.$$ (14.10)

Einstein considered this solution as justifying his conjecture that with sufficiently high density it should be possible to 'close' the Universe. See Reference [46] for the Einstein paper on this topic. In (14.10) we have the radius S of the Universe as given by the matter density ρ_0, with the result that the larger the value of ρ_0, the smaller the value of S. However, if λ is a given universal constant like G, both ρ_0 and S are determined in terms of λ (as well as G and c). How big is λ?

In 1917 very little information was available about ρ_0, from which λ could be determined. The value of

$$S \approx 10^{26} - 10^{27} \, \text{cm}$$

quoted in those days is therefore only of historical interest. If we take $\rho_0 \sim 10^{-31} \, \text{g cm}^{-3}$ as a rough estimate of the mass density in the form of galaxies, we get $S \approx 10^{29} \, \text{cm}$ and $\lambda \approx 10^{-58} \, \text{cm}^{-2}$.

The λ-term introduces a force of repulsion between two bodies that increases in proportion to the distance between them. The attractive force of gravity decreases with distance, whereas the above force of repulsion *increases* with distance. Therefore at a specific distance the two would balance and provide a static universe. Later it turned out that the model was unstable and would either collapse or expand to infinity, depending on which of these two forces dominated. Theoretical objections like this apart, this model did not survive much longer than a decade, for observational reasons discussed in the next section.

Example 14.1.1 Consider a two-body problem in which a small mass m moves under the influence of a large mass M. Ignoring the motion of M and assuming that m is held at rest at a distance r from M, we have the net force on m as

$$F \equiv \left(-\frac{GM}{r^2} + \lambda r c^2 \right) m.$$

For equilibrium F must vanish, thus giving a static distance of separation as

$$r_0 = \left(\frac{GM}{\lambda c^2} \right)^{1/3}.$$

(Note that we have introduced c^2 manifestly to preserve the correct dimensionality.) If, however, the small mass were slightly displaced, F will be non-zero. Writing the displacement away from M as δr, we get

$$\delta F = \left(\frac{2GM}{r^3} + \lambda c^2 \right) m \, \delta r.$$

So, if $\delta r > 0$, $\delta F > 0$, resulting in m moving further away from M. Likewise, if $\delta r < 0$, $\delta F < 0$, thus telling us that m will move towards M. These

movements are indicative of instability, since in neither case does m return
to its original position of rest.

This example gives an intuitive feel for why the Einstein model is
unstable.

14.1.2 The de Sitter Universe

Einstein was initially very satified by this solution, for it reinforced his
belief that the Universe would essentially be unique in having a definitive
spacetime geometry determined by the matter distribution. To this end
his model showed how its radius was determined by its matter density.

However, his expectation that general relativity can yield only such
matter-filled spacetimes as solutions of the field equations was proved
wrong shortly after the publication of his paper in 1917. For, a few
months later in the same year, W. de Sitter [47] published another solution
of the field equations (14.7) with the line element given by

$$ds^2 = c^2 \left(1 - \frac{H^2 R^2}{c^2}\right) dT^2 - \frac{dR^2}{\left(1 - \frac{H^2 R^2}{c^2}\right)} - R^2(d\theta^2 + \sin^2\theta \, d\phi^2),$$

$$(14.11)$$

where H is a constant related to λ by

$$\lambda = \frac{3H^2}{c^2}.$$ $$(14.12)$$

The remarkable feature of the de Sitter universe is that *it is empty*.
Moreover, although the above coordinates give the impression that the
universe is static, it is possible to find a new set of coordinates (t, r, θ, ϕ)
in terms of which the line element (14.11) takes the manifestly dynamic
form

$$ds^2 = c^2 \, dt^2 - e^{2Ht}[dr^2 + r^2(d\theta^2 + \sin^2\theta \, d\phi^2)].$$ $$(14.13)$$

It is easy to verify that test particles with constant values of (r, θ, ϕ)
follow timelike goedesics in this model. Thus the proper separation
between any two particles measured at a given time t increases with time
as e^{Ht}. That is, these particles are all moving apart from one another.

However, these particles have no material status. They have no
masses and they do not influence the geometry of spacetime. In the
dynamic sense the universe is empty, although in the kinematic sense it
is expanding. As Eddington once put it, the de Sitter universe has *motion
without matter*, in contrast to the Einstein universe, which has *matter
without motion*.

14.2 The expanding Universe

Although de Sitter's universe was of academic interest because it contained no matter, it possessed a feature that turned out to have contact with reality, as was discovered a few years later, namely the fact that the Universe is expanding.

In 1929 Edwin Hubble published a paper in the *Proceedings of the National Academy of Sciences* [48] that turned out to be the trend-setter of modern cosmology. Just as Einstein's model marked the beginning of theoretical cosmology, so did Hubble's findings launch *observational cosmology*, the subject dealing with the observational studies of the large-scale Universe.

Hubble's conclusions were based on a long series of observations that had started with V. M. Slipher in 1912 and to which several astronomers had contributed, including Hubble himself and his coworker Milton Humason. These observations typically looked at the spectra of nearby nebulae, which were believed to be galaxies of stars in their own right just like the Milky Way. Barring very few exceptions, which included the great galaxy in Andromeda, the majority of spectra showed absorption lines that were shifted to the red end. This is another instance of a class of astronomical objects showing *redshift*.

Although one may use the relativistic Doppler-shift formula (1.64) derived in Chapter 1, because the redshifts are very small (of the order of a few parts in a thousand) one may use the simpler Newtonian limit of that formula for $|v| \ll 1$, and write the speed of recession of a galaxy of redshift z as

$$v = c \times z. \tag{14.14}$$

Going beyond this result, however, Hubble found that the velocities so computed were increasing in proportion to the distances D of the galaxies from us. Figure 14.1 is based on Hubble's early data.

Hubble's findings can be written in the following form:

$$v = H \times D. \tag{14.15}$$

Thus, in whichever direction we look, we find galaxies moving radially away from us. Does that mean that we are in a special position in the Universe? Rather the opposite! If we sit on any other galaxy and observe the Universe from there, we would see exactly the same picture: namely the other galaxies, including the Milky Way, are receding from us.

How can we express this phenomenon in the language of general relativity? Can we generate models of the Universe that combine de Sitter's notion of expansion with Einstein's notion of non-emptiness?

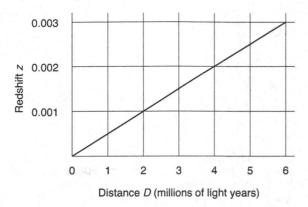

Fig. 14.1. Hubble's
redshift–distance relation
showing that for larger
distances (*D*) the galaxies'
radial velocities (*v*) are
proportionately larger. The
velocity of a galaxy is
proportional to its redshift.

This was the challenge to theoretical cosmologists. For, with this large-scale radial expansion, it became impossible to maintain the myth of a static universe. Models of an expanding universe were needed. The Friedmann models to be discussed shortly do just that, and were in fact obtained by Alexander Friedmann between 1922 and 1924, seven years *before* Hubble's data became well known [49]. Later Abbé Lemaître in 1927 [50] also independently obtained models similar to Friedmann's. However, until the impact of Hubble's observations of 1929, these models remained largely unrecognized.

14.3 Basic assumptions of cosmology

Once we decide to generalize from a static to a non-static model of the Universe, our task becomes more complicated. Figure 14.2(a) shows a spacetime diagram with a swarm of world lines representing particles moving in arbitrary ways. There is no order in this picture, and where two world lines intersect we have colliding particles. It would indeed be very difficult to solve the Einstein field equations for such a mess of gravitating matter. Fortunately, the real Universe does not appear to be so messy.

Hubble's observations indicate that the Universe is (or at least seems to be) an orderly structure in which the galaxies, considered as basic units, are moving apart from one another in a systematic manner. Thus Figure 14.2(b) represents a typical spacetime section of the Universe in which the world lines represent the histories of galaxies. These world lines, unlike those of Figure 14.2(a), are non-intersecting and form a funnel-like structure in which the separation between any two world lines is steadily increasing. One may compare Figure 14.2(b) with the disciplined march of an army unit, and Figure 14.2(a) with a jostling mob after a rowdy football match.

Fig. 14.2. In (a) we have
world lines with no set
pattern, intersecting one
another on some occasions
and in general describing
arbitrary motions of particles.
In (b) we see world lines
showing systematic motion
with each spatial point
identified with a unique
member of the set. The Weyl
postulate stipulates that
large-scale motions of galaxies
come close to (b).

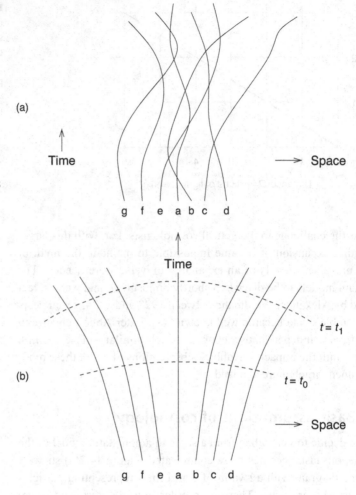

14.3.1 Weyl's postulate

This intuitive picture of regularity is often expressed formally as the *Weyl postulate*, after the early work of the mathematician Hermann Weyl. The postulate states that the world lines of galaxies form a 3-bundle of non-intersecting geodesics orthogonal to a series of spacelike hypersurfaces.

To appreciate the full significance of Weyl's postulate, let us try to express it in terms of the coordinates and metric of spacetime. Accordingly we use three spacelike coordinates x^μ ($\mu = 1, 2, 3$) to label a typical world line in the 3-bundle of galaxy world lines. Further, let the coordinate x^0 label a typical member of the series of spacelike hypersurfaces mentioned above. Thus

$$x^0 = \text{constant}$$

is a typical spacelike hypersurface orthogonal to the typical world line given by

$$x^\mu = \text{constant}.$$

Although in practice the galaxies form a discrete set, we can extend the discrete set (x^μ) to a continuum by the *smooth-fluid approximation*. This approximation is none other than the widely used device of going over from a discrete distribution of particles to a continuum density distribution. In this case we can treat the quantities x^μ as forming a continuum along with x^0 and use them as the four coordinates x^i to describe space and time.

It is worth emphasizing the importance of the non-intersecting world lines. If two galaxy world lines did intersect, our coordinate system above would break down, for we would then have two different values of x^μ specifying the same spacetime point (the point of intersection). In the next chapter we will, however, encounter an exceptional situation in which all world lines intersect at one *singular* point!

Let the metric in terms of these coordinates be given by the tensor g_{ik}. What can we assert about this metric tensor on the basis of the Weyl postulate? The orthogonality condition tells us that

$$g_{0\mu} = 0. \tag{14.16}$$

Further, the fact that the line $x^\mu = $ constant is a *geodesic* tells us that the geodesic equations

$$\frac{\mathrm{d}^2 x^i}{\mathrm{d}x^2} + \Gamma^i_{kl} \frac{\mathrm{d}x^k}{\mathrm{d}s} \frac{\mathrm{d}x^l}{\mathrm{d}s} = 0 \tag{14.17}$$

are satisfied for $x^i = $ constant, $i = 1, 2, 3$. Therefore

$$\Gamma^\mu_{00} = 0, \qquad \mu = 1, 2, 3. \tag{14.18}$$

From (14.16) and (14.18) we therefore get

$$\frac{\partial g_{00}}{\partial x^\mu} = 0, \qquad \mu = 1, 2, 3. \tag{14.19}$$

Thus g_{00} depends on x^0 only. Following the trick used earlier, we can therefore replace x^0 by a suitable function of x^0 to make g_{00} constant. Hence we take, without loss of generality,

$$g_{00} = 1. \tag{14.20}$$

The line element therefore becomes

$$\mathrm{d}s^2 = (\mathrm{d}x^0)^2 + g_{\mu\nu}\, \mathrm{d}x^\mu\, \mathrm{d}x^\nu$$
$$= c^2\, \mathrm{d}t^2 + g_{\mu\nu}\, \mathrm{d}x^\mu\, \mathrm{d}x^\nu, \tag{14.21}$$

where we have put $ct = x^0$. This time coordinate is called the *cosmic time*. It is easily seen that the spacelike hypersurfaces in Weyl's postulate are the surfaces of simultaneity with respect to the cosmic time. Moreover, t is the proper time kept by any galaxy.

Example 14.3.1 *Problem.* Suppose we retain the homogeneity assumption of the Weyl postulate but give up isotropy. This allows $g_{0\mu}$ terms. Show that $g_{0\mu}$ are independent of time ($\mu = 1, 2, 3$).

Solution. We still have $x^\mu =$ constant as a geodesic. So

$$\frac{d^2x^0}{ds^2} + \Gamma^0_{ij} \frac{dx^i}{ds} \frac{dx^j}{ds} = 0, \qquad \frac{d^2x^\mu}{ds^2} + \Gamma^\mu_{ij} \frac{dx^i}{ds} \frac{dx^j}{ds} = 0.$$

Since $x^\mu =$ constant, $dx^\mu/ds = 0$, $d^2x^\mu/ds^2 = 0$. The above equations therefore give

$$\Gamma^i_{00} \left(\frac{dx^0}{ds} \right)^2 = 0.$$

Hence $\Gamma^\mu_{00} = 0 \Rightarrow \Gamma_{\mu|00} = g_{\mu0}\Gamma^0_{00} + g_{\mu\nu}\Gamma^\nu_{00} = g_{\mu0}\Gamma^0_{00} = 0$. This implies $\partial g_{0\mu}/\partial t = 0$.

14.3.2 The cosmological principle

The second important assumption of cosmology is embodied in the *cosmological principle*. This principle states that, at any given cosmic time, the Universe is homogeneous and isotropic. In practical terms it means that, if you are blindfolded and taken to any part of the Universe, then, on the removal of the eye-cover, you would, on the basis of your observations, be able neither to say where you are nor to identify the direction in which you are looking.

We have already come across such spaces in Chapter 6, under the category of *maximally symmetric spaces*. As seen in Equation (6.31), we are able to write the line element of such spaces in the form

$$d\sigma^2 = S^2 \left[\frac{dr^2}{1 - kr^2} + r^2(d\theta^2 + \sin^2\theta \, d\phi^2) \right]. \tag{14.22}$$

The parameter $k = 0, -1,$ or $+1$ and the factor S is spatially constant. It could, however, be a function of cosmic time t without affecting any of the symmetries above. Thus the most general line element satisfying the Weyl postulate and the cosmological principle is given by

$$ds^2 = c^2 \, dt^2 - S^2(t) \left[\frac{dr^2}{1 - kr^2} + r^2(d\theta^2 + \sin^2\theta \, d\phi^2) \right], \tag{14.23}$$

where the 3-spaces $t =$ constant are Euclidean, or flat, for $k = 0$, closed with positive curvature for $k = +1$, and open with negative curvature for $k = -1$. For reasons that will become clearer later, the scale factor $S(t)$ is often called the *expansion factor*.

The line element (14.23) that we have obtained was rigorously derived in the 1930s by H. P. Robertson and A. G. Walker, independently [51, 52]. It is often referred to as the *Robertson–Walker line element*.

The Robertson–Walker line element is sometimes expressed in a slightly different form with the help of the following radial coordinate transformation:

$$\bar{r} = \frac{2r}{1 + \sqrt{1 - kr^2}}. \tag{14.24}$$

We then get the line element as

$$ds^2 = c^2\, dt^2 - \frac{S^2(t)}{\left(1 + \dfrac{kr^2}{4}\right)^2}[d\bar{r}^2 + \bar{r}^2(d\theta^2 + \sin^2\theta\, d\phi^2)]. \tag{14.25}$$

This line element is manifestly isotropic in $\bar{r}, \theta, \phi$. We will, however, continue to use (14.23).

Notice how the simplifying postulates of cosmology have reduced the number of unknowns in the metric tensor from 10 to the single function $S(t)$ (of only one independent variable) and the discrete parameter k that characterize the Robertson–Walker metric. To determine these unknowns we need to solve the Einstein field equations, as was done by Friedmann and Lemaître. We will defer this exercise to the following chapter.

14.4 Hubble's law

Let us first try to understand how the nebular redshift found by Hubble and Humason is accounted for by the Robertson–Walker model. We begin by recalling that the basic units of Weyl's postulate are galaxies with constant coordinates x^μ. We readily identify the x^μ with the (r, θ, ϕ) of Robertson–Walker spacetime. Thus each galaxy has a constant set of coordinates (r, θ, ϕ). This coordinate frame is often referred to as the *cosmological rest frame*. As observers we are located in our Galaxy, which also has constant (r, θ, ϕ) coordinates. Without loss of generality we can take $r = 0$ for our Galaxy. Although this assumption suggests that we are placing ourselves at the centre of the Universe, it does not confer any special status on us. Because of the assumption of homogeneity, *any* galaxy could be chosen to have its radial coordinate $r = 0$. Our particular choice is simply dictated by convenience.

14.4.1 Redshift

Consider a galaxy G_1 at (r_1, θ_1, ϕ_1) emitting light waves towards us. Let us denote by t_0 the present epoch of observation. At what time should a light wave have left G_1 in order to arrive at $r = 0$ at the present time $t = t_0$? To find the answer to this question, we need to know the path of the wave from G_1 to us. Since light travels along null geodesics, we need to calculate the null geodesic from G_1 to us.

From the symmetry of a spacetime we can guess that a null geodesic from $r = 0$ to $r = r_1$ will maintain a constant spatial direction. That is, we expect to have $\theta = \theta_1, \phi = \phi_1$ all along the null geodesic. This guess proves to be correct when we substitute these values into the geodesic equations. Accordingly we will assume that only r and t change along the null geodesic. Next we recall that a first integral of the null geodesic equation is simply $ds = 0$. For the Robertson–Walker line element this gives us

$$c\,dt = \pm \frac{S\,dr}{\sqrt{1 - kr^2}}.$$ (14.26)

Since r decreases as t increases along this null geodesic, we should take the minus sign in the above relation. Suppose the null geodesic left G_1 at time t_1. Then we get from the above relation

$$\int_{t_1}^{t_0} \frac{c\,dt}{S(t)} = \int_0^{r_1} \frac{dr}{\sqrt{1 - kr^2}}.$$ (14.27)

Thus, if we know $S(t)$ and k, we know the answer to our question.

However, consider what happens to successive wave crests emitted by G_1. Suppose the wave crests were emitted at t_1 and $t_1 + \Delta t_1$ and received by us at t_0 and $t_0 + \Delta t_0$, respectively. Then, similarly to (14.27), we have

$$\int_{t_1 + \Delta t_1}^{t_0 + \Delta t_0} \frac{c\,dt}{S(t)} = \int_0^{r_1} \frac{dr}{\sqrt{1 - kr^2}}.$$ (14.28)

If $S(t)$ is a slowly varying function, so that it effectively remains unchanged over the small intervals Δt_0 and Δt_1, we get by subtraction of (14.27) from (14.28)

$$\frac{c\,\Delta t_0}{S(t_0)} - \frac{c\,\Delta t_1}{S(t_1)} = 0,$$

that is,

$$\frac{c\,\Delta t_0}{c\,\Delta t_1} = \frac{S(t_0)}{S(t_1)} \equiv 1 + z.$$ (14.29)

It is not difficult to see that the quantity z defined above is the redshift. The term $c\,\Delta t_1$ is, evidently, the wavelength λ_1 measured by an observer at rest in the galaxy G_1, while $c\,\Delta t_0$ is the wavelength λ_0 measured by an observer at rest in our Galaxy, since in the Robertson–Walker spacetime the cosmic time measures the proper time kept by any galaxy. Thus the wavelength of the light wave increases by a fraction z during the transmission from G_1 to us, provided that $S(t_0) > S(t_1)$. In other words, Hubble's observations of redshift are explained if we assume $S(t)$ to be an increasing function of time. We will refer to this redshift as the *cosmological redshift*. Let us view it in comparison with the two other types of redshifts we have so far encountered.

Our derivation above shows that the cosmological redshift arises from the passage of light through an expanding non-Euclidean space-time. Although in the early days of its discovery it was considered a manifestation of the Doppler effect, the correct general-relativistic treatment shows that it does *not* arise from the Doppler effect, since in our coordinate frame all galaxies have constant (r, θ, ϕ) coordinates. Further, in a non-Euclidean spacetime it is not possible to attach an unambiguous meaning to the relative velocity of two objects separated by a great distance. People are often tempted to relate z to velocity by the special-relativistic relation

$$1 + z = \sqrt{\frac{1 + v/c}{1 - v/c}}. \tag{14.30}$$

Such an interpretation is not valid in our present framework because, as we saw in Chapter 5, special relativity applies only in a locally flat region of spacetime.

It is also necessary to contrast (14.29) with the gravitational redshift described in Chapter 9. The gravitational redshift is characterized by the fact that, if light travelling from object B to object A is redshifted, the light travelling from A to B is blueshifted. In the present case, if light travelling from galaxy A to galaxy B is redshifted, that travelling from B to A will also be redshifted, provided that $S(t)$ is increasing during the transmission of light.

In conclusion, we also consider the oft-expressed confusion at the existence of objects with redshifts greater than 1. Normally the Doppler shift z is interpreted as arising from a source receding from us with the speed $c \times z$. How, then, it is asked, is it that we have objects travelling faster than light in spite of the light-speed limit imposed by special relativity? This way of looking at things is wrong on at least two counts. The formula used to compute velocity is Newtonian and needs to be replaced if one is applying special relativity. The special-relativistic formula above gives $|v| < c$ for all z howsoever large. Secondly, in the cosmological case, the correct formula is not (14.30) but (14.29). The latter simply tells us that the light from the redshifted object left it when the scale factor was $(1 + z)^{-1}$ of its present value. In an expanding Universe this may well be possible. Since we are dealing with curved spacetime, there is no justification in invoking formulae and results from special relativity, which does not apply here.

14.4.2 The velocity–distance relation

With the framework developed so far, we can derive Hubble's law for low-redshift galaxies. The largest redshift in Hubble's 1929 paper was $z \cong 0.003$. At these small redshifts we can use the Taylor expansion to derive a simple linear relation for the distance D_1 of a galaxy G_1 of redshift $z_1 \ll 1$. We define the distance at the present epoch t_0 as

$$D_1 = r_1 S(t_0). \tag{14.31}$$

We also get, by the Taylor expansion of (14.29),

$$\int_0^{r_1} \frac{dr}{\sqrt{1 - kr^2}} \approx r_1, \tag{14.32}$$

$$\int_{t_1}^{t_0} \frac{c \, dt}{S(t)} \approx \frac{c(t_0 - t_1)}{S(t_0)} \tag{14.33}$$

$$S(t_1) \approx S(t_0) - (t_0 - t_1) \cdot \left(\frac{\dot{S}}{S}\right)_{t_0} S(t_0), \tag{14.34}$$

$$S(t_1) = \frac{S(t_0)}{1 + z} \approx S(t_0)(1 - z). \tag{14.35}$$

From these relations we get

$$D_1 \approx r_1 S(t_0) \approx c(t_0 - t_1)$$

$$\approx \left[\left(\frac{\dot{S}}{S}\right)_{t_0}\right]^{-1} cz, \tag{14.36}$$

which can be expressed in the form

$$cz = H_0 D_1, \tag{14.37}$$

with H_0, *the Hubble constant*, given by

$$H_0 = \left(\frac{\dot{S}}{S}\right)_{t=t_0}. \tag{14.38}$$

From a Doppler-shift point of view, cz may be identified with the velocity of recession at small z. In this form (14.37) we have Hubble's *velocity–distance relation*. Expressed as part of the velocity–distance relation, the Hubble constant has the unit of velocity per unit distance, the most common unit in usage being kilometres per second per mega-parsec.[1] In many calculations of observational and physical cosmology we shall use

$$H_0 = h_0 \times 100\,\mathrm{km\ s^{-1}\,Mpc^{-1}}. \tag{14.39}$$

Although Hubble originally obtained $h_0 \sim 5.3$, the present estimate of h_0 is much lower. It is still uncertain, and, until recently, was believed to lie in the range $0.5 \le h_0 \le 1$. Observations with the Hubble Space Telescope (HST) and some ground-based telescopes have narrowed down this range, to around 0.55–0.75. Many cosmologists, however, believe that $h_0 \approx 0.7$.

Another useful way of expressing H_0 is in units of reciprocal time; that is, by expressing

$$\tau_0 = H_0^{-1} \tag{14.40}$$

in units of time. A good time unit for τ_0 is the gigayear (Gyr). The present estimate of τ_0 is in the range of approximately 13–18 Gyr, depending upon the value chosen for H_0. We may refer to τ_0 as the Hubble time scale.

14.5 The luminosity distance

The distance $D_1 = r_1 S(t_0)$ we have defined above may be called the *metric distance*. From the Robertson–Walker metric we deduce that this is the present radius of the sphere centred on us, on which the galaxy is located, the total surface area of the sphere being $4\pi D_1^2$. Using the practice prevailing in galactic astronomy, we may be tempted to argue that, if L is the luminosity of the galaxy G_1, its apparent flux of radiation crossing unit area normally at $r = 0$ will be simply

$$l = \frac{L}{4\pi D_1^2}.$$

This conclusion would, however, be wrong in the expanding Universe. Let us do the calculation correctly.

Let L be the total energy emitted by the galaxy G_1 in unit time at the epoch t_1 when light left it in order to reach us at the present epoch t_0. The redshift z of the galaxy is therefore given by (14.29). It is now necessary to specify the wavelength range of observation. To fix ideas,

---

[1] To those unfamiliar with the *parsec* as a distance unit, we add that it equals 3.0856×10^{18} cm or 3.26 light years. This unit naturally arose from the stellar astronomer's attempts to measure distances of stars using the parallax method.

Fig. 14.3. The distribution of
light emitted by galaxy G_1 is
assumed isotropic, i.e.,
distributed uniformly across
the surface of the sphere
centred on G_1.

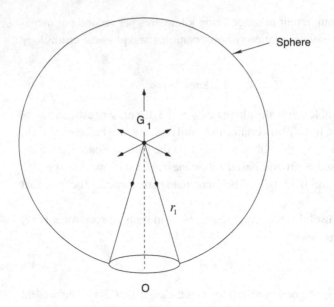

suppose that the intensity distribution of G_1 over wavelengths λ is given
by the normalized function $I(\lambda)$. Thus

$$\mathrm{d}L = L\,I(\lambda)\mathrm{d}\lambda \tag{14.41}$$

is the energy emitted by G_1 per unit time over the bandwidth $(\lambda, \lambda + \mathrm{d}\lambda)$.
If instead of wavelengths we wanted to use frequencies, the correspond-
ing intensity function $J(\nu)$ is related to $I(\lambda)$ by

$$cJ(\nu) = \lambda^2 I(\lambda). \tag{14.42}$$

Both $J(\nu)$ and $I(\lambda)$ are used by the astronomer, depending on convenience.

In the case of isotropic light emission by G_1, by the time its light
reaches us it is distributed uniformly across a sphere of coordinate radius
r_1 centred on G_1 (see Figure 14.3). We have already seen that, in the
Robertson–Walker line element, the area is $4\pi D_1^2$. We now need to
know how much light is received per unit time by us across unit proper
area held perpendicular to the line of sight to G_1, over a bandwidth
$(\lambda_0, \lambda_0 + \Delta\lambda_0)$. Denote this quantity by $\mathcal{F}(\lambda_0)\Delta\lambda_0$. Now two effects
intervene to make the answer different from that expected from Galactic
astronomy.

Note first that because of redshift the arriving light with wavelengths
in the range $(\lambda_0, \lambda_0 + \Delta\lambda_0)$ left G_1 in the wavelength range

$$\left(\frac{\lambda_0}{1+z}, \frac{\lambda_0 + \Delta\lambda_0}{1+z} \right).$$

Now the total amount of energy that leaves G_1 between the epochs t_1 and $t_1 + \Delta t_1$ in the above frequency range is

$$LI\left(\frac{\lambda_0}{1+z}\right) \cdot \frac{\Delta\lambda_0}{1+z} \cdot \Delta t_1.$$

How many photons carry the above quantity of energy? For a small enough bandwidth, we may assume that a typical photon had, at emission, the wavelength $\lambda_0/(1+z)$, a frequency $(1+z)c/\lambda_0$, and hence an energy equal to $(1+z)ch/\lambda_0$, where h is Planck's constant. Therefore the required number of photons is

$$\delta\mathcal{N} = LI\left(\frac{\lambda_0}{1+z}\right)\frac{\Delta\lambda_0}{1+z}\frac{\Delta t_1}{(1+z)ch/\lambda_0}$$

$$= \frac{L\lambda_0}{ch} \cdot \frac{1}{(1+z)^2} \cdot I\left(\frac{\lambda_0}{1+z}\right)\Delta\lambda_0 \, \Delta t_1.$$

At the epoch of reception, these photons are distributed across a surface area of $4\pi r_1^2 S^2(t_0)$ and are received over a time interval $(t_0, t_0 + \Delta t_0)$. Thus the number of photons received by us per unit area held normal to the line of sight and per unit time is given by

$$\frac{L\lambda_0}{ch} \cdot \frac{1}{(1+z)^2} I\left(\frac{\lambda_0}{1+z}\right)\Delta\lambda_0 \cdot \frac{\Delta t_1}{\Delta t_0} \cdot \frac{1}{4\pi \, r_1^2 S^2(t_0)}.$$

At this epoch, because of a scaling down of its frequency by redshift, each photon has been degraded in energy by the factor $(1+z)^{-1}$. Thus each photon now has the energy ch/λ_0. If we multiply the above expression by this factor, we get the quantity we were after:

$$\mathcal{F}(\lambda_0)\Delta\lambda_0 = L\frac{1}{(1+z)^2} \cdot \frac{\Delta t_1}{\Delta t_0} \cdot I\left(\frac{\lambda_0}{1+z}\right) \cdot \frac{1}{4\pi \, r_1^2 S^2(t_0)} \cdot \Delta\lambda_0.$$

However, we note from (14.29) that $\Delta t_1/\Delta t_0$ gives us another factor $(1+z)^{-1}$ in the denominator. Thus finally we get

$$\mathcal{F}(\lambda_0) = \frac{LI(\lambda_0/1+z)}{(1+z)^3 4\pi \, r_1^2 S^2(t_0)}. \tag{14.43}$$

Thus the two effects coming in because of the expansion of the Universe are (1) the reduction by the factor $(1+z)^{-1}$ of energy emitted per quantum and (2) the time-dilatation at the receiving end by the factor $(1+z)$.

In terms of frequencies the result is quoted as *flux density*,

$$S(\nu_0) = \frac{LJ(\nu_0(1+z))}{(1+z)\, 4\pi \, r_1^2 S^2(t_0)}. \tag{14.44}$$

Here $S(\nu_0)\Delta\nu_0$ is the amount of radiation received perpendicular to unit area in unit time across a frequency range $(\nu_0, \nu_0 + \Delta\nu_0)$.

The optical astronomer uses this result in the form (14.43), while the radio astronomer uses it in the form (14.44). The X-ray astronomer uses energies instead of frequencies, so (14.44) is scaled by h. Astronomers have occasion to use these expressions when looking at the various observational tests of cosmology. We will end this section by deriving a few results of interest to optical astronomy.

The expression (14.43) integrated over all wavelengths gives

$$\mathcal{F}_{bol} = \frac{L_{bol}}{4\pi \, r_1^2 S^2(t_0)(1+z)^2}. \tag{14.45}$$

where $L_{bol} (= L)$ is the *absolute bolometric luminosity* of G_1. $\mathcal{F}_{bol}$ is correspondingly the apparent bolometric luminosity of G_1. On the logarithmic scale of magnitudes familiar to the optical astronomer, (14.45) becomes

$$m_{bol} = -2.5 \log\left(\frac{\mathcal{F}_{bol}}{\mathcal{F}_0}\right),$$

$$M_{bol} = -2.5 \log\left(\frac{L_{bol}}{L_\odot}\right) + 4.75, \tag{14.46}$$

$$m_{bol} - M_{bol} = 5 \log D_1 - 5,$$

where

$$\mathcal{F}_0 = 2.48 \times 10^{-5} \, \text{erg} \, \text{cm}^{-2} \, \text{s}^{-1},$$

$$L_\odot = \text{solar luminosity} = 4 \times 10^{33} \, \text{erg} \, \text{s}^{-1}, \tag{14.47}$$

$$D_1 = r_1 S(t_0)(1+z).$$

D_1 is called the *luminosity distance* of G_1. If we are interested in a magnitude defined for a particular waveband around λ_0, say, we may similarly use (14.43) in the logarithmic form with the apparent magnitude defined by

$$m(\lambda_0) = -2.5 \log \mathcal{F}(\lambda_0) + \text{constant},$$

the constant depending on the filter used to select that waveband. It is customary to indicate the filter by a suffix attached to m. Thus m_{pg} stands for photographic magnitude, m_v for visual magnitude, m_b for blue magnitude, and so on.

Note, however, that, when using a specific filter, because of redshift the astronomer has to apply a correction to include the effect of the term $I(\lambda_0/1 + z)$. Thus an astronomer using a red filter may actually be receiving the photons that originated in the blue part of the spectrum of G_1 if $z \approx 1$. This correction, which is crucial to many cosmological observations, is called the *K-correction*.

Example 14.5.1 *Problem.* Calculate the luminosity distance for the de Sitter universe.

Solution. With the notation used in the text we have

$$S(t) = e^{Ht}, \qquad S(t_0)/S(t_1) = e^{H(t_0 - t_1)} = 1 + z.$$

The epochs t_0 and t_1 are related by the result (14.27) with $k = 0$. Thus simple integration yields

$$r_1 = \frac{c}{H}\left(e^{-Ht_1} - e^{-Ht_0}\right).$$

Hence the luminosity distance is

$$D_1 = r_1 S(t_0)(1 + z) = \frac{c}{H}e^{Ht_0}(e^{-Ht_1} - e^{-Ht_0})(1 + z)$$

$$= \frac{c}{H}z(1 + z).$$

14.6 The Olbers paradox

In 1826, Heinrich Olbers, a physician from Germany, carried out a simple calculation which led to a paradoxical answer. His paradox can be phrased as this question: 'Why is the sky dark at night?' What the Olbers calculation shows is that whether we are facing the Sun or not makes no difference: the total radiation received from all stars in the Universe is infinite. This paradox and its resolution have cosmological implications, so it is appropriate to discuss them here.

Olbers assumed that (1) the Universe is homogeneous, isotropic and static, (2) it is infinite in extent and (3) it is filled with radiating objects, each with a constant luminosity. In those days the only geometry recognized was Euclid's, so Olbers did his calculation in its framework.

Take any point in this Universe as the observing post, denoted by the point O. We wish to calculate the total radiation received at O from all the stars in the Universe. To this end, divide the Universe into thin concentric shells centred at O. The volume of a typical shell of radii R and $R + dR$ is

$$4\pi R^2 \, dR$$

and, if the number density of radiating sources is N, the number of such sources in the shell is

$$4\pi R^2 N \, dR.$$

Suppose that each source has luminosity L. At a distance R the source would have a radiation flux of

$$\frac{L}{4\pi R^2},$$

so the total contribution of all radiating sources in the shell is to generate a flux at O of

$$4\pi R^2 \, dR \times N \times \frac{L}{4\pi R^2} = NL \, dR.$$

The total flux from all shells is therefore

$$\mathcal{F} = \int_0^\infty LN \, dR = \infty. \tag{14.48}$$

This was the conclusion of Olbers' calculation. Can its drastic nature be moderated? For clearly the night sky is dark and not infinitely bright.

One possible way out was to note that the typical radiator is of a finite size, so that beyond a certain distance from O the foreground objects would block the radiation of the background population. (In a forest thickly populated with trees we see only the foreground trees.) If a typical radiator is a ball of radius a, then from a distance R it subtends a solid angle at O of

$$\frac{\pi a^2}{R^2}$$

so that the total solid angle subtended by the sources in our shell is

$$\frac{\pi a^2}{R^2} \times 4\pi R^2 N \, dR = 4\pi \times \pi a^2 N \, dR.$$

We integrate this expression to a distance D where it equals the total solid angle 4π of the whole sky. Thus, the whole sky will be covered at a distance

$$D = \frac{1}{\pi a^2 N}. \tag{14.49}$$

If we take our integral only up to this distance, we have a finite answer:

$$\mathcal{F} = NLD \sim \frac{L}{\pi a^2}. \tag{14.50}$$

However, the expression we have arrived at is four times the surface brightness of the typical source. So, if the typical source is like the Sun, the sky should be shining like the solar disc! The above calculation supposes that the solid angles from different shells do not overlap. A more exact calculation can be done taking into account the overlap, but our conclusion is not substantially altered.

To resolve the paradox we can adopt any of the following arguments.

1. The Universe is of finite extent.
2. The Universe is of finite age. If the age is T, say, then we can argue that radiation reaching O today could not have come from distances beyond $c \times T$.

3. Each radiating source lasts only a finite time. This effectively limits the total light reaching O.

4. The Universe is expanding. This is the most dramatic explanation, for it invokes the dimming of radiation by the factor $(1 + z)^{-2}$ as found in formula (14.45). A calculation using this effect usually yields a finite and low value for sky brightness even for infinitely old models.

We will leave the Olbers paradox here as an example of a simple question calling for profound ideas for an answer and turn our attention next to actual models of the real Universe.

Exercises

1. Taking $\rho_0 = 10^{-31}\, \text{g cm}^{-3}$, calculate the radius of the Einstein universe and its total mass in spherical space.

2. By calculating the 3-volume of space within the coordinate region $r = $ constant in the spaces with the spatial line element

$$d\sigma^2 = S^2 \left[\frac{dr^2}{1 - kr^2} + r^2 (d\theta^2 + \sin^2\theta\, d\phi^2) \right], \quad k = 0, 1, -1.$$

develop the three-dimensional analogue of the experiment of covering the surfaces of zero, positive and negative curvature by a plane sheet of paper. (In this experiment the paper exactly covers a surface of zero curvature; it gets wrinkled while covering a surface of positive curvature and it gets torn while covering the surface of negative curvature.)

3. Determine the affine parameter for the radial null geodesic from galaxy G_1 to the origin $r = 0$ in Robertson–Walker spacetime.

4. A particle of mass m is fired today from our galaxy at $t = t_0$ with a linear momentum P_0. Show that the momentum of the particle when it reaches another galaxy at a later epoch t (as measured in the rest frame of that galaxy) is given by

$$P = P_0 \frac{S(t_0)}{S(t)}.$$

Compare this result with the cosmological redshift for photons.

5. Take a galaxy G_1 at (r_1, θ, ϕ) as a fundamental observer and write u_1^k as its velocity vector in the Robertson–Watson frame. Consider parallel propagation of this vector along the null ray connecting the galaxy to the observer O at the origin at the present epoch t_0 of observation. Let this vector be v_1^k at O. This represents the radial velocity of G_1 relative to the cosmological rest frame at O. Use the Doppler effect to work out the redshift for this motion and show that it is none other than z as given by the formula (14.29). You can do this exercise for the Schwarzschild line element and you can show that the gravitational redshift

can also be understood as a Doppler effect for the parallely transported velocity vector of the source along the null geodesic to the observer. Try to generalize these results.

6. In a universe with $S(t) \propto t^{2/3}$ and $k = 0$, a galaxy is observed to have a redshift $z = 1.25$. How long has light taken to travel from that galaxy to us? Express your answer in units of τ_0.

7. Work out the formula (14.45) for the universe with $S \propto t^{2/3}$ and $k = 0$, and compare this with the result for the de Sitter universe. In which model is the galaxy apparently brighter?

8. Why is the 'expanding-universe' solution preferable as a solution to the Olbers paradox, rather than 'a finite universe' or 'a finitely old' universe?

Chapter 15
Friedmann models

15.1 Introduction

The work covered in Chapter 14 did not tell us two important items of information about the Universe: (1) the rate at which it expands as given by the function $S(t)$; and (2) whether its spatial sections $t = \text{constant}$ are open or closed as indicated by the parameter k. To find answers to these questions, it is necessary to go beyond the Weyl postulate and the cosmological principle. We require a dynamical theory that tells us how the scale factor and curvature are determined by the matter/radiation contents of the universe.

A comparison of Newton's law of gravitation with the general theory of relativity shows the latter as enjoying advantages both on the theoretical and on the observational front. General relativity gets round the criticism of Newtonian gravity of violating the light-speed limit. It allows for the permanence of gravitation by identifying its effect with the curvature of spacetime. Observationally it performs better *vis-à-vis* the Solar-System tests and explains the shrinking of binaries through gravitational radiation. It therefore generates greater confidence than Newton's approach does, especially for use in cosmology, where strong gravitational fields are likely to be involved and where distances are so large that the assumption of instantaneous action at a distance would be misleading. Hence we will adopt general relativity as the underlying theory for constructing models of the Universe.

We will now undertake that exercise by constructing the models which Friedmann [49] in 1922–4 and Lemaître [50] in 1927 came up with before Hubble's results became known.

15.2 Setting up the field equations

We begin with the Robertson–Walker line element:

$$ds^2 = c^2\, dt^2 - S^2(t)\left[\frac{dr^2}{1 - kr^2} + r^2(d\theta^2 + \sin^2\theta\, d\phi^2)\right]. \qquad (15.1)$$

We use it first to compute the Einstein tensor and thereby formulate the general-relativistic field equations. To solve them we will next require the energy tensor of the material contents of the Universe.

Accordingly, we set

$$x^0 = ct, \qquad x^1 = r, \qquad x^2 = \theta, \qquad x^3 = \phi \qquad (15.2)$$

so that the non-zero components of g_{ik} and g^{ik} are

$$g_{00} = 1, \qquad g_{11} = -\frac{S^2}{1 - kr^2}, \qquad g_{22} = -S^2 r^2, \qquad g_{33} = -S^2 r^2 \sin^2\theta,$$

$$g^{00} = 1, \qquad g^{11} = -\frac{1 - kr^2}{S^2}, \qquad g^{22} = -\frac{1}{S^2 r^2}, \qquad g^{33} = -\frac{1}{S^2 r^2 \sin^2\theta},$$

$$\sqrt{-g} = \frac{S^3 r^2 \sin\theta}{\sqrt{1 - kr^2}}. \qquad (15.3)$$

The non-zero components of Γ^i_{kl} are then as follows:

$$\Gamma^1_{01} = \Gamma^2_{02} = \Gamma^3_{03} = \frac{1}{c}\frac{\dot{S}}{S},$$

$$\Gamma^0_{11} = \frac{S\dot{S}}{c(1 - kr^2)}, \qquad \Gamma^0_{22} = \frac{S\dot{S}r^2}{c}, \qquad \Gamma^0_{33} = \frac{S\dot{S}r^2 \sin^2\theta}{c},$$

$$\Gamma^1_{11} = \frac{kr}{1 - kr^2}, \qquad \Gamma^2_{12} = \Gamma^3_{13} = \frac{1}{r},$$

$$\Gamma^1_{22} = -r(1 - kr^2), \qquad \Gamma^1_{33} = -r(1 - kr^2)\sin^2\theta,$$

$$\Gamma^2_{33} = -\sin\theta \cos\theta, \qquad \Gamma^3_{23} = \cot\theta.$$

Now we use the expression for the Ricci tensor (*vide* Equation (5.9) in Chapter 5), which may be put in the following form:

$$R_{ik} = \frac{\partial^2 \ln\sqrt{-g}}{\partial x^i \partial x^k} - \frac{\partial \Gamma^l_{ik}}{\partial x^l} + \Gamma^m_{in}\Gamma^n_{km} - \Gamma^l_{ik}\frac{\partial \ln\sqrt{-g}}{\partial x^l}. \qquad (15.4)$$

Straightforward but tedious calculation then gives the following non-zero components of R^i_k:

$$R^0_0 = \frac{3}{c^2}\frac{\ddot{S}}{S}, \qquad (15.5)$$

$$R^1_1 = R^2_2 = R^3_3 = \frac{1}{c^2}\left(\frac{\ddot{S}}{S} + \frac{2\dot{S}^2 + 2kc^2}{S^2}\right). \qquad (15.6)$$

From these we get

$$R = \frac{6}{c^2}\left(\frac{\ddot{S}}{S} + \frac{\dot{S}^2 + kc^2}{S^2}\right), \tag{15.7}$$

and hence

$$G_1^1 \equiv R_1^1 - \frac{1}{2}R = -\frac{1}{c^2}\left(2\frac{\ddot{S}}{S} + \frac{\dot{S}^2 + kc^2}{S^2}\right) = G_2^2 = G_3^3, \tag{15.8}$$

$$G_0^0 \equiv R_0^0 - \frac{1}{2}R = -\frac{3}{c^2}\left(\frac{\dot{S}^2 + kc^2}{S^2}\right). \tag{15.9}$$

We have gone through the details of the calculation to illustrate how techniques of general relativity developed in earlier chapters can be applied to the problem of cosmology. The reader may check that putting $S = \text{constant} = S_0$ and $k = +1$ gives us the formulae (14.3) and (14.4) obtained for the Einstein universe in Chapter 14. As a general comment we remark that, because we have spatial homogeneity, the tensor components above (Equations (15.5)–(15.9)) do not contain any space coordinates. Further, because of isotropy, we have the three space–space components of the Einstein tensor equal. Recalling now the Einstein equations, we get from (15.8) and (15.9) the only non-trivial equations of the set as

$$2\frac{\ddot{S}}{S} + \frac{\dot{S}^2 + kc^2}{S^2} = \frac{8\pi G}{c^2}T_1^1 = \frac{8\pi G}{c^2}T_2^2 = \frac{8\pi G}{c^2}T_3^3, \tag{15.10}$$

$$\frac{\dot{S}^2 + kc^2}{S^2} = \frac{8\pi G}{3c^2}T_0^0. \tag{15.11}$$

We next consider the energy tensor.

15.3 Energy tensors of the Universe

Before we consider specific forms of T_k^i, it is worth noting that two properties must be satisfied by any energy tensor in the present framework of cosmology. The first is obvious from (15.10):

$$T_1^1 = T_2^2 = T_3^3 = -p \tag{15.12}$$

(say). The fact that these three components of T_k^i are equal is hardly surprising since we have already emphasized the condition of isotropy imposed on the Universe. In the light of our discussion of Chapter 7, we identify the quantity p with pressure. We further define the energy density by

$$T_0^0 = \epsilon. \tag{15.13}$$

The second property is not quite so obvious, but is derivable from (15.10) and (15.11). It relates the pressure to the energy density. We note that if we differentiate (15.11) with respect to t we can express the resulting answer as a linear combination of (15.10) and (15.11). The result is in fact equivalent to the following identity:

$$\frac{\mathrm{d}}{\mathrm{d}t}\left[S(\dot{S}^2 + kc^2)\right] \equiv \dot{S}\left[2S\ddot{S} + \dot{S}^2 + kc^2\right],$$

that is,

$$\frac{\mathrm{d}}{\mathrm{d}S}(\epsilon S^3) + 3pS^2 = 0. \tag{15.14}$$

It is not necessary, however, to write down the full field equations (15.10) and (15.11) in order to arrive at (15.14). The above result is a direct consequence of the conservation law implicit in the Einstein equations:

$$T^i_{k;i} = 0. \tag{15.15}$$

Recall that, from the Bianchi identities, (15.15) follows *identically*. We now turn our attention to the specific forms of the energy tensor.

15.3.1 Pressure and random motion of galaxies

We have assumed via the Weyl postulate that the primary unit of the Universe is a galaxy, which may be treated as a point particle. The characteristic length scale of the Universe, now taken as c/H_0, where H_0 is the Hubble constant at present, works out at $\sim 10^{28}$ cm, compared with which the galactic length scale is ~ 30 kpc $\sim 10^{23}$ cm. Thus the galaxy in the Universe is like a bead of diameter 1 cm in a field of size 1 km. We may ideally visualize the 'cosmological fluid' as made of these beads, flowing smoothly with negligible pressure.

The galaxies ideally should follow the Weyl postulate: in reality they do so at best approximately. Random motions of galaxies in clusters, typically of the order of $v \sim 300$ km s^{-1}, provide pressure to the cosmological fluid of the order of $\sim \rho v^2$, ρ being the density of the fluid. That is, the ratio p/ρ is as low as 10^{-6}. So when we write the energy tensor as

$$T^{ik} = (p + \rho)u^i u^k - pg^{ik} \tag{15.16}$$

we may justifiably approximate u^i by $[1, u^\mu]$, and ignore $|u^\mu|$ in comparison with unity. But when is this approximation valid?

Example 15.3.1 *Problem.* How does random motion behave in an expanding Universe?

Solution. Let us take the velocity vector of a typical galaxy as $[1, u^\mu]$, where $|u^\mu| \ll 1$. Since the galaxy follows a geodesic, we get

$$\frac{\mathrm{d}u^\mu}{\mathrm{d}s} + \Gamma^\mu_{ik}\frac{\mathrm{d}x^i}{\mathrm{d}s}\frac{\mathrm{d}x^k}{\mathrm{d}s} = 0.$$

Here we retain only the first-order terms. Using the Christoffel symbols of Section 15.2 we get, for $\mu = 1$, say

$$\frac{du^1}{ds} + 2\Gamma^1_{01} u^1 = 0,$$

i.e.,

$$\frac{du^1}{dt} + 2\frac{\dot{S}}{S} u^1 = 0 \Rightarrow u^1 S^2 = \text{constant}.$$

The same applies to $\mu = 2, 3$. Since we are using comoving coordinates, the physical motion $v^\mu = S u^\mu$. So we get $v^\mu S = \text{constant}$. Thus, as the Universe expands, the random motion decreases as S^{-1}.

The solved example above shows that, in the Robertson–Walker spacetime, the random velocity v^μ varies as $1/S$. Hence, in an expanding Universe, the pressure was more important in the past, is not so important now and will be even less important in the future. As we turn towards the past epoch, we should find the galaxy motions becoming more and more turbulent, since v was larger in the past. Thus, if we use $S \approx 10^{-3} S_0$ (S_0 being the value of S at the present epoch), the p-term would no longer be negligible in this epoch and prior to it.

For such past epochs we have to abandon our simplified picture of cosmology and ask whether galaxies existed as single units then. This question leads us to *cosmogony*, the subject of the origin of the large-scale structure of the Universe. Obviously, galaxies were formed at some stage in the past and, in a proper theory of cosmology and cosmogony, we have to say how and when they were formed. This topic, however, does not fall within the ambit of this text. The reader is referred to [53], which is the companion text to this one.

Returning to our discussion of energy tensors, we see that, if we simply extrapolate $v \propto S^{-1}$ to very low values of S, v becomes comparable to c and our approximation that led us to $v \propto S^{-1}$ breaks down. The correct formula then tells us that the 3-momentum P goes as S^{-1}. In this relativistic domain galaxies have not yet formed and matter is in the form of atomic particles moving very rapidly. Thus we have to use the formula (7.22), and we set

$$p = \frac{1}{3}\epsilon, \tag{15.17}$$

where ϵ denotes the energy density of these fast-moving particles. Thus, we may look upon a typical volume of these early epochs as containing matter particles moving at random relativistically, but any such spherical volume would have a centre of mass of all these particles at rest in the Robertson–Walker frame. In this case the Weyl postulate is not satisfied for a typical particle, but it may still be applied to the centre of mass of a typical spherical volume.

15.3.2 Matter versus radiation domination

So, if S continues to increase from very small values, then (15.17) would hold for the early epochs, just as (15.16), with $p \approx 0$, holds in the present and relatively recent epochs. The transition between the two epochs was through a rather messy phase when neither (15.16) nor (15.17) applied.

If (15.16) holds, then from (15.14) we get, with $p = 0$,

$$\frac{d}{dS}(\rho S^3) = 0, \tag{15.18}$$

which integrates to

$$\rho = \rho_0 \frac{S_0^3}{S^3}, \tag{15.19}$$

ρ_0 and S_0 being the values of ρ and S in the present epoch.

Similarly, substitution of (15.17) into (15.14) leads to

$$\frac{d}{dS}(\epsilon S^4) = 0, \tag{15.20}$$

giving

$$\epsilon \propto S^{-4}. \tag{15.21}$$

We therefore have the following picture. For a distribution of matter (15.21) was applicable when S was very small compared with S_0, while (15.19) holds in the more recent epochs. If, however, on top of matter we also have electromagnetic radiation present in the Universe, it too will contribute to T_k^i. For small S, (15.21) holds uniformly for matter (moving relativistically) and for radiation. However, as S increases we have to be more careful in distinguishing between the contributions of matter and radiation to T_k^i. For, as we shall see later, while matter and radiation were in close interaction at small S, at later epochs they became effectively decoupled from each other. We will go into these details more fully in Chapter 16.

For the present discussion let us assume that, after a certain epoch $t = t_{\text{dec}}$ when S was given by $S = S_{\text{dec}}$, radiation and matter decoupled from each other, each going its own way. Thus we can write

$$T_k^i = T_{k\,|\,\text{matter}}^i + T_{k\,|\,\text{radiation}}^i \tag{15.22}$$

and assume that the divergence of each energy tensor separately vanishes. Since for the radiation energy tensor we have (for $\mu = 1, 2, 3$), say,

$$- T_{\mu\,|\,\text{radiation}}^{\mu} = \frac{1}{3} T_{0\,|\,\text{radiation}}^0 = \frac{1}{3}\epsilon \quad \text{(no sum over } \mu\text{)}, \tag{15.23}$$

we get for $S > S_{dec}$

$$\epsilon = \epsilon_0 \frac{S_0^4}{S^4}. \tag{15.24}$$

What is t_{dec}? Why, if at all, should matter decouple from radiation? What happened prior to $t = t_{dec}$? We defer a discussion of these questions to Chapter 16. There was, however, another important epoch in the past history of the Universe, when the densities of matter and radiation were equal. We will denote it by $t = t_{eq}$, when S was equal to S_{eq}, say. It is easy to estimate this scale as follows.

The present estimates of $\epsilon_0 \approx 4 \times 10^{-13}$ erg cm^{-3} and of $\rho_0 c^2 \geq 3 \times 10^{-10}$ erg cm^{-3} mean that the matter density is about 10^3 times the radiation density. Thus $\epsilon_0 \ll \rho_0 c^2$, and we may ignore the contribution of radiation (in comparison with the contribution of matter) to the field equations (15.10) and (15.11) at the present epoch, and for $S > S_0$. However, for the past epochs with $S < S_0$, we have from (15.19) and (15.21)

$$\frac{\epsilon}{\rho c^2} = \frac{\epsilon_0}{\rho_0 c^2} \cdot \frac{S_0}{S}. \tag{15.25}$$

and we cannot ignore the contribution of radiation for, say, $S_0/S \sim 10^3$. This is the epoch t_{eq}. Indeed, prior to this epoch, that is for $S < S_{eq}$, the relative importance of radiation and matter was inverted: radiation was the more dominant factor in deciding how S should vary with t.

From the above discussion we see that, at $S = S_{eq} \approx 10^{-3} S_0$, we have a transition from a *radiation-dominated* Universe to a *matter-dominated* one. Here we will limit ourselves to the matter-dominated models with negligible pressure, leaving the discussion of the radiation-dominated models to the following chapter. Equations (15.10) and (15.11) are therefore to be solved with

$$T_1^1 = 0, \qquad T_0^0 = \rho_0 c^2 \frac{S_0^3}{S^3}. \tag{15.26}$$

This simplification leads us to the classic models first considered by A. Friedmann in 1922. Basically, these models ignore any contributions of electromagnetic radiation to T_k^i and suppose that the matter in the Universe can be approximated by dust.

We also mention, in passing, that the above analysis ignores the contribution of dark matter. A realistic assessment of dark matter will push up the value of ρ_0 by a factor ~ 6–7.

15.4 The Friedmann models

We will assume that the Universe is (as at present) dust-dominated. For dust models, Equations (15.10) and (15.11) become

$$2\frac{\ddot{S}}{S} + \frac{\dot{S}^2 + kc^2}{S^2} = 0, \tag{15.27}$$

$$\frac{\dot{S}^2 + kc^2}{S^2} = \frac{8\pi G \rho_0}{3} \frac{S_0^3}{S^3}. \tag{15.28}$$

In view of the conservation law given in (15.14), the above two differential equations are not independent, and only one of them is sufficient to determine $S(t)$. Since it is of lower order, we will choose (15.28) for our solution, and consider the three cases $k = 0, 1, -1$ separately.

15.4.1 Euclidean sections ($k = 0$)

This is the simplest case and is also known as the *Einstein–de Sitter model*, since it was given by Einstein and de Sitter in a joint paper [54] in 1932. Equation (15.28) becomes

$$\dot{S}^2 = \frac{8\pi G \rho_0}{3} \frac{S_0^3}{S}. \tag{15.29}$$

We now recall from Chapter 14 that the present value of Hubble's constant is given by

$$\left.\frac{\dot{S}}{S}\right|_{t_0} = H_0. \tag{15.30}$$

Hence, on applying (15.29) to the present epoch, we get

$$\rho_0 = \frac{3H_0^2}{8\pi G} \equiv \rho_c. \tag{15.31}$$

For reasons that will become clear later, ρ_c is often called the *closure density*. With the range of values of H_0 quoted in Chapter 14, we have

$$\rho_c = 2 \times 10^{-29} h_0^2 \, \mathrm{g\,cm}^{-3}. \tag{15.32}$$

The value as estimated by using the current favourite value of $h_0 \sim 0.7$ is considerably higher than the matter density actually observed at present. We will return to this issue in Chapter 16.

Returning to (15.29), it is easy to verify that it has the solution

$$S = S_0 \left(\frac{t}{t_0}\right)^{2/3}. \tag{15.33}$$

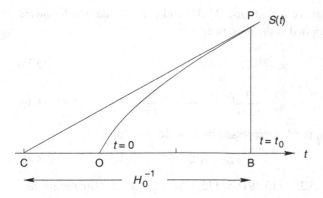

Fig. 15.1. The scale factor of the Einstein–de Sitter model ($k = 0$ in the case of the simplest Friedmann models).

An arbitrary constant that arises from the integration of the differential equation can be set equal to zero by assuming that $S = 0$ at $t = 0$. We also get from Equation (15.30) the *age of the Universe* as the present value of t:

$$t_0 = \frac{2}{3H_0}. \tag{15.34}$$

The constant S_0, the value of the scale factor at the present epoch, is not determined. It has the dimensions of length, and it can be absorbed into the unit of length chosen. Figure 15.1 illustrates this solution.

15.4.2 Closed sections ($k = 1$)

Equations (15.10) and (15.11) now take the form

$$2\frac{\ddot{S}}{S} + \frac{\dot{S}^2 + c^2}{S^2} = 0. \tag{15.35}$$

$$\frac{\dot{S}^2 + c^2}{S^2} - \frac{8\pi G\rho_0 S_0^3}{3S^3} = 0. \tag{15.36}$$

It is convenient to introduce the quantities $q(t)$ and $H(t)$ through the relations

$$\frac{\ddot{S}}{S} = -q(t)[H(t)]^2, \qquad H(t) = \frac{\dot{S}}{S}, \tag{15.37}$$

with their present values denoted by q_0 and H_0. We have already come across H_0, the Hubble constant. The second parameter q_0 is called the *deceleration parameter*, and it is useful for expressing ρ_0 in terms of the closure density.

With the above definitions, (15.35) and (15.36) take the following forms when applied at the present epoch:

$$\frac{c^2}{S_0^2} = (2q_0 - 1) H_0^2,$$
(15.38)

$$\rho_0 = \frac{3}{8\pi G}\left(H_0^2 + \frac{c^2}{S_0^2}\right) = \frac{3H_0^2}{4\pi G}q_0.$$
(15.39)

The density ρ_0 is often expressed in the following form:

$$\rho_0 = \rho_c \Omega_0,$$
(15.40)

so that from (15.38), (15.39) and (15.40) we get the *density parameter*

$$\Omega_0 = 2q_0.$$
(15.41)

Since the left-hand side of (15.38) is positive, we must have

$$q_0 > \frac{1}{2}, \qquad \Omega_0 > 1.$$
(15.42)

Thus our closed model has density *exceeding* the so-called closure density ρ_c. This explains the name 'closure density'. It is the value of the universal density that must be exceeded if the model is to describe a *closed* universe. We mention at this stage the result (to be proved shortly) that for the open models ($k = -1$) the inequalities of (15.42) are reversed.

Using (15.38) and (15.39) to eliminate S_0 and ρ_0, we get the following differential equation:

$$\dot{S}^2 = c^2\left(\frac{\alpha}{S} - 1\right).$$
(15.43)

with α given by

$$\alpha = \frac{2q_0}{(2q_0 - 1)^{3/2}} \frac{c}{H_0}.$$
(15.44)

The parameter α has the dimensions of length. Thus the model is characterized by the parameters H_0 and q_0 (or, alternatively, Ω_0).

Equation (15.43) can be integrated as follows. We get

$$ct = \int \frac{\sqrt{S}\,dS}{\sqrt{\alpha - S}}.$$

Make the substituition in terms of an auxiliary variable Θ:

$$S = \alpha \sin^2\left(\frac{\Theta}{2}\right) = \frac{1}{2}\alpha(1 - \cos \Theta).$$
(15.45)

Then the integral becomes

$$ct = \int \alpha \sin^2\left(\frac{\Theta}{2}\right) d\Theta = \frac{1}{2}\alpha(\Theta - \sin \Theta).$$
(15.46)

Again, as in the case $k = 0$ we have taken $S = 0$ at $t = 0\,(\Theta = 0)$. We therefore get $t = t_0$ by requiring that $S = S_0$. From (15.38) and (15.44) we see that $S = S_0$ at $\Theta = \Theta_0$, where

$$S_0 = \frac{1}{2}\alpha(1 - \cos\Theta_0) = \frac{c}{H_0}(2q_0 - 1)^{-1/2} = \frac{(2q_0 - 1)}{2q_0}\alpha,$$

that is,

$$\cos\Theta_0 = \frac{1 - q_0}{q_0}, \qquad \sin\Theta_0 = \frac{\sqrt{2q_0 - 1}}{q_0}. \tag{15.47}$$

We therefore get the *age of the Universe* as

$$t_0 = \frac{\alpha}{2c}(\Theta_0 - \sin\Theta_0)$$

$$= \frac{q_0}{(2q_0 - 1)^{3/2}}\left[\cos^{-1}\left(\frac{1 - q_0}{q_0}\right) - \sqrt{\frac{2q_0 - 1}{q_0}}\right]\frac{1}{H_0}. \tag{15.48}$$

For example, for $q_0 = 1$ we get

$$t_0 = \left(\frac{\pi}{2} - 1\right)H_0^{-1}. \tag{15.49}$$

Note that S reaches a maximum value at $\Theta = \pi$, when

$$S = S_{\text{max}} = \alpha = \frac{2q_0}{(2q_0 - 1)^{3/2}}\frac{c}{H_0}. \tag{15.50}$$

Thus, for $q_0 = 1$, the Universe expands to twice its present size.

In closed models, therefore, expansion is followed by contraction and S decreases to zero. The value $S = 0$ is reached when $\Theta = 2\pi$; that is, when

$$t = t_{\text{L}} = \frac{\pi\alpha}{c} = \frac{2\pi q_0}{(2q_0 - 1)^{3/2}}\frac{1}{H_0}. \tag{15.51}$$

The quantity t_{L} may be termed the *lifespan* of this universe. For $q_0 = 1$, $t_{\text{L}} = 2\pi H_0^{-1} = 2\pi\tau_o$. Recall that τ_0 is defined by the relation (14.40).

Figure 15.2 illustrates the function $S(t)$ for the closed models for a number of parameter values q_0. All curves have been adjusted to have the same value of H_0 at point P. Notice that the value $S = 0$ is reached sooner in the past as q_0 is increased from just over $1/2$.

15.4.3 Open sections ($k = -1$)

Equations (15.10) and (15.11) become in this case

$$2\frac{\ddot{S}}{S} + \frac{\dot{S}^2 - c^2}{S^2} = 0, \tag{15.52}$$

$$\frac{\dot{S}^2 - c^2}{S^2} - \frac{8\pi G\rho_0 S_0^3}{3S^3} = 0. \tag{15.53}$$

Fig. 15.2. The scale factors
for closed ($k = +1$) Friedmann
models are shown in the
diagram. The life span of the
Universe gets smaller for
larger values of the
deceleration parameter q_0.

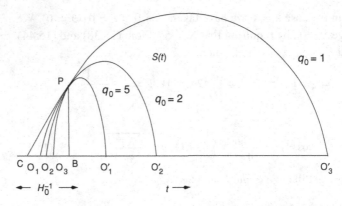

Fig. 15.2. The scale factors for closed ($k = +1$) Friedmann models are shown in the diagram. The life span of the Universe gets smaller for larger values of the deceleration parameter q_0.

We again use the definitions of (15.37) and apply them at the present epoch to get

$$\frac{c^2}{S_0^2} = (1 - 2q_0)H_0^2, \tag{15.54}$$

$$\rho_0 = \frac{3H_0^2}{4\pi G}q_0; \qquad \Omega_0 = 2q_0. \tag{15.55}$$

Thus instead of (15.42) we now have

$$0 \le q_0 \le \frac{1}{2}, \qquad 0 \le \Omega_0 < 1, \tag{15.56}$$

and in place of (15.43) we get

$$\dot{S}^2 = c^2\left(\frac{\beta}{S} + 1\right) \tag{15.57}$$

with

$$\beta = \frac{2q_0}{(1 - 2q_0)^{3/2}}\frac{c}{H_0}. \tag{15.58}$$

As in the $k = +1$ case, the solution of (15.57) may be expressed by a parameter Ψ with

$$S = \frac{1}{2}\beta(\cosh\Psi - 1), \qquad ct = \frac{1}{2}\beta(\sinh\Psi - \Psi). \tag{15.59}$$

The present value of Ψ is given by Ψ_0, where

$$\cosh\Psi_0 = \frac{1 - q_0}{q_0}, \qquad \sinh\Psi_0 = \frac{\sqrt{1 - 2q_0}}{q_0}. \tag{15.60}$$

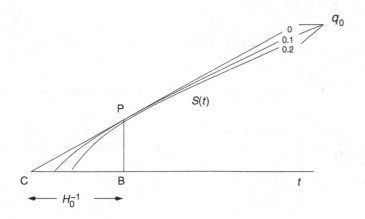

Fig. 15.3. Three cases of the temporal behaviour of the scale factor $S(t)$ for the open Friedmann model corresponding to $q_0 = 0, 0.1$ and 0.2. The age of the Universe is larger for smaller q_0. As shown in the figure, the largest age is $1/H_0$, corresponding to $q_0 = 0$.

We have set $t = 0$ at $S = 0$, as in the two preceding cases. The present value of t is given by

$$t_0 = \frac{\beta}{2c}(\sinh \Psi_0 - \Psi_0)$$

$$= \frac{q_0}{(1 - 2q_0)^{3/2}}\left[\frac{\sqrt{1 - 2q_0}}{q_0} - \ln\left(\frac{1 - q_0 + \sqrt{1 - 2q_0}}{q_0}\right)\right]\frac{1}{H_0}.$$

$$(15.61)$$

Like the Einstein–de Sitter model, these models continue to expand forever. The behaviour of $S(t)$ in these models is illustrated in Figure 15.3.

15.4.4 The Milne model

It is worth pointing out that the model with $k = -1$, $q_0 = 0$, $S(t) = ct$ represents flat spacetime. In fact, by the following coordinate transformation we can change the line element to a manifestly Minkowski form:

$$R = ctr, \qquad T = t\sqrt{1 + r^2},$$
$$ds^2 = c^2\,dT^2 - dR^2 - R^2(d\theta^2 + \sin^2\theta\,d\phi^2).$$

$$(15.62)$$

This model arose naturally in Milne's kinematic relativity [55], which was a cosmological theory with foundations different from those of general relativity. For this reason the above model is sometimes referred to as the *Milne model*.

For a comparison, the three types of Friedmann models ($k = 0, \pm1$) are shown together on the same plot in Figure 15.4. There is a unique

Fig. 15.4. All three types of
Friedmann models shown
together. All share the
property of a singular origin,
as discussed in the text.

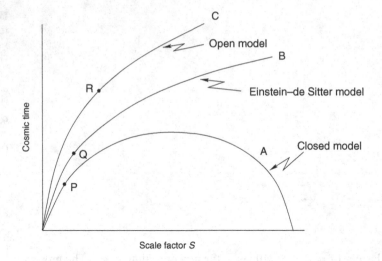

'flat' model (the $k = 0$ case is often so described; but this can be misleading insofar as the 'flatness' refers to the spatial sections $t =$ constant,
not to spacetime as a whole), but a continuous range of $k = \pm 1$ models,
of which two representative ones are shown. The dots P, Q, R lie on a
typical curve $H(t) =$ constant. Thus, for the same Hubble constant, the
open models give a larger age.

Figure 15.4 shows how all the Friedmann models have the common
feature of having $S = 0$ at a certain epoch (which we designate by
$t = 0$). As we approach $S = 0$, the Hubble constant increases rapidly,
becoming infinite at $S = 0$, except in the special case of the Milne model
$k = -1$, $q_0 = 0$. This epoch therefore indicates violent activity and is
given the name *big bang*. It was Fred Hoyle who in the late 1940s gave
this name, largely in a sarcastic vein, as he was, and continued to be,
critical of the big-bang concept. We will discuss the reasons for this *in
extenso* in later chapters. For the time being we simply state that the
name has stuck and has been accepted by a large majority of workers in
cosmology. We also mention that the big bang is the singular state where
all the geodesics of the Weyl postulate meet.

15.4.5 Luminosity distance

A practical result we need is the luminosity distance described in Chapter
14, for it tells us the effective distance of a source at a given redshift,
the distance whose square we divide by in order to estimate the flux of
radiation received from the source normal to a unit area at our location.
If the light left the source at time t_1 to reach us at time t_0, its travel

formula tells us that, for the $k = +1$ model,

$$\int_0^{r_1} \frac{dr}{\sqrt{1 - r^2}} = \int_{t_1}^{t_0} \frac{dt}{S(t)}, \tag{15.63}$$

where the source galaxy is located at $r = r_1$. Using formulae (15.45) and (15.46) we get

$$dt = \frac{1}{2}\alpha[1 - \cos \Theta]d\Theta = S(t)d\Theta,$$

so that the above equation yields the simple solution

$$r_1 = \sin(\Theta_0 - \Theta_1). \tag{15.64}$$

Here we have identified the epoch t_0 with the receiver and t_1 with the source. Using the fact that the redshift of the source is z, formulae (14.29) and (15.45) give

$$1 + z = \frac{1 - \cos \Theta_0}{1 - \cos \Theta_1},$$

which gives

$$\cos \Theta_1 = \frac{z + \cos \Theta_0}{1 + z}. \tag{15.65}$$

On putting together the values of $\cos \Theta_1$ and $\sin \Theta_1$ in Equation (15.64) and using the corresponding values of the trigonometric functions of Θ_0 from Equations (15.47), we get

$$r_1 = \frac{\sqrt{2q_0 - 1}\left[q_0 z + (q_0 - 1)(\sqrt{1 + 2zq_0} - 1)\right]}{q_0^2(1 + z)}. \tag{15.66}$$

The luminosity distance is therefore given by

$$D_1 = r_1 S_0(1 + z)$$

$$= \left(\frac{c}{H_0}\right)\frac{1}{q_0^2}\left[q_0 z + (q_0 - 1)(\sqrt{1 + 2zq_0} - 1)\right]. \tag{15.67}$$

This formula was first derived by Mattig [56] in 1958.

The formula for the open universe can be similarly derived and leads to exactly the same final answer. The case of the Einstein–de Sitter model can be obtained by letting q_0 tend to $1/2$. Of course it is much simpler to derive the result directly from the original formulae for that model. We give the final answer for this case below:

$$D_1 = \frac{2c}{H_0}\left[(1 + z) - (1 + z)^{1/2}\right]. \tag{15.68}$$

Although it is good to be able to derive these formulae analytically, the facility of the computers has made the exercise rather unnecessary. Nevertheless, the above exercise is useful in clarifying the roles of

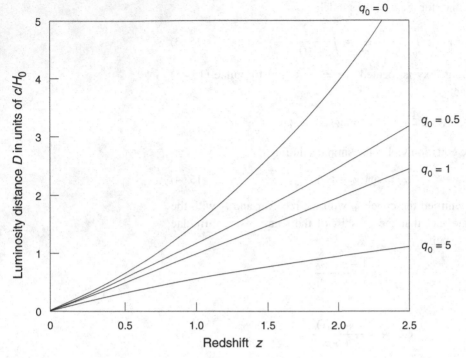

Fig. 15.5. The luminosity function D as a function of redshift plotted for various values of q_0 for the simple Friedmann models (with $\lambda = 0$).

various expressions in cosmology, which otherwise remain hidden in a computer programme. Figure 15.5 shows how D varies with z for different q_0.

15.5 The angular-size–redshift relation

We will next consider an unusual effect that arises because of the non-Euclidean geometry of the typical Robertson–Walker spacetime. In Figure 15.6 we have a spherical galaxy of diameter d at redshift $1 + z$. How will its angular size $\Delta\theta$ depend on redshift? If we associate the

Fig. 15.6. The angular size of an extended source is the angle it subtends at the observer O as shown in the figure.

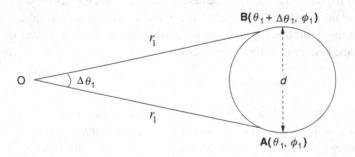

redshift with distance, then we expect distant galaxies to look smaller, i.e., to have progressively smaller $\Delta\theta$.

To decide the answer to this question, consider two neighbouring null geodesics (representing light rays) from the two points A and B at the two extremities of G_1 directed towards our Solar System. Without loss of generality we can choose our angular coordinates such that A has the coordinates (θ_1, ϕ_1), while B has the coordinates $(\theta_1 + \Delta\theta_1, \phi_1)$. (Although we have used homogeneity to take $r = 0$ at our location, we can also use isotropy to choose any particular direction as the polar axis $\theta = 0$, $\theta = \pi$.)

According to the Robertson–Walker line element, the proper distance between A and B is obtained by putting $t = t_1 = $ constant, $r = r_1 = $ constant, $\phi = \phi_1 = $ constant and $\mathrm{d}\theta = \Delta\theta_1$ in (15.1). We then get

$$\mathrm{d}s^2 = -r_1^2 S^2(t_1)(\Delta\theta_1)^2 = -d^2,$$

since in the rest frame of G_1 the spacelike separation $AB = d$. Thus

$$\Delta\theta_1 = \frac{d}{r_1 S(t_1)} = \frac{d(1+z)}{r_1 S(t_0)} \tag{15.69}$$

gives the answer to our question.

Notice that as r_1 increases we are looking at more and more remote galaxies, which must therefore be seen at earlier and earlier epochs t_1. However, in an expanding universe $S(t_1)$ was smaller at earlier epochs t_1, so it is not obvious that $r_1 S(t_1)$ should get progressively larger as we look at more and more remote galaxies. In some cases, therefore, distant objects may look bigger. The effect can be ascribed to 'gravitational bending or lensing' of light as it passes through curved spacetime. (We briefly discussed gravitational lensing in Chapter 12.) Clearly, we need to know how fast $S(t_1)$ decreases as r_1 increases. Although (15.69) provides the answer in an implicit form, we still need to know $S(t)$ in order to be able to perform these integrations.

Let us take the different Friedmann models in this context. It is easy to derive this result for $q_0 = 1/2$. From (15.68) we get

$$\Delta\theta_1 = \frac{d H_0}{2c} \frac{(1+z)^{3/2}}{(1+z)^{1/2} - 1}. \tag{15.70}$$

Straightforward differentiation gives us the result that the minimum value of $\Delta\theta_1$ ($=\theta_{\min}$, say) and the redshift $z = z_{\mathrm{m}}$ at which it occurs are given by

$$\theta_{\min} = 3.375 \frac{d H_0}{c}$$

and

$$z_{\mathrm{m}} = 1.25. \tag{15.71}$$

The cases $q_0 \neq 1/2$ are more involved. We illustrate the case $q_0 > 1/2$. Instead of using D_1 as given by (15.67), it is more convenient to use the parameter Θ introduced in (15.45) and (15.46) and the relations (15.47). We then get

$$\Delta\theta_1 = \frac{d}{r_1 S(t_1)} = \frac{2d}{\alpha}[(1 - \cos\Theta_1)\sin(\Theta_0 - \Theta_1)]^{-1}. \tag{15.72}$$

The constant α is defined by Equation (15.44). Differentiation with respect to Θ_1 tells us that the minimum occurs when

$$\sin\Theta_1 \sin(\Theta_0 - \Theta_1) - (1 - \cos\Theta_1)\cos(\Theta_0 - \Theta_1) = 0,$$

that is,

$$\sin\left(\Theta_0 - \frac{3\Theta_1}{2}\right) = 0,$$

thus giving

$$\Theta_1 = \frac{2\Theta_0}{3}, \qquad 1 + z_{\mathrm{m}} = \frac{1 - \cos\Theta_0}{1 - \cos(2\Theta_0/3)}. \tag{15.73}$$

Using (15.69) we get

$$\Theta_{\mathrm{min}} = \frac{(2q_0 - 1)^{3/2}}{q_0} \frac{1}{\left(1 - \cos\left(\frac{2\Theta_0}{3}\right)\right)\sin\left(\frac{\Theta_0}{3}\right)} \frac{d H_0}{c}. \tag{15.74}$$

The corresponding result for $q_0 < 1/2$ is

$$\Theta_{\mathrm{min}} = \frac{(1 - 2q_0)^{3/2}}{q_0} \frac{1}{\left(\cosh\left(\frac{2\Psi_0}{3}\right) - 1\right)\sinh\left(\frac{\Psi_0}{3}\right)} \frac{d H_0}{c} \tag{15.75}$$

at the redshift z_{m} given by

$$1 + z_{\mathrm{m}} = \frac{\cosh\Psi_0 - 1}{\cosh\left(\frac{2\Psi_0}{3}\right) - 1}. \tag{15.76}$$

Figure 15.7 plots $\Delta\theta_1$ as a function of z for different Friedmann models. Notice how the curves all start with the near-Euclidean result $\Delta\theta_1 \propto z^{-1}$ and then begin to differ from one another at larger z values. In principle this effect might be used to decide which Friedmann model (if any!) comes closest to the actual Universe.

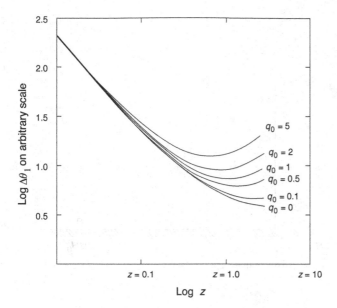

Fig. 15.7. The 'non-Euclidean' behaviour of angular size at large redshifts.

This non-Euclidean effect was first pointed out by R. C. Tolman [57] and a way of using it as a cosmological test was first suggested by Fred Hoyle [58].

15.6 Horizons and the Hubble radius

In cosmological discussions two kinds of horizons often crop up: the *particle horizon* relates to limits on communication in the past, whereas the *event horizon* relates to limits on communication in the future. We will deal with these two concepts in that order.

15.6.1 The particle horizon

It is pertinent to ask the following question. What is the limit on the proper distance up to which we are able to see sources of light? This question is answered as follows. Going back to Equation (14.27) of the preceding chapter, we may have a situation wherein the integral on the left-hand side has a maximum value at the given epoch t_0. This therefore gives a maximum value r_P for the radial coordinate r_1. For a galaxy with $r_1 > r_P$ there is no communication with us in the above fashion.

First we calculate this limiting value r_P of r_1, which for the Friedmann models comes from setting the lower limit for the t-integral at zero. The corresponding limiting proper distance is

$$R_P = S_0 \int_0^{r_P} \frac{dr}{\sqrt{1 - kr^2}}.$$

Fig. 15.8. The particle
horizon of P is contained in
the past null cone (dotted
lines) from P. Thus the particle
B can causally affect P, but not
particle A, which lies outside
the particle horizon of P.

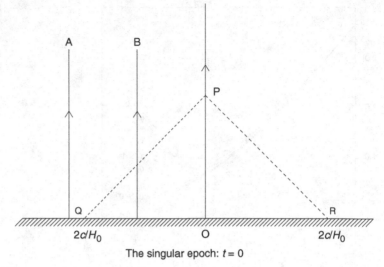

It is then easy to verify that, for the various Friedmann models,

$$
R_{\mathrm{P}} = \frac{c}{H_0} \times
\begin{cases}
2 & (k = 0, \ q_0 = \tfrac{1}{2}) \\[2mm]
\dfrac{2}{\sqrt{2q_0 - 1}} \sin^{-1} \sqrt{\dfrac{2q_0 - 1}{2q_0}} & (k = 1, \ q_0 > \tfrac{1}{2}) \\[4mm]
\dfrac{2}{\sqrt{1 - 2q_0}} \sinh^{-1} \sqrt{\dfrac{1 - 2q_0}{2q_0}} & (k = -1, \ q_0 < \tfrac{1}{2}).
\end{cases}
$$

$$(15.77)$$

The existence of a finite value of R_{P} means that the Universe has a *particle horizon*. Particles with $S(t_0)r_1 > R_{\mathrm{P}}$ are not visible to us at present, no matter how good our techniques of observation are.

Consider, for example, the Einstein–de Sitter model. The result (15.77) gives, in this case, $R_{\mathrm{P}} = 2c/H_0$. This means that at present we are able to see only those galaxies whose proper distance from us happens to be less than $2c/H_0$. See Figure 15.8.

15.6.2 The event horizon

The particle horizon sets a limit to communications from the past. Let us now see how the event horizon sets a limit on communications to the future. Let us ask the following question. A light source at $r = r_1, t = t_0$ sends a light signal to an observer at $r = 0$. Will the signal ever reach its destination? Suppose it does and let t_1 be the time of arrival. Then from (14.27) we get

$$\int_{t_0}^{t_1} \frac{c\,dt}{S(t)} = \int_0^{r_1} \frac{dr}{\sqrt{1 - kr^2}}.$$

This relation determines t_1 for any given r_1, provided that the integral on the left is large enough to match that on the right. Now it may happen that as $t_1 \rightarrow \infty$ the integral on the left converges to a finite value that corresponds to a value of the integral on the right for $r_1 = r_E$, say. In that case it is not possible to satisfy the above relation for $r_1 > r_E$. In other words the signal from the light source at $r_1 > r_E$ will *never* reach the observer at r_0. Thus no two observers can communicate beyond a proper distance

$$R_E = S_0 \int_{t_0}^{\infty} \frac{c\,dt}{S(t)} \tag{15.78}$$

at $t = t_0$.

This limit is called the *event horizon*. It does not exist for Friedmann models but has the value c/H_0 for the de Sitter model, as can be seen in the following calculation.

Example 15.6.1 Consider the de Sitter model described in Chapter 14. Here we have $k = 0$ and $S = e^{Ht}$. Then we get

$$R_E = e^{Ht_0} \int_{t_0}^{\infty} ce^{-Ht}\,dt = \frac{c}{H_0}.$$

That is, if any light source sends a ray of light from beyond this range at time t_0 towards the observer at $r = 0$, it will *never* reach the observer. See Figure 15.9. The reader will immediately notice the similarity of the event horizon here to that for a black hole (see Chapter 13).

Notice that both the event horizon and the particle horizon have radii comparable to c/H_0, which has led to an erroneous conclusion that the length $R_H = c/H_0$ is of the size of the horizon in *any* cosmology. Whether a horizon (particle or event) exists in a cosmological model depends on the scale factor and how the relevant integral (discussed above) behaves. Thus there are cosmological models that do not have any horizon, and for such models the above length does not have any 'signal-limiting' significance. In such cases, it is best to call this length R_H, the *Hubble radius*. The Hubble radius as defined here tells us only the characteristic distance scale of the Universe at $t = t_0$; it does *not* have any causal significance unless it is shown to have horizon properties. We may compare it with the *Hubble time scale* τ_0 defined in the previous chapter.

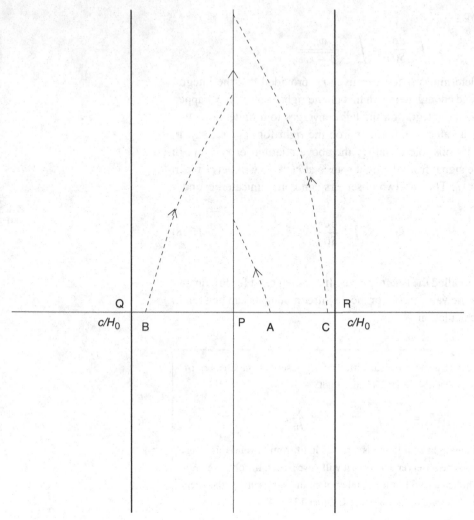

15.7 Source counts

The distribution of discrete luminous sources out to great distances may give indications that spacetime geometry is non-Euclidean. How does the number of galaxies up to coordinate distance r_1 (that is, up to the distance of galaxy G_1) increase with r_1? Let us suppose that at any epoch t there are $n(t)$ galaxies in a *unit comoving coordinate volume* (using the r, θ, ϕ coordinates). The word 'comoving' indicates that, although the galaxies individually retain the same coordinates (r, θ, ϕ), the proper separation between them at any epoch increases with epoch according to the scale factor $S(t_1)$. Thus the proper volume of any region bounded by such galaxies increases as S^3.

When we observe galaxies at radial coordinates between r and $r + dr$, we see them at times in the range $t, t + dt$, where, from (14.27),

$$\int_t^{t_0} \frac{c\, dt'}{S(t')} = \int_0^r \frac{dr'}{\sqrt{1 - kr'^2}}. \tag{15.79}$$

The number of galaxies seen in this shell is therefore

$$dN = \frac{4\pi\, r^2\, dr}{\sqrt{1 - kr^2}} \cdot n(t), \tag{15.80}$$

where t is related to r through (15.79). Thus the required number of galaxies out to $r = r_1$ is given by

$$N(r_1) = \int_0^{r_1} \frac{4\pi r^2 n(t) dr}{\sqrt{1 - kr^2}}. \tag{15.81}$$

If no galaxies are created or destroyed between $r = 0$ and $r = r_1$, we may take $n(t) = $ constant, and the integral can be explicitly evaluated. Clearly, the answer must depend on the parameter k. If we draw a sphere whose surface lies at a proper distance R from the centre in the $k = 0$ (Euclidean) space, its volume will be $4\pi R^3 / 3$. However, a similar sphere drawn in the $k = +1$ (closed) space will have a volume *less* than $4\pi R^3 / 3$, whereas a sphere drawn in the $k = -1$ (open) space will have a volume exceeding this value.

We now apply the above formula to Friedmann models. It is more convenient to use redshift as the distance parameter instead of r or t. As an example, we will work with the case $k = +1$. From (15.64) and the relations that follow it, we have

$$r = \sin(\Theta_0 - \Theta_1),$$

$$\left| \frac{dr}{\sqrt{1 - r^2}} \right| = |d\Theta_1|, \qquad 1 + z = \frac{\sin^2\left(\frac{\Theta_0}{2}\right)}{\sin^2\left(\frac{\Theta_1}{2}\right)},$$

$$\left| \frac{dz}{1 + z} \right| = \cot\left(\frac{\Theta_1}{2}\right) |d\Theta_1| = \sqrt{\frac{1 + 2q_0 z}{2q_0 - 1}} |d\Theta_1|.$$

Therefore the number of astronomical sources with redshifts in the range $(z, z + dz)$ is given by

$$dN = 4\pi \sin^2(\Theta_0 - \Theta_1) \cdot n(t) \cdot \left| \frac{d\Theta_1}{dz} \right| dz.$$

Let us suppose that $n(t)$ is specified as a function $n(z)$ of z. Using (15.65) and some algebraic manipulation, we get

$$dN = 4\pi \cdot n(z) \frac{(2q_0 - 1)^{3/2}}{q_0^4} \frac{[q_0 z + (q_0 - 1)(\sqrt{1 + 2zq_0} - 1)]^2 dz}{\sqrt{1 + 2q_0 z}(1 + z)^3}. \tag{15.82}$$

Suppose $n(z)$ is expressed in a slightly different form. We recall that n was specified as the number of sources per unit *coordinate* volume, in terms of the comoving (r, θ, ϕ) coordinates. What is the relationship between n and the number of sources per unit *proper* volume? Denoting the latter by $\bar{n}$, we have

$$n = \bar{n} S^3 = \frac{\bar{n} S_0^3}{(1+z)^3}. \tag{15.83}$$

From (15.38) we get

$$\frac{\bar{n}}{(1+z)^3} = (2q_0 - 1)^{3/2} \left(\frac{H_0}{c}\right)^3 n. \tag{15.84}$$

Substitution into (15.82) gives

$$dN = 4\pi \left(\frac{c}{H_0}\right)^3 \frac{[q_0 z + (q_0 - 1)(\sqrt{1 + 2zq_0} - 1)]^2 \bar{n} \, dz}{q_0^4 (1+z)^6 \sqrt{1 + 2q_0 z}}. \tag{15.85}$$

In this form (15.85) is applicable to all Friedmann models, even though our derivation assumed $q_0 > 1/2$ and $k = 1$.

This formula played a big role in the early development of observational cosmology. In the 1930s Hubble expected to measure the curvature effects in the counts of galaxies, assuming that galaxies are uniformly distributed. He found out that the effect, if it exists, is too minute to be measurable. Two decades later radio astronomers attempted a similar study using powerful extragalactic radio sources. Here too the curvature effects became dwarfed by other variables such as the spectrum of luminosity of the sources, the evolution of density and luminosity with epoch, possible large-scale inhomogeneity of radio-source distribution, etc., etc.

15.8 Cosmological models with the λ-term

Although our concern in this chapter was mainly with the simplest Friedmann models, we now discuss briefly another class of models given by the modified Einstein equations (14.7) – the equations containing the cosmological constant λ. We have already discussed two special cases of this class of solutions in the last chapter, namely the static Einstein model and the empty de Sitter model. When Hubble's observations established the expanding-Universe picture, Einstein conceded that there was no special need for the λ-term in his equations. In the post-Hubble-law era, he dropped this term from his equations, and the Einstein–de Sitter model discussed in this chapter was the outcome of Einstein's collaboration with de Sitter after abandoning the λ-term.

Nevertheless, in the 1930s eminent cosmologists such as A. S. Eddington and Abbé Lemaître felt that the λ-term introduced certain attractive features into cosmology and that models based on it should also be discussed at length. In modern cosmology the reception given to the λ-term has varied from the hostile to the ecstatic. The term is quietly forgotten if the observational situation does not demand models based on it. It is resurrected if it is found that the standard Friedmann models without this term are being severely constrained by observations. The present compulsion for this term comes partly because of the observational constraints and partly because inputs from particle physics in the very early stages of the Universe have provided a new interpretation for the λ-term, which we shall discuss in Chapter 16.

Putting $\lambda \neq 0$, (15.10) and (15.11) are modified to the following:

$$2\frac{\ddot{S}}{S} + \frac{\dot{S}^2 + kc^2}{S^2} - \lambda c^2 = \frac{8\pi G}{c^2} T_1^1, \tag{15.86}$$

$$\frac{\dot{S}^2 + kc^2}{S^2} - \frac{1}{3}\lambda c^2 = \frac{8\pi G}{3c^2} T_0^0. \tag{15.87}$$

The conservation laws discussed earlier are not affected by the λ-term. If we restrict ourselves to dust only, (15.87) gives us the following differential equation in place of (15.28):

$$\frac{\dot{S}^2 + kc^2}{S^2} - \frac{1}{3}\lambda c^2 = \frac{8\pi G\rho_0}{3}\frac{S_0^3}{S^3}. \tag{15.88}$$

Similarly, (15.86) becomes

$$2\frac{\ddot{S}}{S} + \frac{\dot{S}^2 + kc^2}{S^2} - \lambda c^2 = 0. \tag{15.89}$$

Let us first recover the static model of Einstein. By setting $S = S_0$, $\dot{S} = 0$, $\ddot{S} = 0$ in (15.88) and (15.89), we get

$$\frac{kc^2}{S_0^2} - \frac{1}{3}\lambda c^2 = \frac{8\pi G\rho_0}{3}; \qquad \frac{kc^2}{S_0^2} = \lambda c^2.$$

From these relations it is not difficult to verify that $k = +1$, and we recover the relations obtained in Chapter 14:

$$\lambda = \frac{1}{S_0^2} \equiv \lambda_c, \tag{15.90}$$

$$\rho_0 = \frac{\lambda_c c^2}{4\pi G}. \tag{15.91}$$

We shall denote by $\lambda = \lambda_c$ the critical value of λ for which a static solution is possible. It was pointed out by Eddington that the Einstein universe is unstable. A slight perturbation destroying the equilibrium conditions (15.90) and (15.91) leads to either a collapse to singularity ($S \to 0$) or an expansion to infinity ($S \to \infty$). Eddington and Lemaître instead proposed a model in which λ exceeds λ_c by a small amount. In this case the Universe erupts from $S = 0$ (the big bang) and slows down

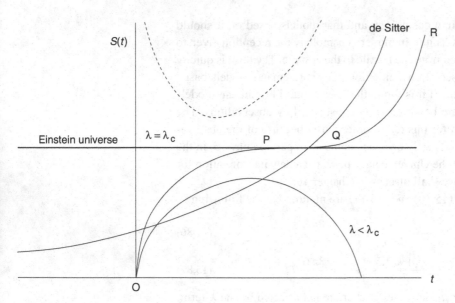

Fig. 15.10. Friedmann models with $k = +1$, $\lambda > 0$ are shown here. The scale factor $S(t)$ is plotted against the cosmic time t.

near $S = S_0$, staying thereabouts for a long time and then expanding away to infinity. It was argued that the quasistationary phase of the Universe would be suitable for the formation of galaxies. This model is illustrated in Figure 15.10, which plots $S(t)$ for a range of values of λ for $k = +1$. The initial (explosive) phase of the Eddington–Lemaître model is shown along the section OP of the curve OPQR, with PQ the quasistationary phase and QR the final accelerated expansion. Notice that for $\lambda < \lambda_c$ the Universe contracts (as in the Friedmann case), whereas for $\lambda > \lambda_c$ it ultimately disperses to infinity, resembling the de Sitter universe.

Figure 15.10 also shows by a dashed line one of another series of models that contract from infinity to a minimum value of $S > 0$ and then expand back to $S \to \infty$. These models are sometimes called *oscillating models of the second kind*, to distinguish them from the models that start from and shrink back to $S = 0$ and are called *oscillating models of the first kind*. This terminology is, however, not quite apt, since there is no repetition of phases in these models as implied by the word 'oscillating'.

The models with $k = 0$ or $k = -1$ do not show these different types of behaviour for $\lambda > 0$. We get from (15.88) a relation of the following type:

$$\dot{S}^2 = -kc^2 + \frac{1}{3}\lambda c^2 S^2 + \frac{8\pi G \rho_0 S_0^3}{3S}, \qquad (15.92)$$

wherein each term on the right-hand side is non-negative. Thus $\dot{S}$ does not change sign, and we get ever-expanding models. For $\lambda < 0$, however,

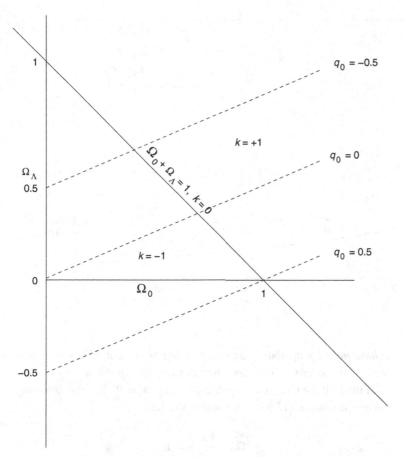

Fig. 15.11. A plot of Ω_Λ against Ω_0 for various Friedmann models is shown for various values of q_0. The 'flat-model' line $\Omega_0 + \Omega_\Lambda = 1$ is of special interest since inflation predicts flat models (see Chapter 16).

we can get universes that expand and then recontract as in the $k = 1$ case for $\lambda < \lambda_c$.

This concludes our discussion of the general dynamical behaviour of the λ-cosmologies. We end this section by writing (15.88) and (15.89) at the present epoch in terms of H_0 and q_0. Thus in place of earlier relations we have

$$H_0^2 + \frac{kc^2}{S_0^2} - \frac{1}{3}\lambda c^2 = H_0^2 \Omega_0, \qquad (15.93)$$

$$(1 - 2q_0)H_0^2 + \frac{kc^2}{S_0^2} - \lambda c^2 = 0. \qquad (15.94)$$

From these we get

$$\Omega_0 = 2q_0 + \frac{2}{3}\lambda \frac{c^2}{H_0^2}. \qquad (15.95)$$

Now there is no unique relationship between q_0 and Ω_0: we have an additional parameter entering the relation. *Note also that it is possible*

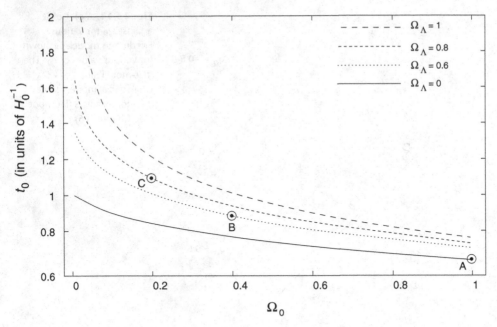

Fig. 15.12. The age of the Universe plotted against Ω_0 for various values of Ω_Λ. The age is seen to increase as Ω_Λ is increased. Thus model A with $\Omega_\Lambda = 0$ has the least age, while the case C for $\Omega_\Lambda = 0.8$ has nearly double that age.

to have negative q_0, that is, an accelerating expansion, if $\lambda > 0$. This is because the λ-term introduces a force of cosmic *repulsion*.

Finally, if the Universe is spatially flat, i.e., $k = 0$, then the following rewrite of relation (15.93) can confirm the fact:

$$\Omega_0 + \frac{1}{3}\frac{\lambda c^2}{H_0^2} = 1.$$

By writing

$$\frac{1}{3}\frac{\lambda c^2}{H_0^2} = \Omega_\Lambda \tag{15.96}$$

the above relation is often expressed in the form

$$\Omega_0 + \Omega_\Lambda = 1. \tag{15.97}$$

See Figure 15.11 showing these relationships in the $\Omega_0 - \Omega_\Lambda$ plane.

A comment is needed here to explain to the reader one reason why the λ-containing models are preferred these days. The measurements of Hubble's constant and the estimates of ages of stars in globular clusters suggest that the ages of the $\lambda = 0$ models are inadequate to accommodate the stellar ages. As Figure 15.12 shows, by having a positive cosmological constant, the age of the Universe can be increased. Therefore, the age constraint can be relaxed if $\lambda > 0$.

With these remarks we wind up our discussion of relativistic cosmological models dominated by dust. We will take up the radiation-dominated models in the following chapter.

Exercises

1. Verify the expressions for the Ricci tensor and the Einstein tensor for the Robertson–Walker line element.

2. Deduce Equation (15.14) from Equation (15.15).

3. Using the Einstein–de Sitter model, estimate the epoch at which the matter and radiation densities in the Universe were equal. For this calculation take $\rho_0 = 10^{-29}$ g cm^{-3} and $\epsilon_0 = 10^{-13}$ erg cm^{-3}, and express your answer as a fraction of the age of the Universe.

4. A galaxy is observed with redshift 0.69. How long did light take to travel from the galaxy to us if we assume that we live in the Einstein–de Sitter universe with Hubble's constant $= 70$ km s^{-1} Mpc^{-1}?

5. In the Friedmann universe with $q_0 = 1$, a galaxy is seen with redshift $z = 1$. How old was the universe at the time this galaxy emitted the light received today? (Take $H_0 = 100$ km s^{-1} Mpc^{-1}.)

6. A light ray is emitted at the present epoch in the closed Friedmann universe. Discuss the possibility of this ray making a round of the universe and coming back to its starting point.

7. Invert the formula (15.67) to express z as a function of $x \equiv D_1 H_0/c$. Show that

$$z = q_0 x - (q_0 - 1) \left(\sqrt{1 + 2x} - 1 \right).$$

Use this relation to show how the linear Hubble velocity–distance relation begins to fan out for cosmological models with different values of q_0.

8. Show why the Friedmann models with $\lambda = 0$ do not have event horizons.

9. The *surface brightness* of an astronomical object is defined as the flux received from the object divided by the angular area subtended by the object at the observation point. How does the surface brightness vary with redshift?

10. Show from first principles that the angular sizes of astronomical objects of fixed linear size will have a minimum at $z = 1.25$ in the Einstein–de Sitter model.

11. If in a Friedmann universe we have a fixed number of sources in a unit comoving coordinate volume and each source emits line radiation of fixed total intensity L_0 at frequency $\bar{\nu}$, show that the radiation background produced by such sources at the present epoch will have the frequency spectrum $S(\nu)d\nu$,

where $S(\nu) = 0$, for $\nu > \bar{\nu}$, whereas for $\nu < \bar{\nu}$

$$S(\nu) = \frac{c}{H_0} \bar{n}_0 L_0 \frac{\nu^{3/2}}{\bar{\nu}^2 \sqrt{2q_0 \bar{\nu} - (2q_0 - 1)\nu}},$$

where n_0 is the proper number density of sources at the present epoch.

12. Given that objects during the quasistationary phase of the Eddington–Lemaître cosmology are now seen with the redshift $z = 2$, what can you say about the value of λ?

13. Deduce that the scale factor in the λ-cosmology with $k = 1$ satisfies the differential equation

$$\dot{S}^2 = c^2 \left(\frac{1}{3} \lambda S^2 - 1 + \frac{\gamma}{S} \right),$$

where

$$\gamma = \frac{2q_0 + \frac{2}{3} \frac{\lambda c^2}{H_0^2}}{\left(2q_0 - 1 + \frac{\lambda c^2}{H_0^2} \right)^{3/2}} \left(\frac{c}{H_0} \right).$$

14. Write down an integral that gives the age of a big-bang universe for $\lambda \neq 0$. Discuss qualitatively how the λ-term may be used to increase the age of the universe.

15. In λ-cosmology, what is the lower limit on the value of λ given the value of q_0?

16. Compute the invariants R, $R_{ik} R^{ik}$ and $R_{iklm} R^{iklm}$ for the Friedmann models and show that they all diverge as $S \to 0$. Is there an exceptional case?

17. Give a general argument to show that, for sufficiently small S, the λ-force is ineffective at preventing the spacetime singularity.

18. Identify the region on the $\Omega_0 - \Omega_\Lambda$ plane corresponding to accelerating models of the Universe.

19. The steady-state model is described by the de Sitter line element and a constant density at all epochs from $t = -\infty$ to $t = +\infty$. Does this model have an infinite sky background as calculated by Olbers? Verify your answer by direct calculation.

20. Show that the line element

$$ds^2 = e^\nu \, dT^2 - e^{-\nu} \, dR^2 - R^2 (d\theta^2 + \sin^2\theta \, d\phi^2),$$

where $e^\nu = 1 - (2GM/R) - (\lambda R^2 / 3)$, describes a spherically symmetric distribution of matter of mass M in an otherwise empty, asymptotically de Sitter, universe. Discuss the effect of the λ-term on the Solar-System tests of general relativity.

Chapter 16
The early Universe

16.1 The radiation-dominated Universe

In the previous chapter we discussed simple cosmological models in which the contents were described as 'dust', i.e., pressure-free matter. We saw that the density ρ of the matter behaves as $\sim S^{-3}$. We also saw briefly that, if radiation were present, its energy density would vary with the scale factor as

$$\epsilon \propto \frac{1}{S^4}.$$

At present the Universe is dust matter-dominated; but if we see the different rates at which ρ and ϵ increased in the past, we find that there was an epoch in the past when the two energy densities were equal. Let us denote this epoch by its redshift z_{eq}. This means that at this epoch the scale factor was a fraction $1/(1 + z_{eq})$ of its present value. Figure 16.1 illustrates the relative variations of matter and radiation densities. We see that, prior to this epoch, radiation dominated over matter in determining the dynamics of the Universe through the Einstein field equations.

What about temperature? During the radiation-dominated era, the temperature was determined by radiation and a simple calculation shows how the temperature also might have been high. This calculation requires the assumption that at present we have a radiation density u_0 that is a relic of an early hot era. With this assumption, the radiation energy density at a past epoch S was given by (15.24):

$$\epsilon = \epsilon_0 \frac{S_0^4}{S^4}. \tag{16.1}$$

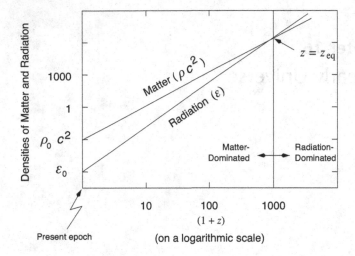

We may therefore assume that in the early epochs the dynamics of
expansion was determined by radiant energy rather than by matter in the
form of dust and that these were high-temperature epochs.

We illustrate the above ideas with a simplified calculation by assuming that the radiation was in blackbody form with temperature T, so
that

$$\epsilon = aT^4, \tag{16.2}$$

where a is the radiation constant. This means that in the early stages of
the big-bang Universe

$$T_0^0 = aT^4, \qquad T_1^1 = T_2^2 = T_3^3 = -\frac{1}{3}aT^4. \tag{16.3}$$

We also anticipate that the space-curvature parameter k will not affect
the dynamics of the early Universe significantly, and set it equal to zero.
Thus, from (15.11),

$$\frac{\dot{S}^2}{S^2} = \frac{8\pi Ga}{3c^2}T^4. \tag{16.4}$$

Further, from (16.2) and (16.1) we get

$$T = \frac{A}{S}, \qquad A = \text{constant}. \tag{16.5}$$

Substituting (16.5) into (16.4) gives a differential equation for S that can
easily be solved. Setting $t = 0$ at $S = 0$, we get

$$S = A\left(\frac{3c^2}{32\pi Ga}\right)^{-1/4} t^{1/2} \tag{16.6}$$

and, more importantly,

$$T = \left(\frac{3c^2}{32\pi Ga}\right)^{1/4} t^{-1/2}. \tag{16.7}$$

Notice that all the quantities inside the parentheses on the right-hand side of the above equation are known physical quantities. Thus, by substituting their values into (16.7), we can express the above result in the following form:

$$T_{\text{kelvin}} = 1.52 \times 10^{10} t_{\text{second}}^{-1/2}. \tag{16.8}$$

In other words, about one second after the big bang the radiation temperature of the Universe was 1.52×10^{10} K. The Universe at this stage was certainly hot enough to have free neutrons and protons around, which, as the Universe expanded, cooled down to facilitate the formation of atomic nuclei. It was George Gamow who appreciated the significance of the early hot era and conjectured that all chemical elements found in the Universe were formed in a primordial nucleosynthesis process [59]. In short, the Universe acted as a fusion reactor.

The idea of a *hot big bang*, as the above picture is called, depends therefore on the assumption that there is relic radiation present today. Later in this chapter we will present the argument that the microwave background discovered in 1965 by Arno Penzias and Robert Wilson is that relic radiation. For the present we will accept this evidence as confirming Gamow's notion of the hot big bang and proceed further.

16.2 Primordial nucleosynthesis

This being a book primarily on general relativity, rather than on cosmology, we will rush through the description of how and when atomic nuclei were synthesized. For details we refer the reader to the companion volume on cosmology [53]. Here we summarize the important steps in this process.

16.2.1 Distribution functions

Assuming an ideal-gas approximation and thermodynamic equilibrium, it is possible to write down the distribution functions of any given species of particles like neutrons, protons, photons, etc. Let us use the symbol A to denote typical species A. Thus $n_A(P)\mathrm{d}P$ denotes the number density of species A in the momentum range $(P, P + \mathrm{d}P)$, where

$$n_A(P) = \frac{g_A}{2\pi^2\hbar^3} P^2 \left[\exp\left(\frac{E_A(P) - \mu_A}{kT}\right) \pm 1\right]^{-1}. \tag{16.9}$$

In the above formula T is the temperature of the distribution, g_A the number of spin states of the species and k the Boltzmann constant, while the equation

$$E_A^2 = c^2 P^2 + m_A^2 c^4 \qquad (16.10)$$

relates the energy to the rest mass m_A and momentum P of a typical particle of species A. Thus for the electron $g_A = 2$, for the neutrino $g_A = 1$, $m_A = 0$, and so on. The $+$ sign in (16.9) applies to particles obeying Fermi–Dirac statistics (these particles are called *fermions*), while the $-$ sign applies to particles obeying Bose–Einstein statistics (particles known as *bosons*). For example, electrons and neutrinos are fermions, whereas photons are bosons. The quantity μ_A is the chemical potential of the species A. The number densities of these species are needed in order to work out their chemical potentials. Since photons can be emitted and absorbed in any amounts, it is normally assumed that $\mu_\gamma = 0$. Present observations suggest that for baryons (B) the ratio

$$\frac{N_B}{N_\gamma} = \frac{\text{Number density of baryons}}{\text{Number density of photons}} \sim 10^{-8} - 10^{-10}$$

is small compared with 1. The smallness of the baryon number density suggests that the number densities of leptons may also be small compared with N_γ, and it is usually assumed that this hypothesis provides a good justification for taking $\mu_A = 0$ for all species. We will assume that $\mu_A = 0$ for all species as a first approximation in our calculations to follow. We will come back to this assumption at a later stage when it may need modification.

We then get the following integrals for the number density (N_A), energy density (ϵ_A), pressure (p_A) and entropy density (s_A) of particle A in thermal equilibrium:

$$N_A = \frac{g_A}{2\pi^2\hbar^3} \int_0^\infty \frac{P^2\, dP}{\exp[E_A(P)/(kT)] \pm 1}, \qquad (16.11)$$

$$\epsilon_A = \frac{g_A}{2\pi^2\hbar^3} \int_0^\infty \frac{P^2 E_A(P) dP}{\exp[E_A(P)/(kT)] \pm 1}, \qquad (16.12)$$

$$p_A = \frac{g_A}{6\pi^2\hbar^3} \int_0^\infty \frac{c^2 P^4 [E_A(P)]^{-1}\, dP}{\exp[E_A(P)/(kT)] \pm 1}, \qquad (16.13)$$

$$s_A = (p_A + \epsilon_A)T. \qquad (16.14)$$

We can deduce a simple relation from these formulae to show that the entropy in a given comoving volume is constant as the Universe expands. Differentiate p_A with respect to T to get

$$\frac{dp_A}{dT} = \frac{g_A}{6\pi^2\hbar^3} \int_0^\infty \frac{c^2 P^4 \exp[E_A(P)/(kT)] dP}{\{\exp[E_A(P)/(kT)] \pm 1\}^2 kT^2}.$$

Now integrate by parts to get for the above integral

$$\frac{g_A}{6\pi^2\hbar^3 T} \int_0^\infty \frac{[3P_A^2 E_A + c^2 P^4 E_A^{-1}]\mathrm{d}P}{\exp[E_A(P)/(kT)] \pm 1} = \frac{p_A + \epsilon_A}{T}.$$

On defining the pressure, energy density and entropy density for a mixture of such gases in thermodynamic equilibrium by

$$p = \sum_A p_A, \qquad \epsilon = \sum_A \epsilon_A, \qquad s = \sum_A s_A, \qquad (16.15)$$

we have the following relation:

$$\frac{\mathrm{d}p}{\mathrm{d}T} = \frac{p + \epsilon}{T}. \qquad (16.16)$$

We will apply this relation next in the expanding Universe. For we shall see that as the Universe expands it cools adiabatically.

We first recall the conservation law satisfied by ϵ and p in the early stages of the expanding Universe, the law given by (15.14),

$$\frac{\mathrm{d}}{\mathrm{d}S}(\epsilon S^3) + 3pS^2 = 0, \qquad (16.17)$$

and use it in conjuction with Equation (16.16). A simple exercise in calculus leads to the conclusion that the entropy in a given volume is constant:

$$\sigma = S^3 \left(\frac{p + \epsilon}{T}\right) = \text{constant.} \qquad (16.18)$$

In general, the above expressions become simplified for particles moving relativistically. In this case, the mean kinetic energy per particle far exceeds the rest-mass energy of the particle, an inequality expressed by

$$T \gg \frac{m_A c^2}{k} \equiv T_A. \qquad (16.19)$$

This is called the *high-temperature approximation*, or the *relativistic limit*.

The thermodynamic details for the various species of interest are given in Table 16.1. The numbers are expressed in units of the quantities for the photon ($g_A = 2$; the symbol for the photon is γ):

$$N_\gamma = \frac{2.404}{\pi^2}\left(\frac{kT}{c\hbar}\right)^3, \qquad \epsilon_\gamma = \frac{\pi^2 (kT)^4}{15\hbar^3 c^3} = 3p_\gamma, \qquad s_\gamma = \frac{4\pi^2 k}{45}\left(\frac{kT}{c\hbar}\right)^3.$$

$$(16.20)$$

Now consider a primordial mixture of bosons and fermions all moving relativistically. The effective energy-density–temperature relationship for this mixture will be

$$\epsilon = \frac{1}{2}gaT^4, \qquad (16.21)$$

Table 16.1. *Thermodynamic quantities for various particle species at* $T \gg T_A$

Particle species A	Symbol	T_A (K)	g_A	N_A/N_γ	$\epsilon_A/\epsilon_\gamma$	s_A/s_γ
Photon	γ	0	2	1	1	1
Electron	e^-	5.93×10^9	2	3/4	7/8	7/8
Positron	e^+		2	3/4	7/8	7/8
Muon	μ^-	1.22×10^{12}	2	3/4	7/8	7/8
Antimuon	μ^+		2	3/4	7/8	7/8
Muon and electron	ν_μ, ν_e	0	1	3/8	7/16	7/16
neutrinos and their	$\bar{\nu}_\mu, \bar{\nu}_e$		1	3/8	7/16	7/16
antineutrinos						
Pions	π^+		1	1/2	1/2	1/2
	π^-	1.6×10^{12}	1	1/2	1/2	1/2
	π^0		1	1/2	1/2	1/2
Proton	p	10^{13}	2	3/4	7/8	7/8
Neutron	n	$T_n - T_p$ $\sim 1.5 \times 10^{10}$	2	3/4	7/8	7/8

where the 'g' factor is related to the total bosonic internal degrees of freedom g_b and fermionic degrees of freedom g_f by

$$g = g_b + \frac{7}{8} g_f. \tag{16.22}$$

The reason becomes clear when we look at the last-but-one column of Table 16.1. The fermionic energy densities carry an extra factor 7/8.

In this approximation consider the electrical potential energy of any two electrons separated by distance r. This is given by

$$U = \frac{e^2}{r}.$$

Now the average inter-electron distance is given by $N_e^{-1/3} \sim c\hbar/(kT)$. Thus the average interaction energy is

$$\langle U \rangle \sim \frac{e^2}{\hbar c} kT.$$

However, kT measures the energy of motion of electrons. Thus the interaction energy is $e^2/(\hbar c) \sim 1/137$ of the energy of motion. Since the fraction is small, we are justified in treating the electrons as free gas.

In contrast, at *low temperatures* $T \leq T_A$ we have for all species with $m_A \neq 0$

$$N_A = \frac{g_A}{\hbar^3} \left(\frac{m_A kT}{2\pi} \right)^3 \exp\left(-\frac{T_A}{T} \right),$$

$$\epsilon_A = m_A N_A, \qquad p_A = N_A kT, \qquad s_A = \frac{m_A N_A}{T} c^2.$$

(16.23)

Notice that with a fall in temperature all these quantities drop off rapidly. We will often refer to this limit as the *non-relativistic approximation*. (For the photon and a zero-rest-mass neutrino $T_A = 0$ and this approximation never applies.)

When applying these results to cosmology, the following considerations usually count. First, the expansion of the Universe is controlled by the species that are in the relativistic limit, for these are the particles that are present in greater abundance. Heavier species are reduced in number because of the exponential damping term of Equation (16.23). Thus, as the temperature of the Universe drops with expansion, the heavier species progressively diminish in dynamical importance.

16.2.2 Decoupling of neutrinos

In general, our understanding of the early epochs of the Universe tells us that there are two processes going on at any given time: the expansion of the Universe with a characteristic rate given by the Hubble constant $H(t) = \dot{S}/S$ and some process involving the interaction of its particle species. If the latter is slower than the former, the process ceases to have any important role in determining the physical properties of the Universe. After the Universe has cooled through to a temperature of $\sim 10^{11}$ K, the first major event to occur because of such a reason is the decoupling of neutrinos.

Using the properties of the weak interactions of physics, one can show that the rate of interaction of neutrinos with leptons (electrons, positrons, muons, neutrinos etc.) is of the order

$$\eta = \mathcal{G}^2 \hbar^{-7} c^{-6} (kT)^5 \exp\left(-\frac{T_\mu}{T} \right).$$

(16.24)

Here $\mathcal{G}$ is the weak-interaction constant. We must now take note of the other rate mentioned earlier, that is relevant to the maintenance of equilibrium of neutrinos – the rate at which a typical volume enclosing them expands. From Einstein's equations we get

$$H^2 = \frac{\dot{S}^2}{S^2} = \frac{8\pi G}{3c^2} \epsilon \approx \frac{16\pi^3 G}{90\hbar^3 c^5} (kT)^4.$$

(16.25)

H, the Hubble constant at the particular epoch, measures the rate of expansion of the volume in question. Thus the ratio of the reaction rate to the expansion rate is given by

$$\frac{\eta}{H} \sim G^{-1/2}\hbar^{-11/2}\mathcal{G}^2 c^{-7/2}(kT)^3 \exp\left(-\frac{T_\mu}{T}\right) \tag{16.26}$$

$$\sim \left(\frac{T}{10^{10}\,\text{K}}\right)^3 \exp\left(-\frac{10^{12}\,\text{K}}{T}\right)$$

$$= T_{10}^3 \exp\left(-\frac{1}{T_{12}}\right). \tag{16.27}$$

Here we have substituted the values of G, $\hbar$, $\mathcal{G}$, c, k and T_μ and arrived at the above numerical expression. Further, we have written the temperatures using the compact notation that T_n indicates temperature expressed in units of 10^n K, i.e.,

$$T_n = \frac{T}{10^n\,\text{K}}.$$

What does (16.27) tell us? As the temperature drops below 10^{12} K, the exponential decreases rapidly. This means that the reactions involving neutrinos run at a slower rate than the expansion rate of the Universe. The neutrinos then cease to interact with the rest of the matter and therefore drop out of thermal equilibrium as temperatures fall appreciably below $T_{12} = 1$. How far below?

The original theory of weak interactions suggested that this temperature may be about $T_{11} = 1.3$. In the late 1960s and early 1970s successful attempts to unify the weak interaction with the electromagnetic interaction led to additional (neutral-current) reactions that keep neutrinos interacting with other matter at even lower temperatures. The outcome of these investigations is that the neutrinos can remain in thermal equilibrium down to temperatures of the order of $T_{10} = 1$.

However, even though neutrinos decouple themselves from the rest of the matter, their distribution function still retains its original form with the temperature dropping as $T \propto S^{-1}$. This is because as the Universe expands the momentum and energy of each neutrino fall as S^{-1} and the number density of neutrinos falls as S^{-3}. Since the temperature of the rest of the mixture also drops as S^{-1} and since the two temperatures were equal when the neutrinos were coupled with the rest of the matter, the two temperatures continue to remain equal even though neutrinos and the rest of the matter are no longer in interaction with one another. These remarks about neutrinos are meant to apply to all four species ν_e, $\bar{\nu}_e$, ν_μ and $\bar{\nu}_\mu$.

16.2.3 Electron–positron annihilation

There is, however, another (later) epoch when the neutrino temperature begins to differ from the temperature of the rest of the matter. First consider the Universe in the temperature range $T_{12} = 1$ to $T_{10} = 1$. In this phase we have the neutrinos, the electron–positron pairs and the photons, each with distribution functions in the high-temperature approximation (see Table 16.1). Thus, referring back to the formula (16.21), we get

$$\epsilon = \frac{9}{2}aT^4. \tag{16.28}$$

Thus in this period the expansion equation is modified from our simplified formula (16.4) (for photons only) to

$$\frac{\dot{S}^2}{S^2} = \frac{12\pi G a}{c^2}T^4 \tag{16.29}$$

and the relation (16.7) is changed to

$$T = \left(\frac{c^2}{48\pi G a}\right)^{1/4}t^{-1/2}, \tag{16.30}$$

which we may rewrite as

$$T_{10} = 1.04\, t_{\text{seconds}}^{-1/2}. \tag{16.31}$$

However, in the next phase the situation becomes complicated, because, with the cooling of the Universe, the electron–positron pairs are no longer relativistic. Thus the high-temperature approximation is no longer valid for them. As they slow down, they annihilate each other. We will not go into the details of this phase but instead jump across to its end, when the pairs have annihilated, leaving only photons (and possibly any excess electrons):

$$e^- + e^+ \rightarrow \gamma + \gamma$$

Thus the energy, originally in $e^\pm$ and photons, is now vested only in photons, raising their number and temperature. How can we evaluate this change? It is here that Equation (16.18), telling us of the constancy of σ, comes to our help.

In the relativistic phase ($T_9 > 5$) of $e^\pm$ we have

$$\sigma = \frac{4S^3}{3T}\left(\epsilon_{e^-} + \epsilon_{e^+} + \epsilon_\gamma\right) = \frac{11}{3}a(ST)^3. \tag{16.32}$$

When the $e^\pm$ have annihilated and left only photons, we have the photon temperature T_γ given by

$$\sigma = \frac{4}{3}\frac{S^3}{T_\gamma}\epsilon_\gamma = \frac{4}{3}a(ST_\gamma)^3. \tag{16.33}$$

We now use the result that the neutrino temperature always declines as S^{-1}. Let us write it as

$$T_\nu = \frac{B}{S}, \quad B = \text{constant}. \tag{16.34}$$

Then (16.32) gives

$$\sigma = \frac{11}{3}aB^3\left(\frac{T}{T_\nu}\right)^3. \tag{16.35}$$

Similarly (16.33) gives

$$\sigma = \frac{4}{3}aB^3\left(\frac{T_\gamma}{T_\nu}\right)^3. \tag{16.36}$$

Now, in the pre-annihilation era $T = T_\nu$, so that (16.35) tells us that $\sigma = (11/3)aB^3$. After annihilation σ must have the same value, so we may equate it to the value given by (16.36). Thus we arrive at the conclusion that the photon temperature at the end of $e^\pm$ annihilation has risen *above* the neutrino temperature by the factor

$$\frac{T_\gamma}{T_\nu} = \left(\frac{11}{4}\right)^{1/3} \cong 1.4. \tag{16.37}$$

So the present-day neutrino temperature is *lower* than the photon temperature by the factor $(1.4)^{-1}$. If we take the latter to be ~ 2.7 K, the former is ~ 1.9 K.

16.2.4 The neutron-to-proton number ratio

We have so far developed a picture of the early Universe that is best expressed in the form of a time–temperature table of events, as shown in Table 16.2 (see also Figure 16.2). We will now be interested in the last entry of Table 16.2.

In our discussion so far we have not paid much attention to baryons – the protons and neutrons that are also present in the mixture. In our approximation of setting the chemical potentials to zero we took the baryon number to be zero. The validity of the approximation depended on the baryon number density being several orders (8 to 10) of magnitude smaller than the photon density. Nevertheless, we must now take note of the existence of baryons, howsoever small their number density; for we need them in order to consider Gamow's idea of nucleosynthesis in the hot Universe. We also emphasize that the baryons at this stage of the Universe (when nucleosynthesis could occur) are not playing any significant role in determining the expansion of the Universe.

Insofar as chemical potentials are concerned, we will take explicit note of them in the following section. However, first notice that the critical temperatures T_n and T_p of Table 16.1 are very high, so the neutron and

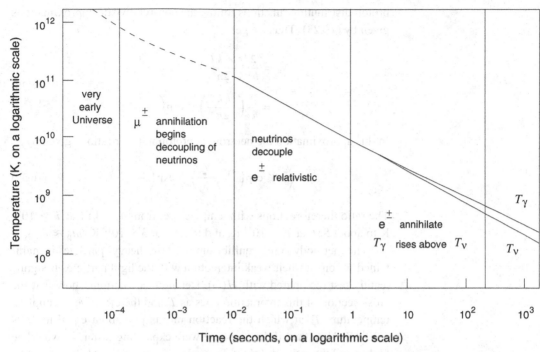

Fig. 16.2. The time–temperature relationship since the Universe was aged about 10^{-4} s, until it became $\sim 10^3$ s old. After the annihilation of $e^{\pm}$ pairs (when the Universe was 1–10 s old) the photon temperature went up above the neutrino temperature by a factor ~ 1.4. That is why the t–T curve splits into two parts.

Table 16.2. *A time–temperature table of events preceding nucleosynthesis in the early Universe*

Time since big bang (s)	Temperature (K)	Events
$\leq 10^{-4}$	$> 10^{12}$	Baryons, mesons, leptons and photons in thermal equilibrium.
10^{-4}–10^{-2}	10^{12}–10^{11}	μ^+ begin to annihilate and disappear from the mixture. Neutrinos begin to decouple from the rest of the matter.
10^{-2}–1	10^{11}–10^{10}	Neutrinos decouple completely. $e^{\pm}$ pairs still relativistic.
1–10	10^{10}–10^{9}	The $e^{\pm}$ pairs annihilate and disappear, raising the photon-gas temperature to ~ 1.4 times the temperature of neutrinos.
10–180	10^{9}–10^{8}	Nucleosynthesis takes place.

proton distribution functions follow the non-relativistic approximations given by (16.23). Thus we get

$$N_p = \frac{2}{\hbar^3} \left(\frac{m_p kT}{2\pi} \right)^{3/2} \exp\left(-\frac{T_p}{T} \right),$$

$$N_n = \frac{2}{\hbar^3} \left(\frac{m_n kT}{2\pi} \right)^{3/2} \exp\left(-\frac{T_n}{T} \right).$$

(16.38)

In this approximation the neutron-to-proton number ratio is given by

$$\frac{N_n}{N_p} \cong \exp\left(\frac{T_p - T_n}{T} \right) = \exp\left(-\frac{1.5}{T_{10}} \right).$$

(16.39)

The ratio therefore drops with temperature, from near $1:1$ at $T \geq 10^{12}$ K to about $5:6$ at $T = 10^{11}$ K, and to $3:5$ at 3×10^{10} K ($m_p \approx m_n$).

The thermodynamic equilibrium of these 'heavy' particles is maintained so long as their weak interaction with the light particles is significantly fast (compared with $H(t)$). Detailed calculations show that the cross section of this interaction goes as T and the effective decoupling temperature T_* at which the reaction rate is just about equal to H is $<10^{10}$ K. Note that, if the Universe were expanding faster, T_* would be higher and the ratio N_n/N_p at decoupling as given by (16.39) would be higher. We will recall this point when relating the helium abundance to the number of neutrino species.

Once the thermodynamic equilibrium ceases to be maintained, the N_n/N_p ratio is given not by (16.39) but by detailed consideration of specific reactions involving the nucleons.

Thus the ratio of neutrons to protons is uniquely determined at the time nucleosynthesis begins, once we know all the parameters of the weak interaction. This is one good aspect of primordial nucleosynthesis theory, which was first pointed out by Chushiro Hayashi in 1950 [60]. We now proceed to discuss its outcome.

16.3 The formation of light nuclei

The process of nucleosynthesis may be considered as a battle between high-speed nuclei flying about in all directions and the strong nuclear force of attraction trying to trap them and bind them together. Clearly, at the higher temperatures the former win, whereas at lower temperatures the latter prevails. This can be seen quantitatively in the following way.

A typical nucleus Q is described by two quantities, A, the atomic mass, and Z, the atomic number, and is written ^A_ZQ. This nucleus has Z protons and $(A - Z)$ neutrons. If m_Q is the mass of the nucleus, its binding energy is given by

$$B_Q = [Zm_p + (A - Z)m_n - m_Q]c^2. \tag{16.40}$$

Let us now consider a unit volume of cosmological medium containing N_N nucleons, bound or free. Since the masses of protons and neutrons are nearly equal, we may denote the typical nucleon mass by m. Thus $m_n \approx m_p = m$. If there are N_n free neutrons and N_p free protons in the mixture, the ratios

$$X_n = \frac{N_n}{N_N}, \qquad X_p = \frac{N_p}{N_N} \tag{16.41}$$

will denote the fractions by mass of free neutrons and free protons. If a typical bound nucleus Q has atomic mass A and there are N_Q of them in our unit volume, we may similarly denote the mass fraction of Q by

$$X_Q = \frac{N_Q A}{N_N}. \tag{16.42}$$

Now, at very high temperatures $(T \gg 10^{10}$ K$)$, the nuclei are expected to be in thermal equilibrium. However, because of their relatively large masses, even at these tempereatures $T \ll T_Q$ and the non-relativistic approximation holds. Further, since we are now concerned with relative number densities, we can no longer ignore the chemical potentials. Thus we have

$$N_Q = g_Q \left(\frac{m_Q kT}{2\pi \hbar^2}\right)^{3/2} \exp\left(\frac{\mu_Q - m_Q c^2}{kT}\right), \tag{16.43}$$

where we have reinstated the chemical potentials μ_Q. Since chemical potentials are conserved in nuclear reactions,

$$\mu_Q = Z\mu_p + (A - Z)\mu_n, \tag{16.44}$$

assuming that the nuclei were built out of neutrons and protons by nuclear reactions.

Using Equation (16.44), the unknown chemical potentials can be eliminated between (16.43) and similar relations for N_p and N_n. The result is expressed in this form:

$$X_Q = \frac{1}{2} g_Q A^{5/2} X_p^Z X_n^{A-Z} \xi^{A-1} \exp\left(\frac{B_Q}{kT}\right), \tag{16.45}$$

where

$$\xi = \frac{1}{2} N_N \left(\frac{mkT}{2\pi \hbar^2}\right)^{-3/2}. \tag{16.46}$$

For an appreciable build-up of complex nuclei, T must drop to a low enough value to make $\exp[B_Q/(kT)]$ large enough to compensate for the smallness of ξ^{A-1}. This happens for nucleus Q when T has dropped down to

$$T_Q \sim \frac{B_Q}{k(A-1)|\ln \xi|}. \tag{16.47}$$

Let us consider what happens when we apply the above formula to the nucleus of ^{4}He. This nucleus is made by fusion of four nucleons, i.e., it requires a four-body encounter. The binding energy of this nucleus is given approximately by 4.3×10^{-5} erg. If we substitute this value into (16.47) and estimate N_N from the currently observed value of nucleon density of about 10^{-6} cm^{-3}, we find that T_Q is as low as $\sim 3 \times 10^9$ K. However, at this low temperature the number densities of participating nucleons are so low that four-body encounters leading to the formation of ^{4}He are extremely rare. Thus we need to proceed in a less ambitious fashion in order to describe the build-up of complex nuclei.

Hence we try using two-body collisions (which are not so rare) to describe the build-up of heavier nuclei. Thus deuterium (^{2}H), tritium (^{3}H) and helium (^{3}He, ^{4}He) are built up in a sequence via reactions like

$$p + n \leftrightarrow {}^2H + \gamma,$$

$$^2H + {}^2H \leftrightarrow {}^3He + n \leftrightarrow {}^3H + p, \tag{16.48}$$

$$^3H + {}^2H \leftrightarrow {}^4He + n.$$

Since formation of deuterium involves only two-body collisions, it quickly reaches its equilibrium abundance as given by

$$X_d = \frac{3}{\sqrt{2}} X_p X_n \xi \exp\left(\frac{B_d}{kT}\right). \tag{16.49}$$

However, the binding energy B_d of deuterium is low, so, unless T drops to less than 10^9 K, X_d is not high enough to start further reactions leading to ^{3}H, ^{3}He and ^{4}He. In fact the reactions given in (16.48), with the exception of the first one, do not proceed fast enough until the temperature has dropped to $\sim 8 \times 10^8$ K.

Although at such temperatures nucleosynthesis does proceed rapidly enough, it cannot go beyond ^{4}He. This is because there are no stable nuclei with $A = 5$ or 8, and nuclei heavier than ^{4}He break up as soon as they are made. Their primordial abundances are extremely small. So the process effectively terminates there. Detailed calculations by several authors have now established this result quite firmly.

So, starting with primordial neutrons and protons, we end up finally with ^{4}He nuclei and free protons. All neutrons have been gobbled up by helium nuclei. Thus, if we consider the fraction by mass of primordial helium, it is very simply related to the quantity X_n – the neutron concentration before nucleosynthesis began. Denoting the helium fraction by mass by the symbol Y, we get

$$Y = 2X_n. \tag{16.50}$$

In Figure 16.3 the cosmic mass fractions of ^{4}He, ^{3}He, ^{2}H and other light nuclei are plotted against a parameter η defined by

$$\eta = \left(\frac{\rho_0}{1.97 \times 10^{-26} \text{g cm}^{-3}} \right) \left(\frac{2.7}{T_0} \right)^3. \tag{16.51}$$

Thus η essentially measures the nucleon density in the early Universe through the formula

$$\rho = \eta T_9^3, \qquad T_9 < 3. \tag{16.52}$$

We will shortly discuss the implications of this parameter for primordial production of light nuclei.

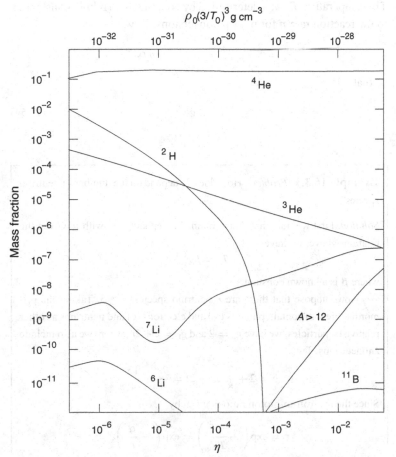

Fig. 16.3. Mass fractions of light nuclei produced in the early Universe for various values of the parameter η related to baryon density. The production of deuterium drops steeply as η approaches a value in excess of 10^{-4}.

16.3.1 Helium abundance and the number of neutrino species

Note that the ^{4}He mass fraction is insensitive to the parameter η. This is because, as we saw just now, it depended only on X_n, which in turn depends more critically on the epoch when the rate of weak interactions fell below the expansion rate. If we go back to (16.39), we see that in the very early stages the neutron-to-proton ratio was determined by the decoupling temperature T_*. A faster expansion rate implies that the ratio became frozen at a higher temperature and so was higher, thus leading to a higher ^{4}He abundance.

To see the effect quantitatively, recall from (16.39) that there was a 'last epoch' of temperature T_* when the neutron-to-proton ratio was determined from considerations of thermodynamic equilibrium:

$$x = \frac{N_n}{N_p} = \exp\left(-\frac{1.5}{T_{*10}}\right). \tag{16.53}$$

The temperature T_* was determined by equating the Hubble constant H to the reaction rate η for n $\leftrightarrow$ p conversions. Now

$$H \propto g^{1/2} T_*^2 \quad \text{and} \quad \eta \propto T_*^4$$

so that

$$T_*^2 \propto g^{1/2}. \tag{16.54}$$

Example 16.3.1 *Problem.* How does T_* depend on the number of neutrino species?

Solution. First we note that T_* is obtained by equating H with η. Given the relations above, we have

$$T_* = \beta g^{1/4},$$

where β is a known constant.

Now suppose that there are r neutrino species, $r \geq 3$. Taking the primordial brew to contain photons (γ) and electrons (e) and neutrinos as other relativistic particles, we have $g_b = 2$ and $g_f = 2 + 2r$, where we also include antineutrinos. Thus

$$g = 2 + \frac{7}{8}(2r + 2) = \frac{7r}{4} + \frac{15}{4}.$$

Since the neutron-to-proton ratio is given by

$$x = \exp\left(-\frac{1.5}{T_{*10}}\right) = \exp\left(-\frac{\alpha}{T_*}\right),$$

say, with $\alpha = 1.5 \times 10^{10}$, the above relations lead to

$$x = \exp\left[-\frac{\alpha}{\beta}\left(\frac{7r+15}{4}\right)^{-1/4}\right].$$

For $r = 3$ we have $x = 1/7$, corresponding to $Y = 1/4$. Numerical calculations for $r = 4$ and $r = 5$ lead to $Y = 0.27$ and 0.29, respectively. Since most estimates of primoridal helium put $Y \lesssim 0.25$, such higher values are ruled out.

The calculation shown above tells us that an increase in the number of neutrino species would result in an increase of Y.

This result is relevant to the question of how many different types of neutrinos exist primordially. The formalisms used by particle physicists allow for three or more neutrino types, ν_e, ν_μ and ν_τ. Having more types of neutrinos existing forces the value of Y upwards. When we look at observations, we discover that the present estimates of helium abundance, $Y \leq 0.25$, rule out the existence of more than three neutrino types.

It is also interesting that the result from particle-accelerator experiments appear to lead to the same conclusion. A series of experiments carried out in 1990 with the large electron–positron collider (LEP) at CERN produced the intermediate Z^0 boson [61] in large numbers. The presence of these particles (which mediate in electro-weak interactions) could be inferred by detecting resonance peaks in the energy-dependent cross sections for producing hadrons and leptons. The width of the peak measures the lifetime of the Z^0 boson, and this in turn can be linked to the number of neutrino species present. The estimate is very close to 3, which is consistent with the above cosmological considerations. This circumstance is considered a notable success of the enterprise of bringing together cosmology and particle physics.

16.3.2 Deuterium abundance and non-baryonic dark matter

In contrast to the behaviour of Y, which does not sensitively depend on the parameter η, the abundances of other light nuclei do depend on η. These abundances are very small compared with Y. The most interesting situation exists for deuterium, whose abundance sharply drops as η rises above 10^{-4} (see Figure 16.3). The present estimate of the deuterium mass fraction is $\sim 2 \times 10^{-5}$. From Figure 16.3, we have $\eta \sim 2 \times 10^{-5}$ to understand the deuterium abundance. For $T_0 = 2.7\,\text{K}$, this value of η corresponds to a present nucleonic density of

$$\rho_0 \sim 4 \times 10^{-31}\,\text{g cm}^{-3}. \tag{16.55}$$

On comparing this with (15.32) and (15.40), we see that $h_0^2 \Omega_0 \lesssim 0.02$ and hence $q_0 \lesssim 0.01$. Therefore, if even such a small amount of deuterium believed to be primordial in origin were found, Friedmann models of the closed variety would be ruled out. There is, however, a loophole in this argument: we can still accommodate non-baryonic matter in the Universe. Such matter does not affect the deuterium abundance, but contributes to Ω_0. Matter of this kind will have to be dark.

To summarize, the process of primordial nucleosynthesis delivers the right abundance of helium, if the parameter η is properly adjusted, and of deuterium, if one allows for a substantial presence of non-baryonic dark matter in the Universe. Apart from marginal production of other light nuclei up to lithium, the pimordial process fails for the production of *all other nuclei*. For these one has to invoke stellar nucleosynthesis, which does produce the right amounts of these remaining (over 200) isotopes [62]. One may be tempted to invoke Occam's razor, and argue that *all* isotopes were produced inside stars. Such an attempt, however, has not succeeded so far: one gets an inadequate quantity of helium in this way, and no deuterium at all. Interestingly, these two failure points of stellar nucleosynthesis are precisely those where the primordial process succeeds.

16.4 The microwave background

The era of nucleosynthesis took place when the temperature was about 10^9 K. The Universe in subsequent phases continued to cool as it expanded, with the radiation temperature dropping as S^{-1}. The presence of nuclei, free protons and electrons did not have much effect on the dynamics of the Universe, which was still radiation-dominated. However, these particles, especially the lightest of them, the electrons, acted as scattering centres for the ambient radiation and kept it thermalized. The Universe was therefore quite opaque to start with.

However, as the Universe cooled, the electron–proton electrical attraction began to assert itself. In detailed calculations performed by P. J. E. Peebles [63], the mixture of electrons and protons and of hydrogen atoms was studied at varying temperatures. Because of Coulomb attraction between the electron and the proton, the hydrogen atom has a certain binding energy B. The problem of determining the relative number densities of free electrons, free protons (that is, ions) and neutral H atoms in thermal equilibrium is therefore analogous to that we considered earlier in deriving (16.45) for the mixture of free and bound nucleons. The only difference is that the binding to be considered now is electrostatic rather than nuclear. Following the same method, we arrive at the formula relating the number densities of electrons (N_e), protons ($N_p = N_e$), and H atoms (N_H) at a given temperature T:

$$\frac{N_e^2}{N_H} = \left(\frac{m_e kT}{2\pi \hbar^2}\right)^{3/2} \exp\left(-\frac{B}{kT}\right), \tag{16.56}$$

where m_e is the electron mass. This equation is a particular case of *Saha's ionization equation*. In about 1920 Meghnad Saha had looked at the problem of ionization in the context of stellar atmospheres and had derived this equation [64].

Writing N_B for the total baryon number density, we may express the fraction of ionization by the ratio

$$x = \frac{N_e}{N_B}. \tag{16.57}$$

Then, since $N_H = N_B - N_e$, we get from (16.56)

$$\frac{x^2}{1-x} = \frac{1}{N_B}\left(\frac{m_e kT}{2\pi \hbar^2}\right)^{3/2} \exp\left(-\frac{B}{kT}\right). \tag{16.58}$$

For the H atom, $B = 13.59$ eV. By substituting for various quantities on the right-hand side of (16.58), we can solve for x as a function of T. The results show that x drops sharply from 1 to near zero in the temperature range of $\sim$5000 K to 2500 K, depending on the value of N_B, that is, on the parameter $\Omega_0 h_0^2$. For example, for $\Omega_0 h_0^2 = 0.01$, $x = 0.003$ at $T = 3000$ K.

Thus by this time most of the free electrons have been removed from the cosmological brew, and as a result the main agent responsible for the scattering of radiation disappears from the scene. The Universe becomes effectively transparent to radiation. This epoch is often called the *recombination epoch*, although the word 'recombination' is inappropriate since the electrons and protons are combining for the first time at this epoch. It is more appropriate to call it the *epoch of last scattering*.

The transparency of the Universe means that a light photon can go a long way ($\sim c/H$) without being absorbed or scattered. Therefore this epoch signifies the beginning of the new phase when matter and radiation became decoupled. This phase has lasted up to the present epoch. During this phase, the frequency of each photon is redshifted according to the rule

$$\nu \propto \frac{1}{S} \tag{16.59}$$

while the number density of photons has fallen as

$$N_\gamma \propto \frac{1}{S^3}. \tag{16.60}$$

It is easy to see that under these conditions the photon distribution function preserves the Planckian form with the temperature dropping as

$$T \propto \frac{1}{S}. \tag{16.61}$$

Does such a Planckian background exist in the Universe today? In 1942 A. McKellar reported that the populated upper levels of the CN molecule in interstellar space led to the conclusion that there was a radiation background of ~2.3 K. This result [65] came during the Second World War when the normal channels of communication between scientists were closed. The result therefore went largely unnoticed. In 1948, Ralph Alpher and Robert Herman, junior colleagues of George Gamow, made the prediction that, since the hot Universe had cooled down, a blackbody radiation background of temperature about 5 K should exist now [66]. They made a guess of the present temperature, since it was not possible to tie down the radiation temperature to the present epoch from the physics of the early Universe. Since cosmology was considered a highly speculative field by the physicists, this important prediction was largely ignored.

This radiation background was subsequently found by Arno Penzias and Robert Wilson, more or less serendipitously [67]. They had planned using a 20-foot horn-shaped reflector antenna to study radiation in the microwave range in the Milky Way. While testing the antenna, they pointed it in various directions and used the wavelength 7.35 cm because it did not attract much Galactic noise. These test measurements contained an unaccounted-for component that was isotropic, i.e., one that could not be ascribed to any specific Galactic or extragalactic source. It was only when they compared notes with the Princeton group that they could identify this radiation with the relic background. For, by 1964, Jim Peebles and Robert Dicke at Princeton had walked along the trodden path to arrive at the same conclusion as Alpher and Herman. To measure the predicted radiation background, Dicke was in fact building a suitable antenna in collaboration with his colleague David Wilkinson.

The Penzias–Wilson measurement at one wavelength, if interpreted as blackbody radiation, gave the temperature 3.5 K. In Figure 16.4 we show the spectrum of the radiation as measured by the Cosmic Background Explorer (COBE) satellite in 1990, with a temperature of 2.735 ± 0.06 K [68]. Besides, the background is extremely homogeneous and isotropic, far more so than the observed distribution of matter in the Universe. The blackbody nature of the intensity–frequency curve has gone a long way towards confirming in most cosmologists' minds the validity of the early hot-Universe picture discussed by Gamow.

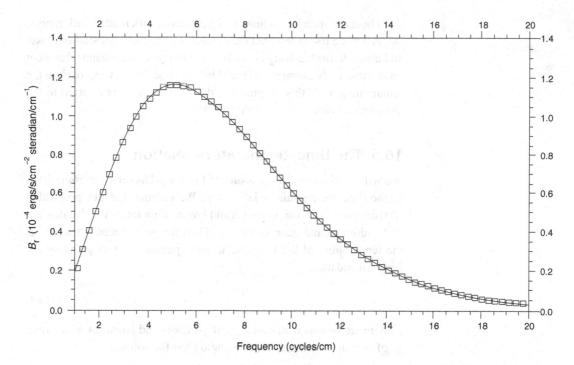

Fig. 16.4. The precise Planckian spectrum for the microwave background obtained by the COBE satellite. The curve shown in the figure corresponds to that of a black body of temperature 2.735 ± 0.06 K.

In the next chapter we will look at some of the observations in cosmology and return to this topic: for the microwave background is currently the most important evidence in favour of the hot-big-bang origin of the Universe. One aspect of the radiation that is still not explained is its present temperature of 2.7 K. This is sometimes stated in the form of the observed ratio of photons to baryons:

$$\frac{N_\gamma}{N_B} = 3.33 \times 10^7 \left(\Omega_0 h_0^2 \right)^{-1} \left(\frac{T_0}{2.7} \right)^3. \qquad (16.62)$$

This ratio has been conserved since the time at which the Universe became essentially transparent, although both N_γ and N_B can be studied theoretically at even earlier epochs. Why the above ratio and no other? Many physicists feel that deeper ideas from particle physics are needed to throw light on this mystery.

There are other fundamental issues to tackle too. Some are stated below.

1. Did the big bang really happen?
2. Why does the Universe exhibit an apparent excess of matter over antimatter?
3. Prior to the neutrons, protons, etc. assumed to be present at primordial nucleosynthesis, what existed in the Universe?
4. How did the large-scale structure of galaxies and clusters evolve from tiny seeds?

There are other issues linked with inflation, dark matter, dark energy, etc. All these issues tempted the cosmologist to push his studies closer and closer to the big-bang epoch in order to try to understand what went on in those early moments. We will briefly describe this approach in the remaining part of this chapter. For details the reader is referred to the companion volume on cosmology [53].

16.5 The time–temperature relation

We will continue our discussions of the early Universe by going back to the time–temperature relationship. We assume that our primordial mixture contained both fermions and bosons with total effective degrees of freedom g_f and g_b, respectively. Then we get the result that relates the temperature of the Universe to its expansion rate as given by the Einstein equation,

$$\frac{\dot{S}^2}{S^2} = \frac{8\pi G}{3}\rho.$$

(16.63)

If there are bosons with a total g_b of g-factors and fermions with a total g_f of g-factors, then the above equation has the solution

$$\rho c^2 = \frac{1}{2}gaT^4$$

(16.64)

with

$$g = g_b + \frac{7}{8}g_f.$$

(16.65)

Thus we have for $g = $ constant

$$S \propto t^{1/2}$$

(16.66)

with

$$t = \left(\frac{3c^2}{16\pi Ga}\right)^{1/2} g^{-1/2}T^{-2}.$$

(16.67)

Here t is the time since the big bang. This relation can be expressed as

$$t_{\text{second}} = 2.4g^{-1/2}T_{\text{MeV}}^{-2} = 2.4 \times 10^{-6}g^{-1/2}T_{\text{GeV}}^{-2},$$

(16.68)

where we have used suitable conversion factors to write the temperature in MeV/GeV units.

This equation gives us at a glance the average particle energy at any given time – the earlier the epoch, the higher the energy. In short, by going closer and closer to the big-bang epoch we get ever higher particle energies.

This circumstance has prompted particle physicists with the idea that collaboration with big-bang cosmologists will be a good venture.

For in particle physics there is a general expectation that at sufficiently high energies all basic physical interactions will be unified under the banner of one master reaction. The unification of the electromagnetic with the weak interaction in the late 1970s showed that the energies required for unification were of the order of 100 GeV. The next step of 'grand unification' of the electroweak theory with the strong interaction appears to require particle energies as high as 10^{15} GeV. Now the particle accelerators at CERN and Fermilab do not go beyond particle energies of the order of 1000 GeV. So it is not really possible to check experimentally the claims of these grand unified theories (GUTs). However, if these theories are considered to apply to the *very early* Universe, then we can identify epochs when that happened. For example, setting $T_{GeV} = 10^{15}$ in Equation (16.68) gives, for $g \cong 100$, $t \cong 2.4 \times 10^{-37}$ s. It is arguable whether any physical meaning can be attached to such a short time scale. But if we do not worry about such operational issues, then in the very early Universe we do have a natural particle accelerator capable of reaching the GUT energies. From the realization that there is much to be gained both by cosmologists and by high-energy particle physicists from collaboration, the subject of *astroparticle physics* was created.

16.6 Some conceptual problems

Going from the present-day cosmological scales to those prevailing in the very early Universe raises some conceptual difficulties, which we highlight first.

16.6.1 The horizon problem

Let us suppose that the initial conditions for the Universe were set fairly early on, at an epoch t in the radiation-dominated phase. From the considerations of Chapter 15 adapted to the scale factor $S \propto t^{1/2}$, we find that the proper radius of the particle horizon at that epoch was

$$R_P = 2ct. \tag{16.69}$$

Whatever physical processes operated at this epoch were limited in range by R_P. Hence we do not expect the homogeneity of physical quantities to extend beyond the diameter $2R_P$, unless we make the somewhat contrived assumption that the Universe was *created* homogeneous. In other words, the causal limitations tell us that no region larger than $2R_P$ in size should be homogeneous.

When the initial conditions were so set, the Universe would expand from them to a much larger size at the present epoch, the factor η by which it would grow being the ratio of scale factors

$$\eta = \frac{S(t_0)}{S(t)}$$

at the present and initial epochs. How do we estimate η?

The simplest method is to compare the temperatures at t and t_0, since $S \propto T^{-1}$. Thus

$$\eta = \frac{T(t)}{T(t_0)},$$

where $T(t)$ is given by (16.67). It is convenient to express T_0 also in GeV:

$$T_0 \,(\text{GeV}) = 2.3 \times 10^{-13} \left(\frac{T_0}{2.7\,\text{K}} \right). \tag{16.70}$$

On combining (16.67) and (16.70) we get the present limit on a homogeneous region as

$$R_{\text{Hom}}(t_0) = 2ct$$

$$= 6.2 \times 10^{17} \times T_{\text{GeV}}^{-1} g^{-1/2} \times \left(\frac{2.7\,\text{K}}{T_0} \right) \text{cm}. \tag{16.71}$$

For $T_{\text{GeV}} \cong 10^{15}$, $g \cong 100$ and $T_0 \cong 2.7\,\text{K}$ we get the surprisingly small value of 62 cm! In other words, we have no reason to expect homogeneity on a scale larger than, say, 1 m. The fact that the relic microwave background is homogeneous on the cosmological scale of $\sim 10^{28}$ cm tells us that there is something seriously wrong with our reasoning above. Yet, the standard model does not provide any loophole out of this so-called *horizon problem*. Notice also that the further we go back in the past (in our attempts to set the initial conditions) the larger will T_{GeV} be and the smaller will the value of $R_{\text{Hom}}(t_0)$ be. Figure 16.5 illustrates the horizon problem.

16.6.2 The flatness problem

When discussing the early and the very early Universe we ignored the kc^2/S^2 term in the field equations. Thus (16.63) should actually have been

$$\frac{\dot{S}^2}{S^2} + \frac{kc^2}{S^2} = \frac{8\pi G\rho}{3}. \tag{16.72}$$

Our justification in ignoring that term was that, as $S \to 0$, $\dot{S}^2 \to \infty$ and, thus, the first term far exceeds the second term on the left-hand side of (16.72). This argument is, however, *scale-dependent*. Thus, if we write $S = At^{1/2}$, then $\dot{S}^2 = A^2/(4t)$. Whether $\dot{S}^2$ exceeds c^2 for $k = \pm 1$ would depend on A. A priori we do not know A, unless we link it with

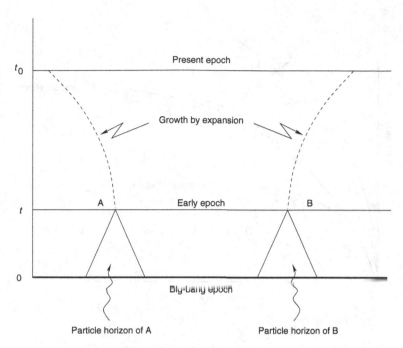

t_0 — Present epoch

Growth by expansion

t — A   Early epoch   B

0 — Big-bang epoch

Particle horizon of A   Particle horizon of B

Fig. 16.5. A and B are two typical observers of the very early Universe far enough apart that their particle horizons (shown by the past light cones) do not overlap. If the homogenization process took place very early in the Universe, then for causal limitation it will be locally limited to the respective light cones of A and B. How then could A and B achieve the same physical conditions around them today? If these horizons did limit the global homogenization process, then today A and B cannot be further apart than ~60 cm. Why then do we find the Universe homogeneous on a scale ~10^{28} cm?

the present size of the Universe. It is more convenient to look at the density parameter Ω instead.

Writing $\rho = \Omega\rho_c$ as in (15.40), we have, at any general epoch when $S \propto t^{1/2}$,

$$\frac{kc^2}{S^2} = (\Omega - 1)\frac{\dot{S}^2}{S^2} = \frac{\Omega - 1}{4t^2}. \qquad (16.73)$$

For the present epoch, on the other hand,

$$\frac{kc^2}{S_0^2} = (\Omega_0 - 1)H_0^2. \qquad (16.74)$$

On dividing (16.73) by (16.74) and using $S \propto T^{-1}$ we get, for $k = \pm 1$,

$$\Omega - 1 = (\Omega_0 - 1) \cdot 4H_0^2 t^2 \cdot \frac{T^2}{T_0^2}.$$

Except for $(\Omega_0 - 1)$ all quantities on the right-hand side are known. Using (16.68) for t and (16.70) for T_0, we get

$$(\Omega - 1) \cong 4.3h_0^2 g^{-1} \times 10^{-21} T_{\mathrm{GeV}}^{-2} \left(\frac{2.7\,\mathrm{K}}{T_0}\right)^2 (\Omega_0 - 1). \qquad (16.75)$$

For $T_{\mathrm{GeV}} = 10^{15}$ and $g \cong 100$, we get for $T_0 \cong 2.7\,\mathrm{K}$

$$\Omega - 1 \cong 4.3h_0^2 \times 10^{-53}(\Omega_0 - 1). \qquad (16.76)$$

Fig. 16.6. The flatness problem described by the curves in the figure shows that, unless the model was initiated at the GUT epoch within the shaded region, today it would not exhibit a matter density comparable to what is observed. In short, it had to be very finely tuned around $\Omega = 1$, to be consistent with modern studies; tuned to the extent of 1 part in 10^{53}.

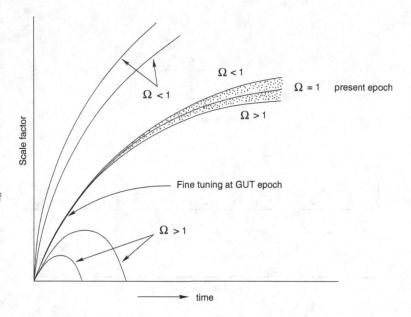

This expression epitomizes what has come to be known as the *flatness problem*. Suppose that the initial conditions including the density parameter Ω were set at the GUT epoch when $T \cong 10^{15}$ GeV. Then the present value of $(\Omega_0 - 1)$ is given by (16.76). Or, to invert the chain of reasoning, suppose that the present observational uncertainty tells us that $|\Omega_0 - 1| \sim \mathcal{O}(1)$. Then, from (16.76), at the GUT epoch Ω differed from unity by a fraction of the order of 10^{-53}. In other words, the departure from the flat value of Ω (=1) at this stage had to be extremely small. Any relaxation of this fine tuning would have led to a far wider range of Ω_0 at present than is permitted by observations.

So our neglect of the curvature term kc^2/S^2 is linked with an extremely fine tuning of the Universe to the flat ($k = 0$) model. If this tuning were not there, the Universe would either have gone into a collapse ($k = 1$) or expanded to infinity ($k = -1$) on time scales of the order of 10^{-35} s that were characteristic of the GUT era.

Figure 16.6 illustrates this conundrum. The shaded region denotes the finely tuned set of Friedmann models that end up today within the observed range $|\Omega_0 - 1| \sim \mathcal{O}(1)$. The curves shown outside this region are the characteristic models with time scales $\sim 10^{-35}$ s which should normally have operated at the GUT stage. What made the Universe get into the shaded region instead?

This problem was first highlighted by R. H. Dicke and P. J. E. Peebles in 1979, who discussed it not at the GUT epoch but at $t \sim 1$ s when the neutrinos had decoupled and pair ($e^{\pm}$) annihilation was to begin. Thus

$T \sim 10^{-3}$ GeV, $g \sim 10$, and we get $\sim 10^{-16}$ instead of 10^{-53} in (16.76). It is clear that the further back in time and closer to $t = 0$ we go the finer is the tuning required. For example, if we were to initialize the problem at the Planck epoch, we would get 10^{-61} for the tuning range instead of 10^{-53}.

16.6.3 The entropy problem

This is a restatement of the flatness problem and the horizon problem in a somewhat different form. The entropy in a given comoving volume stays constant in an adiabatic expansion (see Section 16.2). The present photonic entropy in the observable Universe of characteristic size $R \approx h_0^{-1} \cdot 10^{28}$ cm is given by

$$\Sigma = \frac{4\pi}{3k} a T_0^3 R^3 \approx h_0^{-3} \times 4.4 \times 10^{87} \left(\frac{T_0}{2.7} \right)^3. \tag{16.77}$$

Why such a large value? If the entropy were conserved, we would have $ST = $ constant. However, we found that in the flatness problem this hypothesis led to fine tuning, whereas for the horizon problem it gave an extremely small size of homogeneity. It therefore appears that the trouble lies in $\Sigma = $ constant: it could be resolved if the adiabatic assumption were violated at some stage and Σ boosted to its present value by an enormously large factor.

16.6.4 The monopole problem

In a grand unified theory, whenever there is a breakdown of symmetry of a larger group like SU(5) to a subgroup like SU(3) $\times$ SU(2)$_L$ $\times$ U(1) that contains the U(1) group, there inevitably arise particles that have the characteristics of a magnetic monopole. This is a rigorous mathematical conclusion in gauge field theories. Typically the mass of the monopole (in energy units) is given by $\sim 10^{16}$ GeV. Monopoles are highly stable particles and once created they are not destructible, so they would survive as relics to the present epoch.

At the GUT epoch t, the horizon size being $2ct$, we expect at least one monopole per horizon-size sphere, i.e., a monopole mass density of

$$\frac{10^{16}\ \text{GeV}/c^2}{(4\pi/3)(2ct)^3}.$$

At present this is diluted by the factor $(T_0/T)^3$. For T_0 in GeV units, given by (16.70) and $T = 10^{15}$ GeV, we get the present monopole density as

$$\rho_M \cong 3 \times 10^{-13} \left(\frac{T_0}{2.7\,\text{K}}\right)^3 \text{g\,cm}^{-3}. \tag{16.78}$$

This is far in excess of the closure density $\sim 10^{-29}$ g cm^{-3}, thus making it a very awkard problem for the standard model to solve. Again, as in the earlier cases, the discrepancy grows if, instead of the GUT epoch, we use an even earlier epoch.

16.7 Inflation

The extrapolation to such short time scales has thus brought its problems. Four such problems have just been described. These problems are addressed by the notion of 'inflation', a brainchild of three cosmologists, D. Kazanas, A. Guth and K. Sato, who independently arrived at it while working on astroparticle physics [69, 70, 71]. The scenario of inflation has evolved a lot since its inception in 1980–81. Even today there is no unique commonly accepted fully worked-out model of inflation. Yet its consequences for the big-bang cosmology are attractive enough for most workers to accept on trust a half-baked idea.

When the Universe cools down through the GUT epoch, a phase transition occurs when the single system described by grand unified theory splits into the electroweak and strong interaction. This phase transition can release a lot of energy, which is dumped into the dynamics of the Universe.

An analogy will be in order to illustrate the scenario. Suppose steam is being cooled through the phase-transition temperature of 100 °C. Normally we expect the steam to condense to water at this temperature. However, it is possible to supercool the steam to temperatures below 100 °C, although it is then in an unstable state. The instability sets in when certain parts of the steam condense to droplets of water, which then coalesce, and eventually the condensation goes to completion. In the supercooled state the steam still retains its latent heat, which is released as the droplets form.

In the case of the Universe the 'vacuum' state is identified with the state of lowest energy. However, the meaning of lowest energy changes as the phase transition occurs. The water–steam analogy tells us that the 'true' state of lower energy is that of water and the supercooled steam identifies a 'false' state that has higher energy. When the steam condenses, the false state changes to the true state. Likewise, in the case of the Universe, the phase transition may be delayed, leading to the existence of a false vacuum. When the transition is complete the true vacuum state is attained. In the original Guth version (which had fatal flaws) the switchover to true vacuum was through quantum tunnelling.

Whichever the mechanism of transition from false to true vacuum (and that is specified by a potential function) the dumped energy leads to a rapid expansion of the Universe. Denoting the extra energy density by ϵ_0, we find that it must have dynamical effects via the Einstein equation:

$$\frac{\dot{S}^2 + kc^2}{S^2} = \frac{8\pi G}{3c^2}(\epsilon_0 + \epsilon_r). \tag{16.79}$$

Here $\epsilon_r \propto 1/S^4$ is the energy density of radiation and relativistic particles. Since ϵ_r falls as the Universe expands while ϵ_0 stays constant, the latter clearly dominates. Hence we ignore ϵ_r and solve (16.79). For $k = +1$ we get, for example,

$$S = \left(\frac{3c^4}{8\pi G\epsilon_0}\right)^{1/2} \cosh\left[\left(\frac{8\pi G\epsilon_0}{3c^2}\right)^{1/2} t\right]. \tag{16.80}$$

For $k = -1$ we get a similar expression with 'cosh' replaced by 'sinh'. The main point to note is that for

$$t \gg \left(\frac{3c^2}{8\pi G\epsilon_0}\right)^{1/2} \tag{16.81}$$

either solution approaches closely the $k = 0$ (flat) solution

$$S \propto \exp(at), \quad a = \left(\frac{8\pi G\epsilon_0}{3c^2}\right)^{1/2}. \tag{16.82}$$

This exponential expansion is reminiscent of the de Sitter model. Indeed, the energy tensor of false vacuum simulates the λg_{ik} term of the Einstein equations.

This rapid expansion in an exponential fashion continues until (in the original Guth version) the tunnelling takes place and ϕ attains its true vacuum value. The average time τ for the tunnelling to occur can be computed quantum mechanically. It tells us the factor Z by which the scale factor S increased while inflation lasted. One finds that

$$a\tau \approx 67, \quad Z = \exp(a\tau) \approx 10^{29}. \tag{16.83}$$

In other words, the exponential expansion or *inflation* lasts long enough for the scale factor to blow up by a large multiple $\sim 10^{29}$. Thus if we had started with a curvature term (kc^2/S^2) comparable to the expansion term $(\dot{S}^2/S^2)$ prior to inflation we would end up by having the former reduced by $Z^2 \sim 10^{58}$ while the latter stays constant. This large factor Z not only takes care of the fine tuning in the flatness problem but also resolves the horizon problem (by blowing up the homogeneous region by a factor Z in linear dimensions) and the monopole problem (by reducing the monopole density by the factor Z^3).

For these advantages and more so because inflation initiates the structure formation in the Universe in a way that seems to lead to the right mass distributions at the present epoch, inflation has been accepted by cosmologists as a vital stage in the very early Universe. For the general-relativity purist the model leaves much to be desired. It is an approximate solution with no matching boundary conditions. That is, one does not know how *exactly* the conditions change across the bubble and its surroundings. Neither is the spatial extent of the initial bubble specified. The potential function that distinguishes the true vacuum from the false vacuum is also put in by hand rather than taken from some deep theory.

We conclude our discussion of the early Universe here. At present astroparticle physics is the most active area in theoretical cosmology. Despite its speculative nature, its challenges invite theorists to try their own prescription for how the Universe began. However, in the last analysis, a scientific theory must pass the test of observations. We will discuss a few important observations in the next chapter.

Exercises

1. Substitute the values of c, G and a into (16.7) and verify the numerical coefficient in (16.8).

2. Taking the present-day temperature of the radiation background as 2.73 K and the present baryon density as 10^{-6} cm^{-3}, calculate the number ratio of photons to baryons.

3. A primordial mixture of relativistic bosons and fermions in the early Universe of temperature T has the total energy density given by the formula

$$\epsilon = \frac{\pi^2}{30\hbar^3 c^3} g_*(kT)^4.$$

Show that $g_* = g_b + (7/8)g_f$, where g_b is the total spin degeneracy of all bosons and g_f is the total spin degeneracy of all fermions.

4. The binding energy of the ^{4}He nucleus is $B \cong 4.3 \times 10^{-5}$ erg. Show that for the nucleus $B/[k(A - 1)] \cong 10^{11}$ K. Next assume that the present value of the radiation temperature is 3 K and that of the nucleon density is 10^{-6} cm^{-3}. Using the result that $N_N T^{-3} = $ constant, show that (16.47) gives T_Q for ^{4}He as $\sim 3.2 \times 10^9$ K.

5. If m is the mass of a nucleon and if Ω_0 is the density parameter, show that the present number density of baryons is $3H_0^2\Omega_0/(8\pi Gm)$. Use this formula and the present microwave background temperature $T_0 = 2.7$ K to estimate N_B in (16.58). Solve the Saha equation for $\Omega_0 = 0.1$, $h_0 = 1$ to show that, at 3000 K, $x = 0.003$.

6. Using the Thomson-scattering cross section for the electrons, show that the optical depth of the Universe at the present epoch would be given by $0.08\Omega_0 h_0$ if all electrons in the Universe were free and equal in number to the baryons and there were no non-baryonic matter.

7. Assuming that in the past the electron number density increased as $(1 + z)^3$, use the analysis of Exercise 6 to estimate the smallest redshift at which the Einstein–de Sitter universe was opaque to radiation. (Take $h_0 = 1$.) Comment on the fact that your answer comes out very much lower than $z \sim 1000$.

8. Give arguments to show that the neutrino temperature drops as S^{-1} after neutrinos decouple from the rest of the matter.

9. Why is the present neutrino temperature expected to be lower than the photon temperature? Derive the ratio of the two temperatures from considerations of the early Universe.

10. Suppose we wish to apply flat-space statistical mechanics to the very early Universe at epoch t. The locally flat region may be characterized by a linear size $L \lesssim \alpha ct$, where $\alpha \ll 1$. Estimate the number of relativistic species in this region using a time–temperature relationship. Show that, while this number is $\gg 1$ for the primordial nucleosynthesis era ($t \sim 1 - 200\,\mathrm{s}$), it is < 1 for the GUT era. Can flat-space statistical physics be applied at the GUT era?

Chapter 17
Observational cosmology

We will now take a look at some of the tests of the relativistic cosmological models discussed so far. We will confine ourselves to tests that bring out the general-relativity part of the model, rather than other aspects like astrophysics, particle physics, etc. For a more comprehensive discussion see [53].

17.1 The redshift–magnitude relation

We saw in Chapter 14 that for small redshifts Hubble's law holds. What is the form of this relation when redshifts are not small compared with unity? Formula (15.67) tells us the relationship between luminosity distance D and redshift z for Friedmann models without the cosmological constant.

In the practical form in which this test is often presented, astronomers use apparent magnitudes in place of distances. Thus the D–z relation becomes

$$m - M = 5 \log D - 5$$
$$= 5 \log \left(\frac{c}{H_0 q_0^2} \right) - 5$$
$$+ 5 \log[q_0 z + (q_0 - 1)(\sqrt{1 + 2q_0 z} - 1)]. \qquad (17.1)$$

A. R. Sandage and his colleagues spent a number of years on this cosmological test with the hope that the correct geometry of the Universe would be revealed. Although in the 1960s Sandage often quoted a value of $q_0 \approx 1$, it gradually became clear that a number of uncertainties

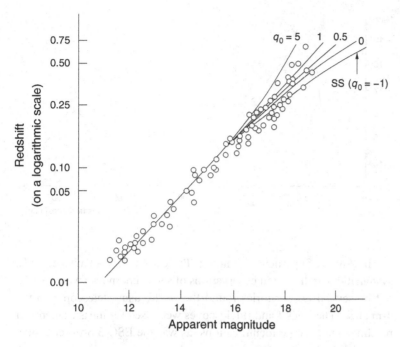

Fig. 17.1. The redshift–magnitude plot for the galaxies which are the brightest in their clusters. The plot is based on the work of Allan Sandage and his colleagues (A. Sandage *et al.* 1978, *Ap. J.*, **221**, 383) who showed that there is very little variation in the luminosity of brightest cluster members. Theoretical curves for various values of q_0 are superposed on the data points.

combine to make this test rather inconclusive. A typical z–m curve obtained by Sandage is shown in Figure 17.1. The various errors and uncertainties that arise in practical applications of this test are many. Some of the issues have been understood and partially resolved; others continue to be difficult to settle. For these reasons this test of spacetime geometry fell into disfavour during the 1980s and early 1990s.

However, in the late 1990s a fresh attempt was made to revive this test when it was realized that Type Ia supernovae can be used to estimate m relatively unambiguously at redshifts as high as unity.

17.1.1 The Hubble diagram using Type Ia supernovae

During the 1980s, it was realized that Type Ia supernovae can serve as standard candles in the following way. The light curve of such a supernova (cf. Figure 17.2) shows an approximately symmetric characteristic rise and fall over $\sim$30 days, followed by a much slower decline. The maximum luminosity of a Type Ia supernova shows an almost uniform value for this population, the dispersion being no more than 0.3 magnitude. Going beyond that, however, we now see that, because of their high peak luminosity, they can be spotted in distant galaxies. Thus they are suitable for determining the z–m relation, out to redshifts of $\sim$1 or even more.

Fig. 17.2. The rise and fall of
light intensity during the
explosion of a supernova of
Type Ia. The peak intensity is
expected to vary from one
supernova to another, but
within a rather narrow range.

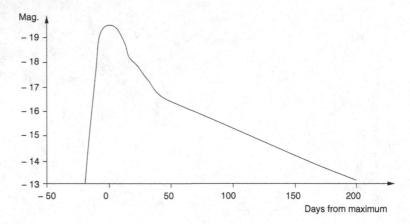

Fig. 17.2. The rise and fall of light intensity during the explosion of a supernova of Type Ia. The peak intensity is expected to vary from one supernova to another, but within a rather narrow range.

In 1988 the Supernova Cosmology Project (SCP) was launched and a systematic search for and observations of such supernovae were carried out by several observatories round the world, using telescopes in the 4-m class. The Keck I and II telescopes were used for measurements of redshifts and spectral identifications, as was the ESO 3.6-m telescope. The database continues to grow.

In 1999 Perlmutter *et al.* [73] used 60 supernovae to draw up the Hubble plot. Of these, 18 came from the work on nearby supernovae by Reiss *et al.* [72]. These were used essentially to set the zero point of the plot, with the remaining 42 coming from the SCP with redshifts starting from 0.18 and going as far as 0.83.

Theoretical Friedmann models with $\lambda = 0$ can be applied to such data using the formulae for $D(z, q_0)$ from Chapter 15. However, their observational fits were not very satisfactory and the parameter space had to be expanded to include the cosmological constant. We briefly discuss the theoretical aspect of these models showing how the m–z relation can be derived numerically. The dimensionless parameters in question are

$$\Omega_0 = \frac{8\pi G \rho_0}{3H_0^2}, \qquad \Omega_\Lambda = \frac{\lambda c^2}{3H_0^2}, \tag{17.2}$$

these being respectively the density parameter and the cosmological-constant parameter.

Using the formulae of Chapter 15, we can write the following relation between the radial Robertson–Walker coordinate r and redshift z:

$$r(z) = \int_{S(t_0)/(1+z)}^{S(t_0)} \frac{c \, dS}{S\dot{S}}. \tag{17.3}$$

It is not difficult to see that using Equation (15.92) we can write the above in the flat case $(k = 0)$ as

$$r(z) = \frac{c}{S_0 H_0} \int_1^{1+z} \frac{dx}{\sqrt{\Omega_\Lambda + \Omega_0 x^3}}. \tag{17.4}$$

Example 17.1.1 Let us reduce (17.3) to (17.4) for the case $k = 0$. From Equation (15.92) we get for $k = 0$

$$\dot{S}^2 = \frac{1}{3}\lambda c^2 S^2 + \frac{8\pi G \rho_0 S_0^3}{3S}.$$

Therefore

$$S^2 \dot{S}^2 = \frac{1}{3}\lambda c^2 S^4 + \frac{8\pi G}{3}\rho_0 S_0^3 S.$$

Using the definitions (15.40) and (15.96), we get

$$\rho_0 = \Omega_0 \frac{3}{8\pi G}\left(\frac{\dot{S}^2}{S^2}\right)_0, \qquad \Omega_\Lambda = \frac{\lambda c^2}{3}\left(\frac{\dot{S}^2}{S^2}\right)_0^{-1}.$$

Writing $H_0 = [\dot{S}/S]_0$, we replace λ and ρ in our equation by Ω_Λ and Ω_0, to get

$$\dot{S}^2 S^2 = \Omega_\Lambda S^4 H_0^2 + \Omega_0 H_0^2 S_0^3 S.$$

Using the relation $S_0/S = x$ we get

$$\dot{S}^2 S^2 = \Omega_\Lambda H_0^2 S_0^4 x^{-4} + \Omega_0 H_0^2 S_0^4 x^{-1}.$$

Hence

$$\frac{c\, dS}{S\dot{S}} = -\frac{c S_0\, dx}{x^2} \cdot \frac{1}{H_0 S_0^2}\left(\Omega_\Lambda x^{-4} + \Omega_0 x^{-1}\right)^{-1/2}$$

$$= -\frac{c}{H_0 S_0}\left(\Omega_\Lambda + \Omega_0 x^3\right)^{-1/2}.$$

The r-integral therefore becomes

$$r(z) = \int_1^{1+z} \frac{c\, dx}{H_0 S_0 \sqrt{\Omega_\Lambda + \Omega_0 x^3}},$$

with the luminosity distance as

$$D(z) = r(z)S_0(1+z) = (1+z)\frac{c}{H_0}\int_1^{1+z} \frac{dx}{\sqrt{\Omega_\Lambda + \Omega_0 x^3}}.$$

Now recall that the luminosity distance $D = r S(t_0)(1 + z)$ and we can write down, in the first approximation, the following z–m relation:

$$m(z) = -2.5 \log L + 5 \log D + \text{constant}. \tag{17.5}$$

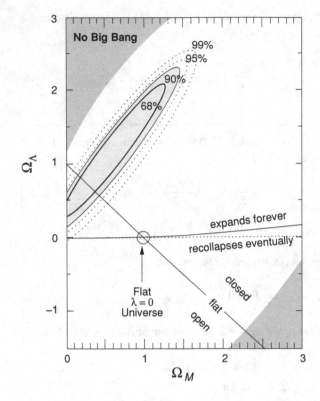

Of course, one has to correct this relation for the K correction and for other possible effects mentioned before. L, as already pointed out, contains a dispersion around the average standard-candle luminosity. In fitting a best-fit curve through the data the dispersions in apparent magnitudes have to be taken into consideration. Figure 17.3 shows how well the various theoretical models match the observations. (Here Ω_M is the same as Ω_0.)

Perlmutter *et al.* found that the simplest Friedmann model, namely the *flat* Einstein–de Sitter model, does not give a statistically satisfactory fit. The flat model, however, does fit well if a non-zero cosmological constant is allowed. That is, consistently with (15.97), i.e.,

$$\Omega_\Lambda + \Omega_0 = 1,$$

the model with $\Omega_0 = 0.28$ gives the best fit. *This implies that a non-zero cosmological constant as high as ≈ 0.7 is needed.* In other words, the Universe has a negative $q_0 \approx -0.6$, if we use the relation (15.95). *Thus the Universe is accelerating.*

Clearly, in view of the profound significance of such a finding, careful follow-up is regularly being done. Several questions arise. How sure are we that there is no evolution in supernovae that would spoil their standard-candle interpretation? Some four or five supernovae in the data have to be left out of the curve-fitting exercise because they lie far off from the best-fit curve: why? Could some other explanation rather than the cosmological constant account for the extra dimming found in the supernovae? The possible effect of dust in standard cosmology has also been invoked by some. They argue that the dust causes the extra dimming that is observed. We will discuss the role of intergalactic dust in the context of the quasi-steady-state cosmology when we return to this test in the final chapter. There we will also discuss how this test plays a complementary role in the parameter space in relation to the cosmic microwave background.

17.2 Number counts of extragalactic objects

The basic idea behind these tests is to find out whether the number counts reveal the non-Euclidean nature of the spacetime geometry of the Universe assumed by most models. We illustrate with a simple example from radio astronomy. Suppose we have a class of radio sources that are (1) uniformly distributed in space and (2) have the same luminosity L. If we further assume that (3) the Universe is of Minkowski type, that is, with Euclidean spatial geometry, the number of sources up to a given distance R will go as

$$N \propto R^3, \tag{17.6}$$

while the flux density from the faintest of the sources up to distance R goes as

$$S \propto R^{-2}. \tag{17.7}$$

By eliminating R between these relations, we get

$$N^2 S^3 = \text{constant}, \quad \text{that is} \quad \frac{D \log N}{d \log S} = -1.5. \tag{17.8}$$

Thus (17.8) tells us how N and S are related under our three assumptions (1), (2) and (3). Under these assumptions N measures the volume and $S^{-1/2}$ the radius of a spherical region centred on the observer, and (17.8) is simply the volume–radius relation in Euclidean geometry.

Given the Robertson–Walker models, we can work out the corresponding relations in non-Euclidean geometries. It is therefore possible, in principle, to test whether the observed relation agrees with one of the various cosmological models. Unfortunately, as with the z–m test, various uncertainties prevent us from drawing a clear-cut conclusion, as we shall see with the counts of galaxies and radio sources below.

17.2.1 Counts of galaxies

In 1936 Hubble attempted number counts of galaxies in order to distinguish between model universes. However, he had to abandon the test because the number of galaxies to be counted is very large, and unless one goes fairly deep in space one cannot detect any significant departures from Euclidean geometry. However, there is one difference from the formula (17.8). Since the optical astronomer measures fluxes in magnitudes, the corresponding relation describing the number N of galaxies brighter than apparent magnitude m becomes

$$\frac{\mathrm{d}\log N}{\mathrm{d}m} = 0.6. \qquad (17.9)$$

Any effect of non-Euclidean geometry will show up as a deviation from this straight-line relation, but these differences become noticeable only at redshifts as high as 0.5, say. By then one needs to identify and count millions of galaxies, even in a small sector of the sky. Although Hubble did not succeed, his programme was revived in recent years by a number of workers, who now have at their disposal many electronic and solid-state devices to facilitate galaxy counts to very faint magnitudes ($m \sim 24$). For example, in 1979 J. A. Tyson and J. F. Jarvis first used techniques of automated detection and classification of galaxies on plates. Their main problem at faint magnitudes was to be able to distinguish stars from galaxies.

Even though one may find ways round these practical problems, the outcomes of such counts are hard to interpret as deviations caused by geometry. Rather, evolutionary effects and inhomogeneities of sample galaxy populations chosen for counting dominate the observations.

17.2.2 Counts of radio sources

In comparison with galaxy counts, counts of radio sources have the advantage that the latter are not as numerous as galaxies. For this reason, after Hubble's galaxy-count programme had come to nothing and as radio astronomy became established during the 1950s, it was felt that the time was ripe to have a go at the radio-source-count test. Radio astronomers also felt that strong radio sources could be seen at much further distances than galaxies, and hence they would provide more stringent tests on the large-scale geometry of the Universe. Also they are much rarer than galaxies, so there are not many of them to count.

M. Ryle at Cambridge, B. Mills at Sydney and J. Bolton at Caltech did pioneering work on the source-count programme. Since the radio astronomer measures S over a specified bandwidth, he tends to plot $\log N$ against $\log S$, where S is the *flux density*, the flux S received over a frequency band divided by the bandwidth. The usual unit for S is the jansky (Jy) (named after Karl G. Jansky, who did pioneering work in radio astronomy in the 1930s), which equals 10^{-26} W m^{-2} Hz^{-1}. Similarly, the *power* of the radio source is defined as luminosity over a unit frequency band per unit solid angle and is expressed in units of watts per hertz per steradian (W Hz^{-1} Sr^{-1}).

The early source counts led to considerable controversy and a lot of discussions, largely because the problem of interpreting the observed source counts was oversimplified. Several factors intervene to make a simple conclusion elusive. Some of them are as follows.

1. Counts are affected by local large-scale inhomogeneities of the matter distribution. For example, a local void near us would make the log N–log S curve steeper than Euclidean.
2. There could be different types of sources, all mixed up in the survey. For example, quasars and radio galaxies mixed together would give misleading answers. These populations need to be counted separately.
3. Evolution in number density or luminosity of the source population will easily mask any geometrical effect that is being looked for.
4. The luminosity function needs to be known before the source count can be reliably made. This last point is important since, unlike the optical astronomer, the radio astronomer cannot measure the redshift of the radio source and so relies on the flux density to estimate its distance.

The surveys today are much more sophisticated and accurate than the early pioneering radio surveys of 1955–65. (Figure 17.4 shows an example.) But they have also brought the realization that distinguishing between different geometries in this way is not possible.

17.3 The variation of angular size with distance

This test was briefly discussed in Chapter 15, where we saw that the angular size of an object of fixed projected linear size does not steadily decrease with its spatial distance from us. Figure 15.7 showed how the angular size changes with the redshift of the object in various Friedmann models. In 1958 F. Hoyle [58] first suggested that this property of non-Euclidean geometries could in principle be tested by radio-astronomical observations.

Early attempts to look for this effect in galaxies at different redshifts failed since it was not possible in the 1960s to carry out measurements of angular sizes of galaxies so far away. After Hoyle's proposal, several radio astronomers took up the challenge. The typical radio source is a linear structure and it is not so difficult to measure the angle subtended by it. In the radio case, however, obtaining redshifts directly is not possible:

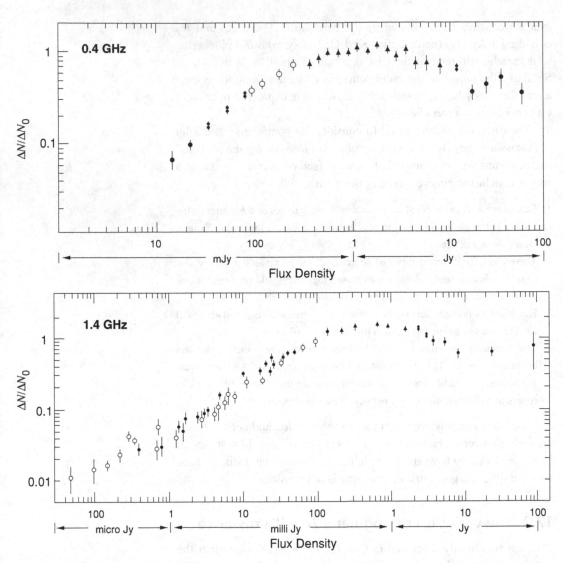

Fig. 17.4. Two curves of source counts, one at 0.4 GHz and the other at 1.4 GHz. ΔN are the numbers in a flux-density range $(S, S + \Delta S)$ while ΔN_0 are the numbers in the same flux-density range in a static Euclidean spacetime. Notice that the counts go down to millijansky (top curve) and microjansky (bottom curve), respectively, compared with the early surveys which reached down to a few jansky. It is, however, not possible to relate these curves to a specific spacetime geometry.

one must optically identify the source and then measure the redshift of the optical counterpart. After several studies in the radio no clear signal emerged since there were other, more dominant, effects whose presence would be significant. These effects included (1) a projection effect while measuring the angle subtended by a linear structure at the observer;

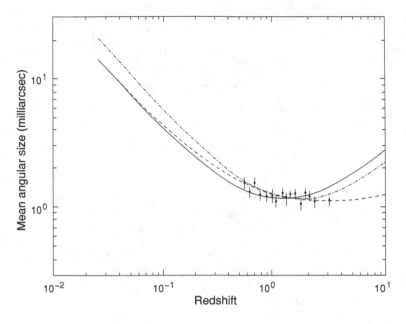

(2) dispersion in linear size, i.e., there is no standard yardstick to refer to; and (3) the ubiquitous possibility of evolution that would introduce z-dependent factors into the answer.

To eliminate or at least to minimize the evolutionary effect of the intergalactic medium, Kellermann in 1993 suggested that the test be applied to the very tiny inner components of quasars which are seen through very-long-baseline interferometry. His preliminary studies gave a result in broad agreement with the Einstein–de Sitter model. However, a more thorough analysis of a sample of 256 ultracompact sources with redshifts in the range 0.5 to 3.8 by J. C. Jackson and Marina Dodgson showed that this model is in fact ruled out and that, for better fits, one needs to invoke the cosmological constant. Figure 17.5 shows the θ–z curves for three types of models fitted by them to the data. The points shown represent median values of data divided into 16 bins.

To sum up, the θ–z test, to begin with, looked a simple and elegant way of checking on spacetime geometry, but observational realities turned out, once again, to be frustrating to the theorist!

17.4 The age of the Universe

The formulae in Chapter 15 give t_0 as the age of the Universe according to the various Friedmann models. These formulae depend on two parameters, H_0 and q_0 (or Ω_0), both of which have been discussed before. Additionally we have the choice of including the cosmological constant. We are now in a position to take a look at the question of whether the

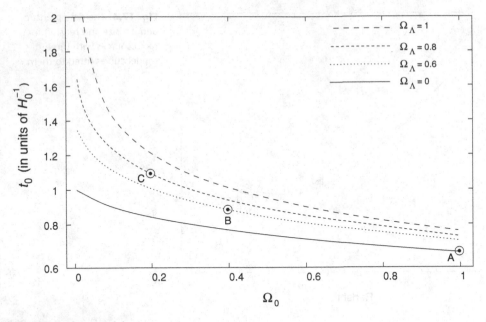

Fig. 17.6. The age of the Universe plotted as a function of the density parameter Ω_0 for various curves of given Ω_Λ. In general the age increases with Ω_Λ.

Friedmann age estimates are consistent with the various astrophysical estimates of the age of the universe. Figure 15.12, reproduced here as Figure 17.6, gives the range of values of the ages of the Friedmann models for various values of these parameters, for purposes of comparison.

At present there are two ways of estimating the ages of galaxies, both of which have been applied to our Galaxy. A primary requirement of consistency is, of course, that the age of the Universe in a Friedmann model must exceed the age of any object in it.

17.4.1 Stellar evolution

This method, applied to globular clusters in our Galaxy, is based on the principle that stars become redder and brighter when they leave the main sequence to become red giants. Since the red giant phase in the star's life lasts a comparatively short time, say up to about 10% of the time the star spends on the main sequence, the turning point from the main sequence to the giant branch provides the cluster age to within 10% uncertainty.

Let the *cluster age*, the time when the stars turn off from the main sequence, be denoted by $t_c \times 10^9$ years, and let Y and Z be the helium and metal abundances in the star at this stage. The calculations of stellar evolution then show that

$$\log t_c = 1.035 + 2.085(0.3 - Y) - 0.03(\log Z + 3). \tag{17.10}$$

Thus the age depends critically on the helium abundance Y. Y can be estimated from a comparison of the time a star spends on the horizontal branch with the time it spends on the red giant branch. If this ratio is R, then calculations show that

$$Y = 0.3 - 0.39 \log\left(\frac{f}{R}\right), \qquad (17.11)$$

where $f = 2$ if the stellar model takes account of semiconvection and certain other effects, whereas $f = 1$ if these effects are not taken into account. R can be estimated from the observed ratio of horizontal-branch stars and red giant stars in the cluster.

Cluster ages deduced by this method fall in the range from $\sim 13 \times 10^9$ to $\sim 18 \times 10^9$ years.

17.4.2 Nuclear cosmochronology

In 1960 F. Hoyle and W. A. Fowler first demonstrated how the relative abundances of radioactive nuclei of long lifetimes can lead to estimates of the age of our Galaxy. The method had already been used for estimating the age of the Solar System. For example, current observations of the abundance ratio $^{87}Sr/^{86}Sr$ plotted against $^{87}Rb/^{86}Sr$ in various Solar-System materials (such as meteorites) give its age accurately as $t_S \simeq 4.54 \times 10^9$ years.

As illustrated in Figure 17.7, the method of nuclear cosmochronology attempts to estimate the time elapsed before the Solar System was formed. According to this method, we start our nuclear clock at $t = 0$ with the birth of the Galaxy. The stars evolve and the more massive ones become supernovae, which manufacture long-lived radioactive nuclei in the so-called *r-process* (the rapid absorption of neutrons by heavy nuclei). The rate at which this process goes on is denoted by a function

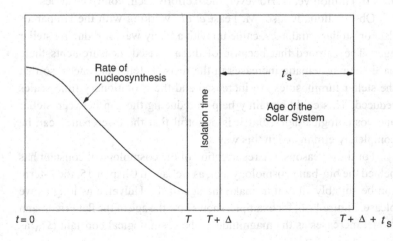

Fig. 17.7. Three time spans need to be estimated in order to estimate the age of the Galaxy. First we need the time T spent in the r-process to manufacture long-lived radioactive nuclei. Then one needs the isolation time Δ, which can be estimated (together with T) from data on abundances of radioactive isotopes. Finally t_S is estimated from radioactivity data of Solar-System material.

$p(t)$, which declines to negligible value at $t = T$. Between this epoch and the formation of the Solar System there occurs a short time gap Δ, known as the *isolation time*, during which we may ignore nucleosynthesis, in particular the r-process. Thus the total nuclear age of the Galaxy is

$$t_G = T + \Delta + t_S. \tag{17.12}$$

By techniques using data on radioactive isotopes and the observed abundances of certain long-lived nuclei, one can estimate T and Δ.

The nuclear age so estimated lies in the range between 6 and 20 billion years, the width of this range indicating the span of uncertainties in the various quantities used for determining the various time intervals.

It is clear nevertheless, when these age estimates and the estimates from globular clusters are compared with those of Figure 17.6, that models with $h_0 = 1$ and $\Omega_0 \geq 1$ will find it very difficult to accommodate the above astrophysical estimates of the age of our Galaxy. In particular, the original inflationary model without λ is ruled out because it predicts $\Omega_0 = 1$ unequivocally. One needs the cosmological constant.

To make the problem easier for the conventional point of view, attempts are being made to see whether the stellar and radioactive ages can be brought down significantly. For example, if significant mass loss occurs during the main-sequence stage of stellar evolution then the time spent by the star on the main sequence is reduced. (For it started with higher mass and evolved faster.) By arguing in this way it may be possible to reduce the ages of globular clusters to values as low as $(7–10) \times 10^9$ years. Likewise W. A. Fowler and C. C. Meisl have recalculated the nuclear age of the Galaxy using a time-dependent model for nucleosynthesis in which an early 'spike' is followed by a uniform rate of synthesis. They claim that the age then comes down to 11 ± 1.6 (1σ) billion years. However, these efforts seem contrived at best.

Observationally also, M. Feast *et al.*, working with the Hipparcos data on stellar parallaxes, came up with a likely way of reducing stellar ages. They argued that because of these revised measurements there have been systematic increases in the revised stellar distances, so that the stellar luminosities are increased and the evolutionary time scales reduced. This could certainly help in reducing the gap between stellar and cosmological ages, but it is doubtful that the discrepancy can be completely eliminated in this way.

For these reasons, the resurrection of the cosmological constant has helped the big-bang cosmology. For, as we saw in Chapter 15, the λ-term can be suitably chosen to make the age of the Universe as long as we please. Figure 17.6, for example, shows how the age of the flat Friedmann model increases as the magnitude of the cosmological constant (Ω_Λ as defined in Chapter 15) is increased. However, the introduction of this constant increases the cosmological distances and thereby increases the probability of a distant light source being gravitationally lensed. On the basis of the frequency of lensed objects, upper limits have been placed on the dimensionless parameter Ω_Λ: it is generally agreed that Ω_Λ cannot much exceed 0.75.

17.5 Abundances of light nuclei

It is generally recognized that nuclei with relative atomic masses $A \geq 12$ are synthesized in stars through various processes discussed in theories of stellar evolution. The nuclei ^{6}Li, ^{9}Be, ^{10}B and possibly ^{11}B could be produced in galactic cosmic rays by the break-up of heavy nuclei as they travel through the interstellar gas. It is the lighter nuclei, in particular ^{2}H, ^{3}He, ^{4}He and ^{7}Li, that appear to pose difficulties of production in stars in the amounts observed. Further, their abundances are such that they could have been produced in the big-bang nucleosynthesis. We will discuss here briefly the data on ^{4}He and ^{2}H, and what constraints they place on standard cosmology.

17.5.1 ^{4}He

The observed helium abundances (always denoted by mass fraction Y) in the Universe are quoted as lying in the broad range $0.13 \leq Y \leq 0.34$. The scatter is wide because of the uncertainties of various observational estimates. Further, the estimate of primordial helium in the Sun at the time the Solar System formed $\sim$4.54 $\times$ 10^9 years ago depends on the solar model and hence cannot be uniquely fixed. M. Peimbert, S. Torres Peimbert and J. F. Rayo have suggested that the break-down of Y at any location is as follows:

$$Y = Y_0 + \Delta Y,$$
$$Y_0 = 0.23 \pm 0.02, \tag{17.13}$$
$$\Delta Y \cong (2.5 \pm 0.5) \times Z,$$

where Y_0 is the primordial helium abundance, ΔY the stellar helium abundance and Z the abundance of heavy elements made by stars. Since $Z \leq 0.02$, $\Delta Y \leq 0.06$.

There are occasional reports of low values of helium abundance and these need to be probed more deeply. For helium, once produced, is difficult to destroy. In this sense, the theoretical primordial value of Y mentioned in Chapter 16 is expected to be the lower bound of Y found today.

We note that in the primordial picture Y_0 is relatively insensitive to h_0 and Ω_0. However, the introduction of new light leptons would push up the neutron/proton ratio and hence the value of Y_0. The following formula due to R. V. Wagoner summarizes this result for the fraction η defined in (16.51) exceeding $\sim 10^{-5}$:

$$Y_0 = 0.333 + 0.0195 \log \eta + 0.380 \log \xi. \tag{17.14}$$

Here the fraction $\xi = 1$ if no new particles except those considered in Chapter 16 are assumed to be present in the early Universe. In terms of our notation of Chapter 16, this implies $g = 9$. If there are more particles, $g \rightarrow g + \Delta g$, where $\Delta g = \Delta g_b + (7/8)\Delta g_f$, and

$$\xi^2 = 1 + \frac{\Delta g}{g}. \tag{17.15}$$

For $Y_0 \leq 0.25$ and $\Omega_0 = 0.01$, only one new neutrino is allowed, the so-called τ-neutrino. If, however, Y_0 were as high as 0.28, up to four new leptons would be permitted by this constraint, whereas a value as low as 0.21 would land the standard big-bang model in real trouble. The smallest value of Y_0 allowed by the standard model is close to 0.236. Results from the accelerator experiments based on measuring the decay width of the Z^0 boson suggest that the number of neutrino species is 3.01 ± 0.10. Thus there is a broad consistency between cosmology and particle physics.

Observations of Y_0 are therefore of great importance and they continue to be reported, as observers sharpen their spectroscopic diagnostics. These estimates may be seen as indicative only in placing constraints on the parameters of the standard cosmology.

17.5.2 ^{2}H

The deuterium abundance, which we will denote here by $X(^2H)$, was first measured in 1973, mainly from the Lyman-series absorption lines in the ultraviolet spectra of the bright stars observed with the Copernicus satellite. There have been several measurements of this important fraction. It is found that generally

$$9 \times 10^{-6} \leq X(^2H) \leq 3.5 \times 10^{-5}.$$

Although a mean interstellar value of $X(^2H) \simeq 2 \times 10^{-5}$ is often quoted, there is considerable variation in its value from cloud to cloud. It is not clear whether these variations are due to partial destruction of primordial deuterium through various processes. It has to be destruction, since so far no satisfactory stellar scenario for production of deuterium is known. Thus the primordial value would correspond to the upper end of the range of observations. At least we expect it to exceed $\sim 2 \times 10^{-5}$. (Contrast this situation with that for ^{4}He, for which there is no destruction mechanism but processes of production exist in stars.)

Referring back to Figure 16.3, we see that a primordial abundance $X(^2H) \geq 2 \times 10^{-5}$ implies that the baryonic density at present cannot exceed 4×10^{-31} g cm^{-3}, which in turn sets an upper limit on the present baryon-density parameter $(\Omega_B)_0$:

$$h_0^2(\Omega_B)_0 \leq 0.02. \tag{17.16}$$

It is interesting to note that in 1996, from measurements of deuterium abundance in clouds around a high-redshift quasar, Tytler, Fan and Burles placed limits on the baryon-density parameter:

$$h_0^2(\Omega_B)_0 \approx 0.024 \pm 0.006.$$

Thus, if matter in the Universe is predominantly made of baryons, the Universe must be open. However, in modern thinking influenced by inflation the condition $\Omega = 1$ must be satisfied. So one invokes non-baryonic dark matter and dark energy (a euphemism for the λ-term). If all dark matter were baryonic, say distributed as black holes or burnt-out cores of stars or brown dwarfs, etc., the currently favoured cosmological model fails, as mentioned in the concluding section of this chapter.

There are, however, fine tunings involved here. For the restriction on baryon density implies a relatively tight relation between density and temperature, i.e., the constant of proportionality in the relation $\rho_B \propto T^3$ in the relation (16.52) has to be correctly chosen for the model to give the right answer. One may therefore question whether this can be claimed as a deductive success of the big-bang cosmology.

17.6 The microwave background radiation

Measurements of the microwave background radiation (MBR) occupy the centre stage of observational cosmology today. As mentioned in Chapter 16, following the finding of the radiation background by Penzias and Wilson, the background has been assumed to be a relic of the early Universe. Following this interpretation, it has been probed by various teams of observers to obtain its spectrum, polarization, anisotropy and angular power spectrum. These observational details and the subtleties of their interpretation within the framework of a 'standard' model of the Universe are not appropriate in a text on general relativity. We will therefore only briefly summarize some relevant studies.

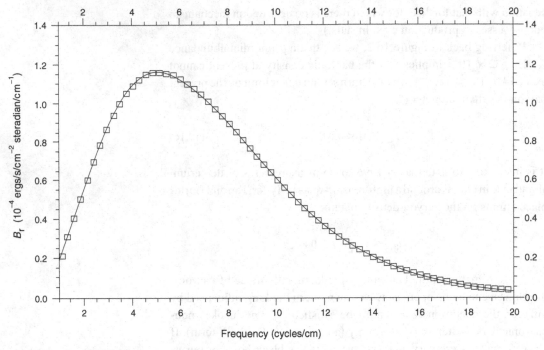

Fig. 17.8. The COBE plot of the spectrum of the cosmic microwave background radiation.

17.6.1 The spectrum

To check the true blackbody character of the radiation, it is necessary to have detectors above the Earth's atmosphere, since ground-based measurements do not reach the peak wavelengths of the expected blackbody curve due to atmospheric absorption. There were several early attempts using balloons and rockets. Some of these reported departures from the Planckian spectrum later turned out to be false alarms. The most accurate and exhaustive study was carried out in 1990 by the satellite COBE.

This satellite was launched in 1989 and obtained a beautiful spectrum as shown in Figure 17.8. See Reference [68]. The COBE measurements gave a very precise Planckian spectrum with a blackbody temperature of

$$T_0 = 2.735 \pm 0.06\,\text{K}. \tag{17.17}$$

The overall sensitivity and accuracy of the experiment made it clear that some of the earlier claims of significant departures from the Planckian spectrum at high frequencies were erroneous. Indeed, even laboratory experiments are not known to produce a Planckian spectrum of this level of accuracy.

17.6.2 The angular power spectrum

When we look at the distribution of a physical quantity across the celestial sphere, its anisotropies can be best described with the help of spherical harmonics. The physical quantity describing the anisotrophy of the MBR is its temperature $T(\theta, \phi)$, written as a function of two spherical polar coordinates (such as declination and right ascension). We may accordingly write

$$\frac{\Delta T(\theta, \phi)}{T} = \left[\sum_{l=1}^{\infty} \sum_{m=-l}^{m=l} a_{lm} Y_{lm}(\theta, \phi) \right]. \tag{17.18}$$

The sum over l begins with 1 instead of zero, for the zeroth perturbation is isotropic over the whole sky, and can be absorbed into T. The $l = 1$ term is the so-called *dipole-anisotropy* term, which, as we shall see, arises from the motion of the Earth relative to the rest frame of the MBR. Henceforth we will not include this term also in the above series. The next, $l = 2$, mode is the *quadrupole* mode.

The *angular power spectrum* is specified by quantities C_l defined by

$$C_l \equiv \langle |a_{lm}|^2 \rangle, \tag{17.19}$$

where the averaging is with respect to all realizations of the sky and summed over all m. Thus each C_l tells us the relative strength of the lth harmonic in the overall distribution.

In general we will be interested in looking at $\Delta T/T$ over a certain angular scale ϑ. Thus, if we take two directions denoted by unit vectors $\mathbf{e}_1$ and $\mathbf{e}_2$ enclosing an angle ϑ between them, we get

$$\mathbf{e}_1 \cdot \mathbf{e}_2 = \cos \vartheta. \tag{17.20}$$

Now we define the *autocovariance function* which tells us how the temperature fluctuations compare over directions separated by the angle ϑ:

$$C(\vartheta) = \left\langle \frac{\Delta T(\mathbf{e}_1)}{T}, \frac{\Delta T(\mathbf{e}_2)}{T} \right\rangle, \tag{17.21}$$

which, for stationary fluctuations, can be expressed in the form

$$C(\vartheta) = \frac{1}{4\pi} \sum_{l=2}^{\infty} (2l + 1) C_l P_l(\cos \vartheta). \tag{17.22}$$

Suppose that, from observations of a single sky, we have obtained the estimate of the autocovariance function as $C(\vartheta)$:

$$\hat{C}(\vartheta) = \frac{1}{4\pi} \sum_{l=2}^{\infty} |\hat{a}_{lm}|^2 P_l(\cos \vartheta), \tag{17.23}$$

Fig. 17.9. The COBE map of small-scale anisotropy of the cosmic microwave background radiation.

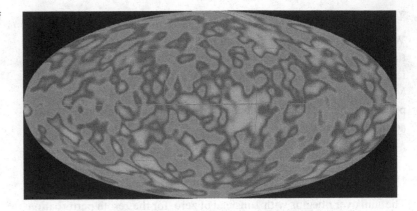

where the $\hat{a}_{lm}$ are determined from a single observation of the sky. In this case one needs to estimate the *cosmic variance* of the quantity $C(\vartheta)$. This can be shown to be

$$\langle |\hat{C}(\vartheta) - C(\vartheta)|^2 \rangle = \left(\frac{1}{4\pi}\right)^2 \sum_{l=2}^{\infty} (2l+1) C_l^2 P_l^2(\cos \vartheta). \qquad (17.24)$$

The first evidence of small-scale anisotropy came with the COBE satellite in 1992. (See Reference [74].) The COBE map of the sky is shown in Figure 17.9.

Later a more detailed angular power spectrum along the above lines was obtained by the Wilkinson Microwave Anisotropy Probe (WMAP) satellite. A simplified WMAP power spectrum is shown in Figure 17.10.

In practice the details are considerably intricate when one attempts to extract the signal from the actual sky data. We will not go into those details here. We point out, however, that the Legendre polynomials

Fig. 17.10. The power spectrum of anisotropies of the radiation background.

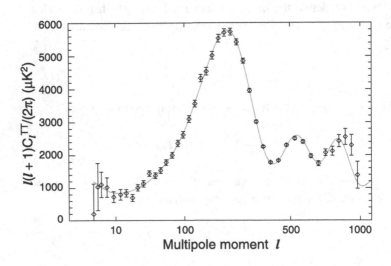

$P_l(\cos \vartheta)$ contain the following information: the typical angular scale of anisotropy corresponding to the index l is of the order of $180°/(\pi l)$.

The value of these measurements lies in constraining the theories of structure formation and through them the cosmological parameters. It is too early to say what the final picture is going to be like. We will content ourselves with listing a few possible *causes* of anisotropies of the MBR so that their signals may be looked for in such measurements. The smallness of angles implies that we are looking at higher harmonics l in the range ~ 10 to $\sim 10^4$.

The *Sachs–Wolfe effect* measures the metric fluctuations near the last-scattering surface. For example, if there is inhomogeneity of matter (clumping/voids) in a given region, this would lead to fluctuation of g_{ik} from the homogeneous Robertson–Walker form. In Newtonian terms we may argue that the photons making up the radiation background come out of different potential (φ) wells, and this would produce a change of energy, and hence of T, given by

$$\left. \frac{\Delta T}{T} \right|_{\text{energy}} = \frac{\delta \varphi}{c^2}. \tag{17.25}$$

In addition to this there is time dilatation, so that the photons emerging from a potential well are delayed in relation to surface photons and therefore encounter the scale factor S at a later epoch. For the Einstein–de Sitter universe $S \propto t^{2/3}$ and the fluctuation in T is given by

$$\left. \frac{\Delta}{T} \right|_{\text{time delay}} = -\frac{\delta S}{S} = -\frac{2}{3}\frac{\delta t}{t} = -\frac{2}{3}\frac{\delta \varphi}{c^2}, \tag{17.26}$$

because the gravitational redshift produces the above time delay. On adding the two effects we get

$$\frac{\Delta T}{T} = \left. \frac{\Delta T}{T} \right|_{\text{energy}} + \left. \frac{\Delta T}{T} \right|_{\text{time delay}} = \frac{1}{3}\frac{\delta \varphi}{c^2}. \tag{17.27}$$

In addition to this there can be *tensor* fluctuations, which will produce small contributions to $\Delta T / T$. Since these fluctuations are associated with time-dependent changes in the metric tensor they are essentially caused by gravitational waves. Some inflationary models predict gravitational-wave-type fluctuations, which are potentially detectable.

The *Sunyaev–Zel'dovich effect* suggests that the photons of the MBR entering a cluster with hot gas will be 'kicked upstairs' to higher (X-ray) energy by the Thomson scattering from high-energy electrons. Thus, if observed in the direction of the cluster, we should find a drop in the intensity of radiation. A crude approximation modelling the cluster as an isothermal sphere of radius R_c gives a fractional drop in the MBR temperature as

$$\frac{\Delta T}{T} = -\frac{4R_c n_e k T_e \sigma_T}{m_e c^2},$$ (17.28)

where n_e is the electron density in the cluster, T_e the electron temperature, m_e the electron mass and σ_T the Thomson-scattering cross section. So far there have been several claims of positive detection of this effect, in clusters ranging up to redshifts $\gtrsim 2$. This not only shows that the MBR extends that far, but also, by giving an estimate of R_c, enables a determination of Hubble's constant. The values of H_0 determined in this way are of the order of ≤ 40 km s^{-1} Mpc^{-1}, i.e., much lower than their standard measurements.

Clearly this effect, though not directly linked with the formation of large-scale structure, is nevertheless a useful tool for cosmologists.

Sakharov oscillations constitute another measurable effect of MBR anisotropy, through velocity effects from acoustic oscillations of perturbations inside the horizon at the last scattering surface. These material oscillations lead to fluctuations of photons and their temperature, both being related to the wavelength of oscillations. So one may see periodic behaviour, showing a peak in the C_l coefficients of the power spectrum estimated at

$$l_{peak} \approx 200 \Omega_0^{-\frac{1}{2}}.$$ (17.29)

The announcement of the detection of such a peak (incongruously called the 'Doppler peak', since oscillations of matter rather than velocities are responsible for the effect) was made in 2000 by the 'BOOMERANG' (Balloon Observations Of Millimetric Extragalactic Radiation ANd Geomagnetics) group of experimentalists. They found a peak amplitude $\Delta T_{200} = (69 \pm 8)$ µK at $l_{peak} = (197 \pm 6)$. The group in fact measured the angular power spectrum at $l = 50$ to 600.

Following COBE and WMAP, a more ambitious statellite-borne experiment, viz. the PLANCK, has been in preparation. The ESA's Planck Surveyor will measure the radiation at frequencies in the range 30–100 GHz with the Low Frequency Instrument and in the range 100–190 GHz with the High Frequency Instrument. The expected resolution is ~10 arcmin with sensitivity for $\Delta T/T \sim 2 \times 10^{-6}$.

The interest of cosmologists has now shifted towards understanding more (smaller) peaks of the power-spectrum curve occurring at higher frequencies, as well as the small degree of polarization found in the radiation. The WMAP team was the first to report this feature and it is hoped to measure it more accurately through later surveys.

To summarize, the MBR is being looked upon as a mine of information by big-bang cosmologists. Since the MBR is regarded as a relic of the early Universe, at least dating back to the last-scattering surface, its spectrum and anisotropies should contain valuable information about the past developments of the Universe, much as the archaeological remains at a site contain information about its past history.

17.7 Dark matter and dark energy

17.7.1 Spiral galaxies

The best handle on the mass contained in a typical spiral is given by its rotation curve. Figure 17.11 illustrates the principle by means of a flat, disc-shaped object representing a circular distribution of stars moving round a common centre C. The rotation velocity v of a star S at a distance r from C is related (in an equilibrium distribution) to the gravitational force F_r acting on S towards the centre:

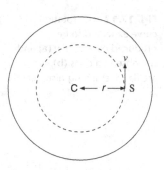

Fig. 17.11. A disc-like distribution as shown in the figure fails to produce a flat rotation curve as seen in Figure 17.12(b).

$$m\frac{v^2}{r} = F_r. \tag{17.30}$$

Therefore, if we have v as a function of r, we get F_r as a function of r. Then, by Newton's law of gravitation (which is applicable here because the gravitational fields are weak), we can determine the mass distribution. For example, if most of the mass were concentrated in the nuclear region around C, we would have $F_r \propto r^{-2}$ and $v \propto r^{-1/2}$. The light distribution across a spiral galaxy does suggest the above to be a good approximation. However, in actual fact the rotation curve – the function $v(r)$ – is flat for most galaxies. That is, after rising sharply outside the nuclear region, v first declines slightly and then remains constant, equal to v_0 (say). Moreover, this relation extends well beyond the visible disc. Figure 17.12 shows some examples.

The implication of this result is either that there is more mass in the outer parts of the galaxy than is indicated by its luminosity distribution, or that Newton's laws of motion and the inverse-square law of gravitation might not be valid over the Galactic distance range (a few kiloparsecs). Taking the former (and less radical) view, astronomers have estimated the masses of spirals. S. M. Faber and J. S. Gallagher have listed the rotation velocities and masses contained within the Holmberg radius (where the surface brightness drops to $\sim 26.5 m_{\text{pg}}$ arcsec^{-2}) for 39 spirals. Since the luminosities are also known, we can estimate the mean value of η (the mass-to-light ratio in units of $M_{\odot}/L_{\odot}$) for this sample. The result is

$$\eta \cong (9 \pm 1)h_0. \tag{17.31}$$

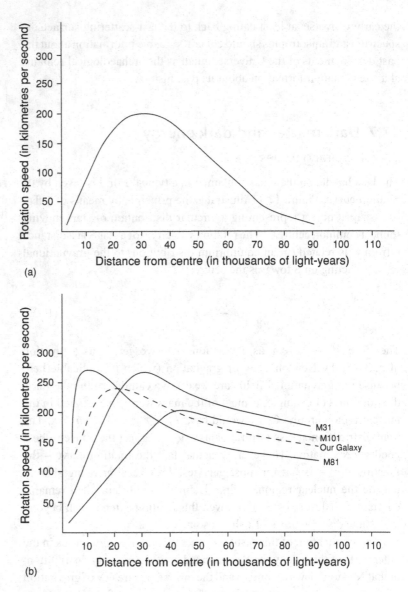

Fig. 17.12. The rotation
curve expected to be
produced by a galaxy (a) and
the observed ones (b), which
are flat over a long distance.

17.7.2 Clusters of galaxies

As early as 1933 F. Zwicky had pointed out what has now become well
known as the *missing-mass problem* in clusters. The problem can be
briefly stated as follows. If we estimate the mass of galaxies moving
in one another's gravitational field in a cluster, then the virial theorem
gives the mass of the cluster in terms of the velocity dispersion and the
effective mean radius:

$$M = \langle v^2 \rangle \frac{R}{G}. \tag{17.32}$$

From observations of the velocity dispersion $\langle v^2 \rangle^{1/2}$ we can therefore estimate the total mass M in the cluster. This value comes out considerably higher than that estimated on the basis of mass/light ratios η_G of individual galaxies. That is, if we see n galaxies in the cluster and if the total luminosity in the cluster is L, then the mass in the cluster is $L\eta_G$. Zwicky was the first to point out that

$$L\eta_G \ll M. \tag{17.33}$$

For the Coma cluster, for example, $M/(L\eta_G) \sim 300$ (see Exercise 7).

Typically one arrives at a cluster mass in the neighbourhood of $10^{15} h_0^{-1} M_\odot$. Observations suggest that there are about 4000 large clusters within a 'local' sphere of radius $600 h_0^{-1}$ Mpc. This leads to a mean density of matter in clusters of

$$\rho_{0cl} \approx 4 \times 10^{-31} h_0^2 \text{ g cm}^{-3}. \tag{17.34}$$

The density estimated for galaxies is of the same order, although not all galaxies reside in clusters. The clusters have proportionately higher mass than the galaxies contained in them because the M/L ratio for them is as high as $\sim 300 h_0 \times M_\odot/L_\odot$, about ten times higher than that for galaxies. This is why the clusters appear to require greater amounts of dark matter for their virial equilibrium.

Observations of X-rays from clusters have indicated that the emission is through bremsstrahlung from hot gas, and the amount of baryonic matter in the Coma cluster is not sufficient to account for the missing mass estimated by application of the virial theorem. If the ratio of baryonic to total gravitating matter in the Coma is representative of the universal value, then the total density parameter Ω_0 is constrained by the inequality

$$\Omega_0 \le \frac{0.15 h_0^{-1/2}}{1 + 0.55 h_0^{3/2}}. \tag{17.35}$$

With this type of inequality, it is clear (i) that, if the deuterium in the Universe were made primordially, then we cannot have the density parameter attain the upper limit with baryons alone; and (ii) that the Universe is open ($k = -1$), unless there is a large quantity of dark matter residing *outside the clusters*. Already at this stage the *known* baryonic content of the cluster mass (M_B) as a fraction of the total cluster mass (M_{tot}) threatens a contradiction with observations of deuterium abundance. For example, for the Coma cluster, we have

$$\frac{M_B}{M_{tot}} \approx 0.01 + 0.05 h_0^{-3/2}. \tag{17.36}$$

If this ratio were universal, it would lead to a conflict with the deuterium-abundance constraint for $\Omega_0 = 1$. In fact, if $h_0 = 0.65$, say, then setting this ratio equal to $0.01h_0^{-2}$, for consistency with the deuterium abundance, gives $\Omega_0 \approx 0.23$. Thus, if the universal value of Ω_0 were claimed to be unity (as originally required by inflation), then the conclusion has to be that baryons are selectively located in clusters, while the non-baryonic matter fills the intercluster space. This epicyclic statement could be avoided, if one admits to a low-density Universe.

17.7.3 Dark energy

We have already referred to the Type Ia supernovae in high-redshift galaxies providing a test of the Hubble relation. In section 17.1.1 we found that Einstein's cosmological constant λ had to be resurrected in order to explain the observed redshift–magnitude relation. Later it became apparent that this remedy was too simple and rather unsatisfactory on two counts. Firstly, if one assumes (as is natural) that the value of λ observed today is a relic of the inflationary era, then a major difficulty arises. The inflation was driven by an effective λ, some 108 orders of magnitude higher than the λ observed today. In short, we want today a finely tuned relic λ of magnitude $\sim 10^{-108}$ of the primordial inflationary λ. Secondly, a *constant* λ turns out to be inadequate to explain the z–m curve for supernovae up to redshifts ~ 1.6. One needs an epoch-dependent λ, so an elaborate theoretical structure that goes well beyond Einstein's simple modification needs to be created.

Today this extra force is popularly known as *dark energy* and various theoretical models for it are being investigated.

17.8 The standard model of cosmology

The studies during the first decade of the twenty-first century have concentrated largely on the microwave background and the observation of redshifts and apparent magnitudes of Type Ia supernovae. These studies, together with the constraints imposed by structure-formation scenarios, the age of the Universe, the abundances of light nuclei, the density of dark matter, etc., have led to the following breakdown of the matter–energy contents of the most favoured or 'standard' model of cosmology:

$$\Omega_{\mathrm{m}} = 0.04, \qquad \Omega_{\mathrm{NBDM}} = 0.23, \qquad \Omega_\Lambda = 0.73. \qquad (17.37)$$

The fact that these parameters can be quoted with very small error bars has led to the adjective 'precision cosmology' being applied to the standard model. Also, because several constraints are satisfied by these parameters, this approach is referred to as 'concordance cosmology'.

Complimentary though these adjectives are, the present confidence of cosmologists in the standard model may turn out to be illusory. For, to begin with, these 'omega' values are not determined directly. Neither are the theoreticians able to identify the non-baryonic dark matter in terms of any known particle. The dark-energy part also rests on highly speculative physics, having moved away from the original cosmological constant of Einstein. In fact, if we take the 'last-scattering-surface' interpretation of the microwave background, we find that no event in the Universe, prior to this epoch of redshift ~ 1000, could be directly observed. Thus we have to rely on indirect evidence in the form of relics of those early epochs – and interpretation of relics can very often be controversial. Certainly it is not unique.

Likewise, the high-energy physics on which the properties of the early Universe rest has not been tested beyond energies of the order of ~ 1000 GeV. Thus there is a big gap between what has been tested and verified and what is uncritically assumed, the latter being twelve orders of magnitude in energy above the former. A key feature of the standard model, inflation, also rests in the speculative era. As mentioned earlier, the inflationary model has not yet been obtained as an exact solution of the field equations with matched boundary conditions, like the Kerr and Schwarzswchild solutions are. When stellar astrophysicists were faced with the possibility of stars existing as very compact balls filled with neutrons, the so-called *neutron stars*, they spent considerable research effort trying to understand the state of matter at densities $\sim 10^{15}$g cm^{-3}. Big-bang cosmologists have not spent any part of their time worrying about matter densities as high as $\sim 10^{50}$ g cm^{-3}, that existed at the GUT epoch.

These are some of the reasons why one needs to be cautious about any conclusions drawn from such early epochs. Before modelling the Universe in such extreme conditions, there is a need to examine the theoretical foundations of relativity, to get a feel for the quantum theory of gravity and to clarify the uncertainties existing in some of the phenomena observed in extragalactic astronomy. In the following, final, chapter we discuss briefly some of these frontier areas.

Exercises

1. Suppose the intergalactic medium produces an absorption cross section $\kappa(\lambda)$ per unit mass at wavelength λ. Show that the increase in apparent magnitude of a galaxy of redshift z_1 in the steady-state Universe due to this process is given by

$$\Delta m = 2.5 \log_{10} e \cdot \tau(z_1, \lambda_0),$$

where λ_0 is the wavelength of observation and

$$\tau(z_1, \lambda_0) = \frac{c\rho_0}{H} \int_0^{z_1} \kappa\left(\frac{\lambda_0}{1+z}\right) \frac{dz}{1+z},$$

where ρ_0 is the density of absorbing material. (For the steady-state Universe assume the de Sitter line element and a constant density of matter.)

2. If the luminosity of a galaxy seen at the epoch t is related to the present epoch t_0 by the formula

$$L(t) = L(t_0)\left(\frac{t_0}{t}\right)^a,$$

where $a = $ constant, calculate the change in the apparent magnitude produced by this effect for a galaxy of redshift z in the Einstein–de Sitter cosmology.

3. In a globular cluster the metal content $Z \sim 10^{-3}$ and the ratio of horizontal-branch stars to red giants is 0.9. Show that in the $f = 1$ model the age of the globular cluster is about 11.9×10^9 years, whereas in the $f = 2$ model it is increased to around 2.0×10^{10} years.

4. Show that in a disc-shaped galaxy with surface density $\sigma(r) \propto r^{-1}$ one gets flat rotation curves

$$v^2 = 2\pi G r \sigma(r) = \text{constant}.$$

5. Suppose $I(\lambda) \propto \lambda^2$ in the range $2500\,\text{Å} < \lambda < 5000\,\text{Å}$. A galaxy of redshift 0.5 is being observed in a wavelength band centred on $5000\,\text{Å}$. Another galaxy of redshift 0.7 is also observed at $5000\,\text{Å}$. Show that the K-terms for the two galaxies will differ by $\sim 0.41^m$.

6. A radio galaxy of redshift $z = 0.1$ has a spectral function $\propto \nu^{-1}$ and a luminosity of 10^{44} erg s^{-1} over the frequency range $150\,\text{MHz} \le \nu \le 1500\,\text{MHz}$. For $h_0 = 1$ show that the flux density of the galaxy is ~ 350 Jy at $1000\,\text{MHz}$ and ~ 1750 Jy at $200\,\text{MHz}$. (Neglect any cosmological effects.)

7. In the Coma cluster of galaxies the observed velocity dispersion is ~ 861 km s^{-1}, while the radius of the cluster is $\sim 4.6h_0^{-1}$ Mpc. Show that the cluster mass given by the virial theorem is $\sim 1.5 \times 10^{15} h_0^{-1} M_\odot$. The total luminosity of the cluster is estimated at $\sim 7.5 \times 10^{12} h_0^{-2} L_\odot$. Show that the mass/light-ratio parameter η for the cluster is $\sim 300 h_0$.

8. Let $f(L)\,dL$ denote the number of radio sources per unit volume in the luminosity range $(L, L + dL)$. Suppose that for small redshifts the plot of $\log z$ against $\log L$ follows a straight line of slope $1/2$. Also assume that the number of points in equal intervals of $\log z$ is found to be constant. Using Euclidean geometry with distance $\propto z$, deduce from these observations that $f(L) \propto L^{-2.5}$. The survey is limited to sources with flux density exceeding S_0.

9. The nucleus ^{87}Rb decays to ^{87}Sr with a half-life of $\tau = 4.7 \times 10^{10}$ years. Let $X(t)$ and $Y(t)$ denote the numbers of these nuclei in a meteorite at any time

t, so that the quantity $X(t) + Y(t)$ is conserved. Let t_0 denote the epoch when the Solar System was formed. Show that a plot of relative abundances $X(t)/Z$ against $Y(t)/Z$, where Z is the number of ^{86}Sr nuclei (which remain unchanged), leads to a straight line whose slope is given by

$$\exp(\lambda t_0) - 1,$$

where $\lambda = \tau^{-1} \ln 2$.

10. The ratio of occupied levels for $J = 1$ and $J = 0$ states for the CN molecule in the star ζ-Ophiuchi is 0.55 ± 0.05 and in the star ζ-Persei it is 0.48 ± 0.15. The energy difference between the two levels is equal to kT, $T = 5.47\,K$ and the occupation weights are $g_1/g_0 = 3$. Deduce that the temperatures of the incident radiation lie in the respective ranges $3.22 \pm 0.15\,K$ and $3.00 \pm 0.6\,K$.

11. Suppose nuclear physics tells us that the age of a galaxy of redshift $z = 0.5$ is 10^{10} years. Use this information to set a limit on a function of H_0 and q_0. If $H_0^{-1} = 1.8 \times 10^{10}$ years, is $q_0 = 1$ possible?

12. A radio source shows an angular separation of $1'$ of arc from a galaxy of redshift $z = 0.44$. Using the Einstein–de Sitter cosmology, estimate the linear separation of the radio source from the galaxy, assuming that source and galaxy are at the same redshift. (Use $H_0^{-1} = 1.8 \times 10^{10}$ years.)

13. Let $\sigma(r)$ denote the surface mass density at a point P located at distance r from the centre of a thin, disc-shaped galaxy. Show that the gravitational force F_r at P is directed towards the centre of the galaxy and is given by

$$F_r = G \int_0^\infty \sigma(rx)x \, dx \int_0^{2\pi} \frac{(1 - x\cos\theta)d\theta}{(1 - 2x\cos\theta + x^2)^{3/2}}.$$

Chapter 18
Beyond relativity

We have come to the end of our account of the theories of relativity: special and general. While the former was briefly reviewed in the first chapter, we spent 16 chapters presenting the general theory from scratch. After preparing the background of vectors and tensors in the curved spacetime, we introduced the notions of parallel propagation, covariant differentiation, spacetime curvature and symmetries of motion. We then introduced physics through the notions of the action principle and energy-momentum tensors.

This was the appropriate stage to introduce the basics of general relativity: the principle of equivalence, Einstein's field equations and their Newtonian limit. Following these notions, we introduced the Schwarzschild solution and the various tests of general relativity, largely within the Solar System. We also discussed the budding field of gravitational radiation and the attempts to detect it coming from cosmic sources. Our next topic was relativistic astrophysics, which deals with compact massive objects such as supermassive stars and black holes. We also briefly touched upon the very interesting topic of gravitational lensing. This was followed by a discussion of some highlights of relativistic cosmology.

This presentation is indicative of the scope of general relativity. While it has created a niche for itself in theoretical physics as a remarkable intellectual exercise, it has also justified its status as the most effective physical theory of gravitation by explaining and predicting several gravitational phenomena. At the same time we need to look ahead and ask whether the search for the ideal theory of gravitation ends here or whether there is scope for further improvement in its framework. Certainly,

Fig. 18.1. The Foucault pendulum at the Inter-University Centre for Astronomy and Astrophysics takes 75 hours to complete one rotation round the vertical axis. The suspended ball slowly changes its direction of oscillation as seen against the background of the floral design below.

despite the successes of general relativity, we are still a long way from understanding gravity. When Newton was asked deep questions about the nature of gravity he replied *Non fingo hypotheses*.[1] Here we discuss a few assorted ideas inspired by general relativity, or attempting to take it further. They neither present the last word nor claim to be exhaustive.

18.1 Mach's principle

There are two ways of measuring the Earth's spin about its polar axis. By observing the rising and setting of stars the astronomer can determine the period of one revolution of the Earth around its axis: the period of $23^h56^m4^s.1$. The second method employs a Foucault pendulum whose plane gradually rotates around a vertical axis as the pendulum swings (see Figure 18.1). Knowing the latitude of the place of the pendulum, it is possible to calculate the Earth's spin period. The two methods give the same answer.

At first sight this does not seem surprising. Since we are measuring the same quantity, we should get the same answer regardless of the method used. Closer examination, however, reveals why the issue is non-trivial. The two methods are based on different assumptions. The first method measures the Earth's spin period against a background of distant stars, whereas the second employs standard Newtonian mechanics in a spinning frame of reference. In the latter case, we take note of how Newton's laws of motion are modified when their consequences are measured in a frame of reference spinning relative to the 'absolute space' in which these laws were assumed, by Newton, to hold.

[1] I do not frame hypotheses.

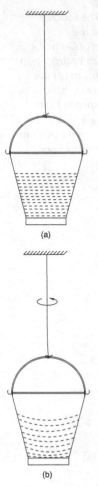

Fig. 18.2. A schematic description of Newton's bucket experiment. The stationary bucket (a) hanging by a thread has the water level in it flat (and horizontal). However, if the bucket is twisted round the thread and let go, the twisted thread unwinds and makes the bucket spin. As the bucket spins rapidly, the water level in it becomes curved (b), rising at the rim and dipping at the centre. Newton argued that this experiment demonstrated rotation relative to absolute space.

Thus, implicit in the assumption that equates the two methods is the coincidence of absolute space with the background of distant stars. It was Ernst Mach in the last century who pointed out that this coincidence is non-trivial. He read something deeper in it, arguing that the postulate of absolute space that allows one to write down the laws of motion and arrive at the concept of inertia is somehow intimately related to the background of distant parts of the Universe. This reasoning is known as 'Mach's principle' and we will analyse it further.

When expressed in the framework of the absolute space, Newton's second law of motion takes the familar form

$$\mathbf{P} = m\mathbf{f}. \tag{18.1}$$

This law states that a body of mass m subjected to an external force $\mathbf{P}$ experiences an acceleration $\mathbf{f}$. Let us denote by Σ the coordinate system in which $\mathbf{P}$ and $\mathbf{f}$ are measured. This frame represents Newton's absolute space.

Newton was well aware that his second law has the simple form (18.1) only with respect to Σ and those frames that are in uniform motion relative to Σ. If we choose another frame Σ' that has an acceleration $\mathbf{a}$ relative to Σ, the law of motion measured in Σ' becomes

$$\mathbf{P}' \equiv \mathbf{P} - m\mathbf{a} = m\mathbf{f}'. \tag{18.2}$$

Although (18.2) outwardly looks the same as (18.1), with $\mathbf{f}'$ the acceleration of the body in Σ', something new has entered into the force term. This is the term $-m\mathbf{a}$, which has nothing to do with the external force but depends solely on the mass m of the body and the acceleration $\mathbf{a}$ of the reference frame relative to the absolute space. Realizing this aspect of the additional force in (18.2), Newton termed it 'inertial force'. As this name implies, the additional force is proportional to the inertial mass of the body. Newton discusses this force at length in his *Principia*, citing the example of a rotating water-filled bucket (see Figure 18.2).

According to Mach, the Newtonian discussion was incomplete in the sense that the existence of the absolute space was postulated arbitrarily and in an abstract manner with no reference to the distant stellar background. Why does Σ have a special status in that it does not require the inertial force? How can one physically identify Σ without recourse to the second law of motion, which is based on it?

To Mach the answers to these questions were contained in the observation of distant parts of the Universe. It is the Universe that provides a background reference frame that can be identified with Newton's frame Σ. Instead of saying that it is an accident that Earth's rotation velocity relative to Σ agrees with that relative to the distant parts of the Universe, Mach took it as proof that the distant parts of the Universe must somehow enter into the formulation of local laws of mechanics.

One way this could happen is by a direct connection between the property of inertia and the existence of the universal background. To see this point of view, imagine a single body in an otherwise empty Universe. In the absence of any forces (18.1) becomes

$$m\mathbf{f} = \mathbf{0}. \tag{18.3}$$

What does this equation imply? Following Newton we would conclude from (18.3) that $\mathbf{f} = \mathbf{0}$, that is, the body moves with uniform velocity. But we now no longer have a background against which to measure velocities. Thus $\mathbf{f} = \mathbf{0}$ has no operational significance. Rather, the lack of any tangible background for measuring motion suggests that $\mathbf{f}$ should be completely indeterminate. It is not difficult to see that such a result follows naturally, provided that we come to the remarkable conclusion, also possible from (18.3), that

$$m = 0. \tag{18.4}$$

In other words, the measure of inertia depends on the existence of the background in such a way that in the absence of the background the measure vanishes! This aspect introduces a new feature into mechanics not considered by Newton. The Newtonian view that inertia is a property of matter has to be augmented to the statement that inertia is a property of matter as well as of the background provided by the rest of the Universe. This general idea can be identified with *Mach's principle*.

Such a Machian viewpoint not only modifies local mechanics but also introduces new elements into cosmology. For there is no basis now for assuming that particle masses would necessarily stay fixed in an evolving Universe. This is the reason for considering cosmological models anew from the Machian viewpoint. Presented here are some instances of how various physicists have given quantitative expression to Mach's principle and arrived at new cosmological models.

Although Einstein himself was initially impressed by Mach's arguments, he later came to discount them because they suggested action at a distance. For a historical review of Mach's principle see the collection of articles edited by Barbour and Pfister [75].

Kurt Gödel demonstrated in 1949 that spinning universes in general relativity do not subscribe to Mach's principle. Gödel's model had the universe spinning so that the observer at rest in the local inertial frame of such a universe would see the distant parts of the universe rotating. This counter-example demonstrated that the basic argument on which Mach's principle is formulated cannot itself be guaranteed by general relativity. Other such 'anti-Machian' solutions later emerged from relativistic cosmology and it became clear that one has to go beyond relativity to incorporate Mach's principle.

We briefly recall the twin paradox of Chapter 1. If twins A and B argue as to which of them is the inertial observer, we can now suggest a practical way of resolving the argument. The one who remains unaccelerated relative to the frame provided by the distant parts of the universe is the inertial observer.

18.2 The Brans–Dicke theory

There have been attempts by later scientists such as Sciama [76], Brans and Dicke [77] and Hoyle and Narlikar [78, 79], which modified general relativity and hence cosmology to give explicit quantitative expression to Mach's ideas. Of these we will refer to the Hoyle–Narlikar approach in Section 18.4.1. The Brans–Dicke theory played a very interesting role in offering alternative predictions of the Solar-System tests of gravity, which prompted an upsurge of experimental techniques to make accurate measurements for distinguishing between the predictions of this theory and general relativity. The action principle of this theory is given by replacing the Hilbert term in general relativity by

$$\mathcal{A} = \frac{1}{16\pi} \int_{\mathcal{V}} (\phi R + \omega \phi^{-1} \phi^k \phi_k) \sqrt{-g} \, \mathrm{d}^4 x.$$

The parameter ω distinguishes the Brans–Dicke theory from general relativity, with the scalar field ϕ playing the role of G^{-1}. By appropriate scaling, one can show that this theory approaches general relativity as $\omega \to \infty$. The Solar-System tests have placed a lower limit of the order of ~ 3000 on this parameter.

Nevertheless, the cosmological models emerging from the Brans–Dicke theory can still be significantly different from standard cosmology sufficiently early in the Universe. For example, the inflationary regime can be different because of the additional terms in the action. The idea seemed to solve some of the conceptual problems of the original inflationary model but ran into trouble because the distortions it produced in the cosmic microwave background were unacceptably high. Undeterred by these setbacks, the inflation enthusiasts explored a variation on the Brans–Dicke theme by adding higher-order couplings of the scalar field with gravity, which led to the notion of 'hyper-extended inflation'. (See for example the paper by Mathiazhagan and Johri [80].) However, none of these ideas seem to have received much following in later years.

To summarize, considerations of the early and very early Universe could possibly probe the differences between general relativity and the Brans–Dicke theory further. Insofar as observations of relatively recent epochs are concerned, however, because of the largeness of ω, for most practical purposes the differences between the Brans–Dicke theory and general relativity are insignificant.

18.3 Spacetime singularity and matter creation

When the Friedmann models were finally recognized as providing the simplest models of the expanding Universe, one aspect of these models was somewhat disturbing – their origin in a spacetime singularity. At the beginning this was considered an anomaly; the choice of an exceptional symmetry of spacetime (the Weyl postulate and the cosmological principle) was held responsible for the singularity. Thus, it was argued, the introduction of anisotropy in the form of shear and spin in the Universe would remove the singularity. In this context, A. K. Raychaudhuri obtained a simple but elegant result that was to have a far-reaching effect on the issue of spacetime singularity [81]. Raychaudhuri showed, with the help of an equation determining the evolution of a volume element, that the introduction of spin goes towards removing the singularity, whereas shear has the opposite effect. The irony was that one could obtain solutions with shear and no spin, but not with spin and no shear. So a demonstration of the avoidance of singularity remained an unattainable goal.

The Raychaudhuri equation arises in relativistic cosmology when we look at the bundle of timelike geodesics defined by the Weyl postulate. If u^i is the unit tangent to the geodesic, we define the spin-vorticity 3-tensor for the cosmic fluid by $\omega_{\mu\nu} = \frac{1}{2}(u_{\mu;\nu} - u_{\nu;\mu})$.

Writing the line element in the form

$$ds^2 = dt^2 + 2g_{0\mu}\, dt\, dx^\mu + g_{\mu\nu}\, dx^\mu\, dx^\nu, \tag{18.5}$$

where the geodesics are specified by $x^\mu = \text{constant}$ and t is the cosmic time, the (0, 0) component of field equations in the case of dust of density ρ then becomes

$$\frac{\ddot{Q}}{Q} = \frac{1}{3}(2\omega^2 - 4\pi\, G\rho - \phi^2), \tag{18.6}$$

where $Q^6 = -g$ and

$$2\omega^2 = -g^{\lambda\mu} g^{\sigma\tau} \omega_{\lambda\sigma} \omega_{\mu\tau},$$

$$\phi^2 = \frac{1}{4} g^{\mu\nu} \dot{g}_{\nu\sigma} g^{\sigma\lambda} \dot{g}_{\lambda\mu} - \frac{1}{3}\left(\frac{\partial}{\partial t} \ln \sqrt{-g}\right)^2. \tag{18.7}$$

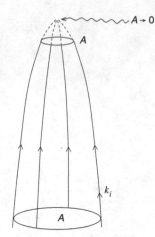

Fig. 18.3. The bundle of geodesics focusses in the future with its cross section A decreasing to zero. This effect was discussed in the context of spacetime singularity by A. K. Raychaudhuri.

The ϕ term is identified with shear and it goes the opposite way (to the spin term) through promoting singularity by helping the scale of the cosmic volume, Q, approach zero. It vanishes when the expansion is isotropic.

The Raychaudhuri equation can be stated in a slightly different form as a *focussing theorem*. In this form it describes the effect of gravity on a bundle of null geodesics spanning a finite cross section. Denoting the cross section by A, we write the equation of the surface spanning the geodesics as $f =$ constant. Define the normal to the cross-sectional surface by $k_i = \partial f / \partial x^i$. Figure 18.3 shows the geometry of the bundle.

By invoking the analogue of the hydrodynamic conservation law, we deduce

$$k^l A_{,l} = [k^l_{,l}]A. \tag{18.8}$$

Additionally we also have from the null geodesic condition

$$k^l k_{i;l} = 0. \tag{18.9}$$

Using a calculation similar to that which led to the geodetic deviation equation in Chapter 5, we get the focussing equation as

$$\frac{1}{\sqrt{A}} \frac{d^2 \sqrt{A}}{d\lambda^2} = \frac{1}{2} R_{im} k^i k^m - |\sigma|^2, \tag{18.10}$$

where

$$|\sigma|^2 = \frac{1}{2} k_{i;m} k^{i;m} - \frac{1}{4} [k^n_{;n}]^2. \tag{18.11}$$

Equation (18.10) is similar to the Raychaudhuri equation with $|\sigma|^2$ being the square of the magnitude of shear. With Einstein's equations, we can rewrite (18.10) as

$$\frac{1}{\sqrt{A}} \frac{d^2 \sqrt{A}}{d\lambda^2} = -4\pi G \left(T_{im} - \frac{1}{2} g_{im} T \right) k^i k^m - |\sigma|^2. \tag{18.12}$$

For focussing of the bundle of rays we need $A \to 0$, so the right-hand side should be negative. This is helped by the shear term in the above equation, just as Raychaudhuri had found. The first term on the right-hand side of the focussing equation also has this property if

$$\left(T_{im} - \frac{1}{2} g_{im} T \right) k^i k^m \geq 0.$$

For dust we have $T_{im} = \rho u_i u_m$ and this condition is satisfied with the left-hand side equalling $\rho (u_i k^i)^2$. (Remember that k_i is a null vector, so $g_{im} k^i k^m = 0$.) Thus the normal tendency of matter is to focus light rays by gravity.

The *singularity* theorems of Penrose and Hawking [82] use this basic feature to state conditions that inevitably lead to a spacetime singularity. The condition of the positivity of the T_{ik} term in the equation above plays a crucial role in general. We will not go into these details except to highlight this work as a field deserving further research. In particular, the positive-energy condition suggests that there may be non-singular spacetimes if it is violated and there are negative energy fields. We will now describe a line of thinking in which such fields are used to avoid the initial (or *any*) singularity.

18.4 The quasi-steady-state cosmology

In the late 1940s, H. Bondi and T. Gold [83] and F. Hoyle [84] independently proposed the steady-state cosmology as an alternative to the standard cosmology. The cosmology envisaged the Universe as described by the Robertson–Walker line element, with $k = 0$ and $S(t) = \exp(Ht)$, where the Hubble constant H is strictly a constant. In fact the name 'steady state' implies that the spacetime has a timelike Killing vector, and that physical conditions at any epoch t are the same. One consequence of this requirement is that as the Universe expands there is creation of matter to keep its density ρ constant, the rate of creation per unit volume being $3H\rho$. The cosmology thus has no singular epoch and no hot past. Bondi and Gold believed that the entire dynamics and physics of the Universe should follow from a single principle which they enunciated as the *perfect cosmological principle*. This principle takes the usual cosmological principle a stage further by additionally requiring homogeneity of the Universe with time. For, the authors argued, without such an invariance being guaranteed, one cannot be confident that the laws of physics known today had the same form at all times past and present. Without such a guarantee, one cannot interpret observations of the distant Universe unambiguously. Bondi and Gold called this model the steady-state model. Hoyle arrived at the same model by modifying Einstein's field equations by adding terms that allowed for creation of matter. His approach had been more physical than philosophical and dictated by the requirement to understand the origin of all the matter observed in the Universe today.

In the 1950s and early 1960s, the steady-state cosmology provided a stimulus to observers to stretch the limits of their observing technology to test the predictions of this model and to distinguish it from the standard cosmology. In the end most cosmological tests involving discrete source populations turned out to be inconclusive, as it became clear that one first needs to understand the various sources of observational errors as well as the physical properties of the sources used for the tests before drawing unequivocal conclusions. Nevertheless, the steady-state theory failed on two important counts, namely providing a setting for the origin of light nuclei (especially deuterium and helium) and explaining the origin of the microwave background.

The theory, which had been abandoned in the 1970s and 1980s, was revived in a new form by F. Hoyle, G. Burbidge and J. V. Narlikar in 1993 [85] and developed to some level of detail in a number of papers. These details include the basic rationale and genesis of the idea, its astrophysical and observational consequences, a formal theoretical structure, cosmological models and a model for structure formation. (For these details in one place, see Reference [86].) We briefly summarize and assess this quasi-steady-state-cosmology (QSSC) model, since we feel that, although it has not been studied in anything like the detail one finds for the standard model, at present it is the only available alternative to which the same observational and theoretical criteria for a viable cosmology can be applied.

18.4.1 Broad features of the QSSC model

The theoretical structure of this cosmology and its relationship to observations are summarized below.

(1) The cosmology is based on the Machian theory of gravitation first proposed by Hoyle and Narlikar in 1964 [78, 79]. The theory of Hoyle and Narlikar starts with the premise that the inertial mass of any particle is determined by the surrounding Universe. In field-theoretical language, the inertia is a scalar field whose behaviour is determined by an action principle. As shown later by Hoyle et al. [87], the theory permits broken particle world lines, i.e., creation and destruction of matter. In the cosmological approximation of a well-filled Universe, the field equations become.

$$R_{ik} - \tfrac{1}{2}g_{ik}R + \lambda g_{ik} = -\frac{8\pi G}{c^4}[T_{ik} - f(C_i C_k - \tfrac{1}{4}g_{ik}C^l C_l)], \quad (18.13)$$

where C is the scalar field representing the inertial effect associated with the creation of a new particle, and a consequence of Mach's principle is that the constants in these equations can be related to the fundamental constants of microphysics and the large-scale features of the Universe. Thus, restoring c for the sake of units, we have

$$G = \frac{3\hbar c}{4\pi m_{\mathrm{P}}^2}, \qquad \lambda = -3\left(\frac{m_{\mathrm{P}}}{\mathcal{N}}\right)^2, \qquad f = \frac{2}{3}\hbar c.$$

Here m_{P} is the mass of the basic particle created and $\mathcal{N}$ the number of such particles in the observable Universe. From the above it is easy to identify m_{P} with the Planck mass, which makes $\mathcal{N}$ of the order of 10^{60}

and λ of the order of $10^{-56}\,\text{cm}^{-2}$. Notice that its sign is negative, i.e., it represents an *attractive* rather than a repulsive force. The coupling constant f is positive, thus requiring the C-field stress and energy to act repulsively on matter and space because of the explicit minus sign in the stress tensor. It is assumed that the creation of a particle of mass m_{P} is possible, provided that a 'creation threshold' is attained by the ambient C-field, namely, $C_l C^l = m_{\text{P}}^2$. At the time of creation the momentum of the new particle is balanced by C_i. In such cases, we may have situations with $T^{ik}_{;k} \neq 0$, although the divergence of the right-hand side overall is zero.

(2) The cosmological models in this theory are driven by the creation process, and it is argued that the creation process does not occur uniformly everywhere, but *preferentially* near massive objects that have collapsed to something close to the state of a black hole. This is because the gravitational field in the neighbourhood of such an object is high and permits the local value of $C_l C^l$ to rise high enough to reach the creation threshold. The Planck particle so created is assumed to be unstable, however, and decays, within a time scale of the order of 10^{-43} s, into baryons, leptons, pions, etc. along with the release of a substantial amount of energy. The creation of matter is compensated for by the creation of the C-field, and, as the strength of the field rises, its repulsive effect makes the space expand rapidly (as in the inflationary scenario), thus causing an explosive ejection of matter and energy. The origin and outpouring of very high energy in quasars, active galactic nuclei, etc. are claimed by the QSSC to be phenomena representing minicreation events like these.

In a typical minicreation event, the central object itself may break up as its gravitational binding is loosened by the growth of the negatively coupled C-field. Thus it may also happen that the central object may eject a coherent piece along the line of least resistance. The QSSC authors argue that some of the 'anomalous redshift' cases (see [88, 89]) can be explained by invoking this phenomenon. What are these cases? Typically in such a case one sees two objects, e.g. a quasar and a galaxy, say, very close to each other but with very different redshifts. The probability of their being projected close to each other by chance is very low. Are they near neighbours? If so, their different redshifts violate Hubble's law. Two cases of such anomalous redshifts are shown in Figure 18.4.

(3) The cosmological solutions are driven by the minicreation events, each of which produce local expansions of space. The averaged effect of a large number of such events over a cosmological volume can be approximated by a homogeneous and isotropic solution of the field equations. As in the standard cosmology, the Robertson–Walker line element can be used to describe such a spacetime. The work of Sachs *et al.* [90] has shown that the generic solution for all three cases,

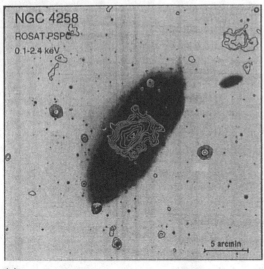

NGC 4258
ROSAT PSPC
0.1-2.4 keV

5 arcmin

(a)

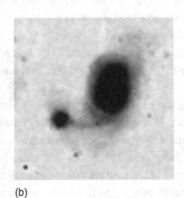

(b)

Fig. 18.4. Two typical cases of anomalous redshifts. In (a) we have two quasars of redshifts 0.4 and 0.65 aligned across an NGC galaxy of redshift 0.002. The precise alignment and close proximity suggest ejection of quasars, which are X-ray sources, by the galaxy which also houses an X-ray source. In (b) we have a big galaxy, NGC 7603, apparently connected by a filament to a companion galaxy. The redshifts of the two galaxies are 0.029 and 0.056, respectively. In either of cases (a) and (b), for maintaining consistency with Hubble's law one has to assume all these configurations to be projection effects with probabilities as low as 10^{-4}.

$k = +1, 0, -1$, has a long-term steady expansion interspersed with short-term oscillations. For example, the scale factor for $k = 0$ is given by

$$S(t) = \exp(t/P)[1 + \eta \cos \tau(t)],$$

where $0 < \eta < 1$, so that S oscillates between two finite values and $\tau(t)$ is almost like t during most of the oscillatory cycle, differing from it mostly during the stage when S is close to the minimum value. The period of oscillation Q is small compared with P. The QSSC is therefore characterized by the following parameters: P, Q, η and z_{max}, the maximum redshift seen by the present observer in the current cycle. Sachs *et al.* [90] took $P = 20Q$, $Q = 4.4 \times 10^{10}$ years, $\eta = 0.8$ and $z_{max} = 5$ as an indicative set of values. The QSSC workers have argued that the cosmology is by no means tightly constrained around these values by the various cosmological tests. Figure 18.5 illustrates one such case.

(4) How is the cosmic microwave background (CMB) produced in this model? The QSSC oscillations are finite, with the maximum redshift observable in the present cycle at ~5–6. Thus each cycle is matter-dominated. The radiation background is, however, maintained from one cycle to next. Thus, from the minimum scale phase of one cycle to next, its energy density is expected to fall by a factor $\exp(-4Q/P)$. This drop is made up by the thermalization of starlight produced during the cycle. Thus, if ϵ is the energy density of starlight generated in a cycle and u_{max} is the energy density of the CMB at the start of a cycle, then $\epsilon \cong 4u_{max}Q/P$. If the cycle minimum occurred at redshift z_{max},

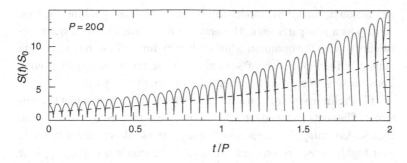

Fig. 18.5. The scale factor of a typical model of the quasi-steady-state cosmology. See the text for details.

then the present CMB energy density would be $P\epsilon/[4Q(1+z_{max})^4]$. By substituting the values of ϵ, P, z_{max} and Q we can estimate the present-day energy density of the CMB and the result agrees well with the observed value of $\sim 4 \times 10^{-13}$ erg cm^{-3} corresponding to a temperature ~ 2.7 K.

How is the starlight thermalized? Consider the following scenario. The cooling of metallic vapours produces whisker-like particles of lengths ~ 0.5–1.0 mm, which convert optical radiation into millimetre-wave radiation. Such whiskers typically form in the neighbourhood of supernovae (which synthesize and eject metals), and are subsequently pushed out of the galaxy through the pressure of shock waves. It can be shown that a density of $\sim 10^{-35}$ g cm^{-3} of such whiskers close to the minimum of the oscillatory phase would suffice for thermalization of starlight.

While the thermalized radiation from previous cycles will be very smoothly distributed, a tiny fraction ($\sim 10^{-5}$) will reflect anisotropies on the scales of rich clusters of galaxies in the present cycle. The angular scales for this anisotropy will be of the order of $\sim 1/100$, $-1/250$ for clusters and superclusters, corresponding to l-values ~ 100–200. A recent comparison with the WMAP data shows an acceptable fit to observations of the power spectrum of CMB fluctuations [91].

(5) In a recent paper Burbidge and Hoyle [92] argued that a case may be made for *all* isotopes having been made in stars, including the light ones generally assumed to be of primordial origin. They showed that possible stellar scenarios exist for production of these nuclei.

(6) The QSSC has been applied to the redshift–magnitude relation obtained by using Type Ia supernovae. Narlikar *et al.* [93] have reexamined the problem in the context of the QSSC for the data used for fitting the standard models, with or without the cosmological constant. As we have seen, the QSSC requires intergalactic dust in the form of metallic whiskers. This whisker population acts to produce further absorption in the light from distant galaxies and supernovae therein. Taking this effect

into account, the QSSC model can be fitted to data by taking the dust density as a free parameter. The optimized fit turns out to be quite satisfactory. Also the optimum whisker density turns out to be in the right range for thermalization of starlight into the microwave background. Thus there is an overall consistency in the parameters used.

(7) Preliminary work on structure formation has shown that the pattern of filaments and voids for clusters can be generated by minicreation events. Assuming that creation of new galaxies takes place selectively near highly dense regions, and that too at the maximum density phase of a typical QSSC cycle, one can simulate the resulting distribution for 10^5–10^6 galaxies on a computer. It is observed that an initial random distribution changes over into a supercluster–void distribution after a few cycles. The two-point correlation function of the galaxies created also tends to a power-law form with the index -1.8, as observed. See Reference [86].

While the various physical and astrophysical aspects of the QSSC have not been studied in anything like the depth to which the standard cosmology has been probed, these preliminary studies suggest that the cosmology, certainly as an alternative to the currently favoured option, deserves more critical attention than it has so far received.

18.5 Quantum gravity

Experience in the rest of physics (except gravity) shows that the classical equations of fields and particles break down at the microscopic level, to be replaced by the notions of quantum theory. When does one make a transition from the classical to the quantum version? A 'rule of thumb' is to evaluate the action $\mathcal{A}$ over the characteristic 4-volume for the problem and compare with $\hbar$. If the ratio $\mathcal{A}/\hbar$ is much larger than unity then the problem can be adequately handled by classical physics. If the ratio is comparable to unity then we need the quantum version to solve the problem.

How does this prescription work for gravity? A look at the action principle (8.7) shows that the limit sought above can be obtained by equating the gravitational action

$$\mathcal{A}_g = \frac{c^3}{16\pi G} \int_\mathcal{V} R\sqrt{-g}\, \mathrm{d}^4 x \qquad (18.14)$$

to Planck's constant. For $\mathcal{A}_g \gg \hbar$ we can trust our classical description of spacetime geometry, whereas for $\mathcal{A}_g \ll \hbar$ a quantum description of cosmology is indispensable. But to evaluate $\mathcal{A}_g$ we need $\mathcal{V}$, the 4-volume of the spacetime manifold.

In the big-bang model we take $\mathcal{V}$ as the 4-volume enclosed by the particle horizon and bounded by the time span of the Universe. Thus at any epoch t for $k = 0$, $S \propto t^{1/2}$, the particle horizon is defined by

$$rS = 2ct.$$

For $S \propto t^{1/2}$, $R = 0$ and so $\mathcal{A}_g = 0$. However, this happens because the trace of T_k^i is zero in the early Universe. As an order of magnitude estimate we may take R_0^0 instead of R in the computation of $\mathcal{A}_g$: R_0^0 gives us an idea of how the geometrical part of the action changes with time. For $S \propto t^{1/2}$, $R_0^0 = 3/4(c^2 t^2)$. Thus up to the epoch t

$$\mathcal{A}_g \sim \frac{c^4}{16\pi G} \int_0^t \frac{3}{4c^2 t_1^2} \frac{4\pi}{3} (2ct_1)^3 \, dt_1 = \frac{c^5}{4G} t^2.$$

By equating $\mathcal{A}_g$ to $\hbar$ we get

$$t = 2t_P = 2\sqrt{\frac{G\hbar}{c^5}} \cong 10^{-43} \text{ s}. \tag{18.15}$$

This time span is called the *Planck time. No classical discussion of gravity can be pushed to time scales $t < t_P$.* We have already encountered very short time scales of the order of 10^{-38} s in Chapter 16 when GUTs operated. The above quantum-gravity time scale corresponds to an even higher energy of $E \sim 10^{19}$ GeV. This energy, as seen from (18.15), is simply $\sim \hbar/t_P$.

Thus the present discussions of GUTs and cosmology already take us right up to the Planck epoch. Whether the Universe did indeed have a spacetime singularity at $t = 0$ should be determined not by classical general relativity but by an appropriate theory of quantum gravity.

There are several conceptual and operational problems on the way to a quantum theory of gravity, if we are to look for a quantized version of general relativity. To begin with, the non-linearity of relativity makes the methods which work for standard 'flat-space field theories' inapplicable here. Secondly, in relativity spacetime geometry and gravity are inextricably mixed and so one is not sure what is to be quantized. Thirdly, in flat-space quantizations, inclusion of the dynamical nature of geometry is not required: here it is an essential feature of the problem.

It is not surprising therefore that the quantized version of general relativity has not yet emerged. At present the goal of having a working theory of quantum gravity seems far away. The different approaches that have been tried in order to quantize gravity do not agree on the answer to the following question: did the Universe have a singular epoch? A simple approach based on conformal fluctuations suggests that, if we include quantum fluctuations of homogeneous and isotropic universes, then the spacetime singularity would 'most probably' be averted. The probability here is in the sense of quantum mechanics. An event is most probable if the quantum probability of its *not* happening has measure zero. The result can in fact be stated in a more general form proved by this author, namely that, if one considers most general quantum conformal fluctuations of a classical singular cosmological solution, then, most probably, singularity is not present in these fluctuations [94].

18.5.1 Radiating black holes

The quantum theory of gravity being recognized as a long-term project, work has been proceeding in the meantime on a simpler notion, that of field-theory quantization in curved spacetime; and it is producing some interesting (and unexpected) results. Here we present in brief the original example of this approach applied to black-hole physics.

As the name 'black hole' implies, we do not expect any radiation to come out of such an object. For a spherical object of mass M, the black-hole condition is reached when its surface area equals $4\pi R_s^2$, where R_s, the Schwarzschild radius, is given by

$$R_s = \frac{2GM}{c^2}. \tag{18.16}$$

No material particle or light signal emitted from $R \leq R_s$ can go into the region $R > R_s$: at least, this is what classical general relativity tells us.

We saw in Chapter 13 that the behaviour of black holes is in many ways analogous to thermodynamics. Thus the area of the horizon is like entropy and surface gravity like temperature. Can this analogy be pushed further, closer to becoming reality? If so, temperature implies radiation and the black hole is expected to radiate. This seemed a very unlikely conclusion given the physical nature of black holes.

Nevertheless, in 1974 Stephen Hawking [95] made the remarkable suggestion that a black hole can radiate. Hawking's calculation went beyond classical physics: it considered what happens when any field (for example, the electromagnetic field) is *quantized* in the spacetime containing a black hole. As we have already seen, the quantum-mechanical description of vacuum is much more involved than the classical description, which simply states that a vacuum is empty. According to quantum field theory, the vacuum is seething with virtual particles and antiparticles whose presence cannot be detected directly. Their interference with physical processes in spacetime can, however, lead to detectable results. Hawking found that one such result when considered in the spacetime outside a black hole is that an observer at infinity sees a flux of particles coming out from the vicinity of a black hole. We will not go through

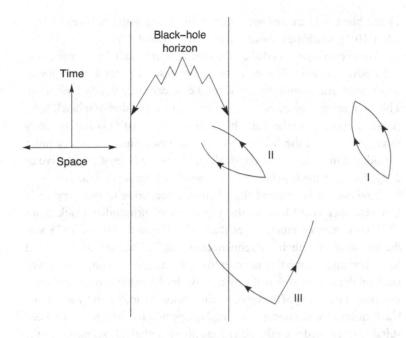

Fig. 18.6. The event horizon of a black hole is shown in the midst of the vacuum containing virtual particle–antiparticle pairs. In some cases one member of the pair with negative energy (case III) is gobbled up by the black hole. The other member with positive energy is set free and gives the impression that the black hole has emitted it.

the calculations leading to this result; we will simply study the consequences of such a process in the early Universe. Figure 18.6 provides a qualitative description of how the Hawking process operates. Not all aspects of the Hawking process have been worked out yet. An important issue still unresolved, for example, is that of back reaction: how the emission of particles by the black hole affects and alters the geometry of spacetime outside and what effect this change has on the process of radiation by the black hole.

The idea we shall use here is that a spherical black hole of mass M ejects particles in a thermal spectrum of temperature T given by

$$kT = \frac{\hbar c^3}{8\pi GM} \sim 10^{26} M_g^{-1},$$ (18.17)

where $M_g = M$ expressed in grams. The emission of particles by the black hole as per the rules of blackbody radiation leads to a mass-loss rate given by

$$\frac{dM_g}{dt} \sim -10^{26} M_g^{-2}\, \text{s}^{-1}.$$ (18.18)

The $\sim$ implies that a numerical constant of the order of 1 appears on the right-hand side to take account of the number of particle species emitted. If we integrate (18.18) we find that the entire mass of the black hole is radiated away in a time τ given by

$$\tau \sim 3 \times 10^{-27} M_g^3\, \text{s}.$$ (18.19)

Thus a black hole created soon after the big bang with a mass exceeding $\sim 5 \times 10^{14}$g would just about last until the present day.

The process described above is slow to start with, when a black hole is massive and cold. However, as M decreases T rises and the mass-loss rate increases until finally it reaches a catastrophically high level. This final stage is often called evaporation or explosion of a black hole. As seen above, a stellar-mass black hole ($M_g \geq 10^{33}$) is hardly likely to explode within the lifetime of the Universe! Since the black holes considered in various astrophysical scenarios are at least as massive as $2M_\odot$, for them the Hawking process is only of academic interest.

However, it is claimed that there are scenarios in the very early Universe that could lead to the formation of primordial black holes (PBHs) of masses much lower than $M_\odot$. Bernard Carr in 1975 was the first to discuss their consequences at length. Carr investigated PBH formation and evaporation in order to see whether the currently observed nucleon density as well as the microwave background can be explained in terms of emission of baryons, leptons, photons and so on by low-mass black holes. These concepts are highly speculative, and have not been suitably integrated with the other (equally speculative!) scenarios of the very early Universe.

The interesting aspect of this approach is that PBHs act as sources of various particles that need somehow to be created in the Universe. The suggestion that PBHs evaporating today might account for the observed γ-ray bursts, however, does not seem to be correct, since the spectrum of γ-rays emitted by a PBH is not like the spectrum observed in burst events.

There are several loose ends still to be sorted out in the PBH scenario. At the deepest level one has to understand how they can form in the first place, since the usual process of gravitational collapse that is supposed to lead to stellar or more-massive black holes cannot apply here. Next one needs to express the concepts of thermodynamics and statistical mechanics in highly curved spacetime in order to give precise meaning to the notions of temperature and blackbody spectrum: the formulae (18.17) and (18.18) merely use a naive extrapolation of flat-spacetime thermodynamics. Further, the problem of back reaction still remains unresolved. Finally, on the observational front, this bizarre concept still awaits a befitting application in the real Universe.

18.6 Concluding remarks

This brings us to the end of this chapter as well as this book. We have tried to cover the theory of relativity at an elementary level. The present chapter gives some glimpses into concepts not covered in the book.

Since the 1960s general relativity has added several new feathers to its cap, both in applications to observations (e.g., relativistic astrophysics, gravitational radiation, gravitational lensing) and purely in theory (e.g., spacetime singularity, field quantization in curved spacetime). It has inspired further intellectual developments such as the loop theory of quantum gravity [96] and string theory [97]. We have kept away from both these descriptions since definitive and observable conclusions have still to emerge from these very interesting approaches.

In the end, we close by quoting a short verse by Jack C. Rossetter that was inspired in 1950 by the popular belief that general relativity is a very difficult-to-understand type of theory:

> To Einstein, hair and violin
> We give our final nod,
> Though understood by just two folks,
> Himself – and sometimes God.
> [From *The Mathematics Teacher*,
> November 1950, p. 341]

Most of that mystique round general relativity has by now dissipated and it is seen today as an intellectual achievement *par excellence*, enriching, rather than isolated from, the rest of physics.

Exercises

1. In Newtonian gravity an oblate Sun will generate a gravitational potential

$$\phi = \frac{GM_\odot}{r}\left[1 - J\left(\frac{R_\odot}{r}\right)^2 P_2(\cos\theta)\right],$$

where J is the quadrupole-moment parameter and P_2 is the second Legendre polynomial. Show that the orbit of a planet precesses because of the above gravitational effect at the rate $3\pi R_\odot^2 J/l^2$, where l is the semi latus rectum of the orbit. Estimate the precession rate for Mercury for $J = 2.5 \times 10^{-5}$. What significance does this calculation have for the Brans–Dicke theory?

2. The Brans–Dicke theory can be re-expressed as a theory in which $G = $ constant but the particle masses change with epoch. Show that this is achieved by a conformal transformation

$$\bar{g}_{ik} = \frac{\phi}{\bar\phi}g_{ik}, \quad \bar\phi = \text{constant}.$$

The field equations then become (in the new metric)

$$\bar{R}_{ik} - \frac{1}{2}\bar{g}_{ik}R = -\kappa \bar{T}_{ik},$$

where κ is constant. Although these look like Einstein's equations, the $\bar{T}_{ik}$ contain ϕ and its derivatives. Show from the new field equations that

$$\Box \ln \phi = \frac{8\pi G}{(2\omega + 3)c^4} \bar{T}$$

with $G =$ constant. This form of the theory was obtained by Dicke in 1962. The particle masses in this version vary as

$$\bar{m} = m \sqrt{\frac{\bar{\phi}}{\phi}}, \quad m = \text{constant}.$$

3. Show that the deceleration parameter for the steady-state universe is equal to -1 at all epochs.

4. Discuss the validity of the following statement: 'Of the various ways of resolving Olbers' paradox, the only way open to the steady-state model is that of the expansion of the Universe'.

5. Write down the expression for the angle subtended at the observer by a spherical cluster of radius R at $z = z_{max}$ in the QSSC. Relate this expression to the angular scale of anisotropy of the microwave background in the QSSC.

6. Explain why the QSSC does not have an Olbers-type problem of darkness of the night sky.

7. Assuming that our Galaxy has been radiating at the rate of 4×10^{43} erg s^{-1} for a time 3×10^{17} s and that this energy is derived from conversion of hydrogen to helium, estimate how much helium is formed in this way. (Energy of 6×10^{18} erg g^{-1} is released when hydrogen is converted to helium.) Comment on this answer in relation to the primordial mass fraction of helium obtained in Chapter 16.

8. Compute $\mathcal{A}_g(t)$ for the closed Friedmann model with given values of q_0 and h_0, taking the time interval as $(0, t)$ and the spatial extent covering the whole (spherical) space. Estimate the epoch at which $\mathcal{A}_g = \hbar$. Why do you get an answer different from t_P?

9. Show that, at the Planck epoch, the Schwarzschild radius of a primordial black hole just filling the particle horizon is of the same order as the Compton wavelength of the black hole.

References

[1] Michelson, A. A. and Morley, E. W. 1887, On the relative motion of the earth and the luminiferous aether, *Phil. Mag.*, S5, **24**, 449–463

[2] Einstein, A. 1905, Zur Elektrodynamik bewegter Körper, *Ann. der Phys.*, **17**, 891–921

[3] Minkowski, H. 1908, Die Grundgleichungen für die electromagnetischen Vorgänge in bewegten Körpern, *Nachr. Königl. Gesell. Wiss. Göttingen*, 53–111

[4] Synge, J. L. 1965, *Relativity: The Special Theory* (Amsterdam: North Holland)

[5] Stephenson, G. and Kilmister, C. W. 1958, *Special Relativity for Physicists* (Englewood Cliffs, NJ: Prentice Hall)

[6] Prokhovnik, S. J. 1967, *The Logic of Special Relativity* (Cambridge: Cambridge University Press)

[7] Eisenhart, L. P. 1926, *Riemannian Geometry* (Princeton, MA: Princeton University Press)

[8] Einstein, A. 1915, Zur allgemeinen Relativitätstheorie, *Preuβ. Akad. Wiss. Berlin Sitzber.*, 844

[9] Hilbert, D. 1915, Die Grundlagen der Physik, *Königl. Gesell. Wiss. Göttingen, Nachr. Math.-Phys. Kl.*, 395

[10] Schwarzschild, K. 1916, Über das Gravitationsfeld eines Maßpunktes nach der Einsteinschen Theorie, *Sitzber. Deutsch. Akad. Wiss. Berlin, Kl. Math.-Phys. Tech.*, 189

[11] Pound, R. V. and Rebka, G. A. 1960, Apparent weight of photons, *Phys. Rev. Lett.*, **4**, 337

[12] Taylor, J. H., Fowler, L. A. and McCullach, R. M. 1979, Measurements of general relativistic effects in the binary pulsar PSR 1913+16, *Nature*, **277**, 437

[13] Coles, P. 2001, Einstein, Eddington and the 1919 eclipse, in *Proceedings of International School on 'The Historical Development of Modern Cosmology'*, ASP Conference Series, Eds. V. J. Martinez, V. Trimble and M. J. Pons-Borderia

[14] Counselman, C. C. III, Kent, S. M., Knight, C. A., Shapiro, I. I. and Clark, T. A. 1974, Solar gravitational deflection of radio waves measured by very long baseline interferometry, *Phys. Rev. Lett.*, **33**, 1621

[15] Fomalont, E. B. and Sramek, R. A. 1975, A confirmation of Einstein's general theory of relativity by measuring the bending of microwave radiation in the gravitational field of the Sun, *Ap. J.*, **199**, 749

[16] Anderson, J. D., Esposito, W., Martin, C. L. and Muhleman, D. O. 1975,
      Experimental test of general relativity time-delay data from Mariner 6 and 7,
      Ap. J., **200**, 221

[17] Bertotti, B., Iess, L. and Tortora, P. 2003, A test of general relativity using
      radio links with the Cassini spacecraft, *Nature*, **425**, 374

[18] Will, C. M. 1998, The Confrontation between General Relativity and
      Experiment, Lecture Notes, SLAC Summer School on Particle Physics,
      gr-qc/9811036

[19] Møller, C. 1955, *Theory of Relativity* (Oxford: Clarendon Press)

[20] Landau, L. D. and Lifshitz, E. M. 1971, *The Classical Theory of Fields*, 4th
      edition (London: Pergamon)

[21] Misner, C. W., Thorne, K. S. and Wheeler, J. A. 1970, *Gravitation*,
      (San Francisco, CA: W. H. Freeman)

[22] Taylor, J. H. and Weisberg, J. M. 2003, The relativistic binary pulsar
      B1913+16, *Binary Radio Quasars*, ASP Conference Series, Eds. M. Bailes,
      D. J. Nice and S. E. Thorsett

[23] Hoyle, F. and Fowler, W. A. 1963, On the nature of strong radio sources,
      Nature, **197**, 533

[24] Chandrasekhar, S. 1939, *An Introduction to the Study of Stellar Structure*
      (Chicago, IL: University of Chicago Press)

[25] Bowers, R. and Deeming, T. 1984, *Astrophysics* (Boston, MA: Jones and
      Bartlett)

[26] Fowler, W. A. 1964, Massive stars, relativistic polytropes and gravitational
      radiation, *Rev. Mod. Phys.*, **36**, 545 and 1104

[27] Chandrasekhar, S. 1931, The maximum mass of ideal white dwarfs, *Ap. J.*,
      74, 81

[28] Chandrasekhar, S. 1964, Dynamical instability of gaseous masses
      approaching the Schwarzschild limit in general relativity, *Phys. Rev. Lett.*, **12**,
      114 and 437; also in *Ap. J.*, **140**, 417

[29] Zwicky, F. 1937, Nebulae as gravitational lenses, *Phys. Rev.*, **51**, 290

[30] Zwicky, F. 1937, On the probability of detecting nebulae which act as
      gravitational lenses, *Phys. Rev.*, **51**, 679

[31] Schneider, P., Ehlers, J. and Falco, E. E. 1992, *Gravitational Lenses* (Berlin:
      Springer-Verlag)

[32] Chitre, S. M. and Narlikar, J. V. 1979, On the apparent superluminal
      separation of radio source components, *M.N.R.A.S.*, **187**, 655

[33] Walsh, D., Carswell, R. F. and Weymann, R. J. 1979, 0957+561 AB: twin
      quasistellar objects or gravitational lens?, *Nature*, **279**, 381

[34] Datt, B. 1938, Über eine Klasse von Lösungen der Gravitationsgleichungen
      der Relativität, *Z. Phys.*, **108**, 314

[35] Oppenheimer, J. R. and Snyder, H. 1939, On continued gravitational
      contraction, *Phys. Rev.*, **56**, 455

[36] Eddington, A. S. 1924, A comparison of Whitehead's and Einstein's formulas,
      Nature, **113**, 192

[37] Kruskal, M. D. 1960, Maximal extension of Schwarzschild metric, *Phys. Rev.*,
      119, 1743

[38] Szekeres, G. 1960, On the singularities of a Riemannian manifold, *Publ. Mat. Debrecen*, **7**, 285

[39] Reissner, H. 1916, Über die Eigengravitation des elektrischen Feldes nach der Einsteinschen Theorie, *Ann. der Phys.*, **50**, 106

[40] Nordström, G. 1913, Zur Theorie der Gravitation vom Standpunkte des Relativitätsprinzips, *Ann. der Phys.*, **42**, 533

[41] Kerr, R. P. 1963, Gravitational field of a spinning mass as an example of algebraically special metrics, *Phys. Rev. Lett.*, **11**, 237

[42] Bardeen, J. M., Carter, B. and Hawking, S. W. 1973, The four laws of black-hole mechanics, *Commun. Math. Phys.*, **31**, 161

[43] Hawking, S. W. 1972, Black holes in general relativity, *Commun. Math. Phys.*, **25**, 152

[44] Bowyer, S., Byram, E. T., Chubb, T. A. and Friedman, H. 1965, Cosmic X-ray sources, *Science*, **147**, 394

[45] Narlikar, J. V., Appa Rao, K. M. V. and Dadhich, N. K. 1974, High energy radiation from white holes, *Nature*, **251**, 590

[46] Einstein, A. 1917, Kosmologische Betrachtungen zur allgemeine Relativitätstheorie, *Preuß. Akad. Wiss. Berlin Sitzber.*, 142

[47] De Sitter, W. 1917, On the relativity of inertia: remarks concerning Einstein's latest hypothesis, *Proc. Koninkl. Akad. Wetensch. Amsterdam*, **19**, 1217

[48] Hubble, E. P. 1929, A relation between distance and radial velocity among extragalactic nebulae, *Proc. Nat. Acad. Sci. (USA)*, **15**, 168

[49] Friedmann, A. 1922 and 1924, Über die Krummung des Raumes, *Z. Phys.*, **10**, 377 and **21**, 326

[50] Lemaître, Abbé 1931, A homogeneous universe of constant mass and increasing radius accounting for the radial velocity of extragalactic nebulae, *M.N.R.A.S.*, **91**, 483 (translated from the 1927 original in French)

[51] Robertson, H. P. 1935, Kinematics and world structure, *Ap. J.*, **82**, 248

[52] Walker, A. G. 1936, On Milne's theory of world-structure, *Proc. Lond. Math. Soc. (2)*, **42**, 90

[53] Narlikar, J. V. 2002, *An Introduction to Cosmology* (Cambridge: Cambridge University Press)

[54] Einstein, A. and de Sitter, W. 1932, On the relation between the expansion and the mean density of the universe, *Proc. Nat. Acad. Sci. (USA)*, **18**, 213

[55] Milne, E. A. 1935, *Relativity, Gravitation and World Structure* (Oxford: Clarendon Press)

[56] Mattig, W. 1958, Über den Zusammenhang zwischen Rotverschiebung und scheinbarer Helligkeit, *Astron. Nachr.*, **284**, 109

[57] Tolman, R. C. 1933, *Relativity, Thermodynamics and Cosmology* (Oxford: Oxford University Press)

[58] Hoyle, F. 1959, The relationship of radioastronomy and cosmology, in *The Paris Symposium on Radio Astronomy*, p. 529, Ed. R. N. Bracewell (Stanford, CA: Stanford University Press)

[59] Gamow, G. 1946, Expanding universe and the origin of elements, *Phys. Rev.*, **70**, 572

[60] Hayashi, C. 1950, Proton–neutron concentration ratio in the expanding universe at the stages preceding the formation of the elements, *Prog. Theor. Phys. (Japan)*, **5**, 224

[61] Yang, J., Schramm, D., Steigman, G. and Rood, R. T. 1979, Constraints on cosmology and neutrino physics from big bang nucleosynthesis, *Ap. J.*, **227**, 697

[62] Burbidge, E. M., Burbidge, G. R., Fowler, W. A. and Hoyle, F. 1957, Synthesis of the elements in stars, *Rev. Mod. Phys.*, **29**, 547

[63] Peebles, P. J. E. 1971, *Physical Cosmology* (Princeton, MA: Princeton University Press)

[64] Saha, M. N. 1920, Ionization in the Solar chromosphere, *Phil. Mag.*, **40**, 472

[65] McKeller, A. 1941, Molecular lines from the lowest states of diatomic molecules composed of atoms probably present in interstellar space, *Pub. Dom. Astrophys. Obs., Victoria, B. C.*, **7**, 251

[66] Alpher, R. A. and Herman, R. C. 1948, Evolution of the Universe, *Nature*, **162**, 774

[67] Penzias, A. A. and Wilson, R. W. 1965, Measurement of excess antenna temperature at 4080 Mc/s, *Ap. J.*, **142**, 419

[68] Mather, J. C., Cheng, E. S., Esplic, R. E. Jr. *et al.* 1990, A preliminary measurement of the cosmic microwave background spectrum by the Cosmic Background Explorer (COBE) satellite, *Ap. J. Lett.*, **354**, L37

[69] Kazanas, D. 1980, Dynamics of the universe and spontaneous symmetry breaking, *Ap. J.*, **241**, L59

[70] Guth, A. H. 1981, Inflationary universe: a possible solution to the horizon and flatness problems, *Phys. Rev.*, **D23**, 347

[71] Sato, K. 1981, First order phase transition of a vacuum and the expansion of the universe, *M.N.R.A.S.*, **195**, 467

[72] Reiss, A. G., Filippenko, A. V., Challis, P. *et al.* 1998, Observational evidence from supernovae for an accelerating universe and a cosmological constant, *A. J.*, **116**, 1009

[73] Perlmutter, S., Aldering, G., Goldhaber, G. *et al.* 1999, Measurement of Ω and Λ from 42 high redshift supernovae, *Ap. J*, **517**, 565

[74] Smoot, G. F., Bennett, C. L., Kogut, A. *et al.* 1992, Structure in the COBE differential microwave radiometer first-year maps, *Ap. J. Lett.*, **396**, L1

[75] Barbour, J. and Pfister, H. 1995, *Mach's Principle: From Newton's Bucket to Quantum Gravity* (Boston, MA: Birkhäuser)

[76] Sciama, D. W. 1953, On the origin of inertia, *M.N.R.A.S.*, **113**, 34

[77] Brans, C. and Dicke, R. H. 1962, Mach's principle and a relativistic theory of gravitation, *Phys. Rev.*, **124**, 125

[78] Hoyle, F. and Narlikar, J. V. 1964, A new theory of gravitation, *Proc. Roy. Soc.*, **A282**, 191

[79] Hoyle, F. and Narlikar, J. V. 1966, A conformal theory of gravitation, *Proc. Roy. Soc.*, **A294**, 138

[80] Mathiazhagan, C. and Johri, V. B. 1984, An inflationary universe in Brans–Dicke theory, *Class. Quant. Grav.*, **1**, L29

[81] Raychaudhuri, A. K. 1955, Relativistic cosmology I, *Phys. Rev.*, **98**, 1123

[82] Hawking, S. W. and Ellis, G. F. R. 1973, *The Large Scale Structure of Space-time* (Cambridge: Cambridge University Press)

[83] Bondi, H. and Gold, T. 1948, The steady state theory of the expanding universe, *M.N.R.A.S.*, **108**, 252

[84] Hoyle, F. 1948, A new model for the expanding universe, *M.N.R.A.S.*, **108**, 372

[85] Hoyle, F., Burbidge, G. and Narlikar, J. V. 1993, A quasi-steady state cosmological model with creation of matter, *Ap. J.*, **410**, 437

[86] Hoyle, F., Burbidge, G. and Narlikar, J. V. 2000, *A Different Approach to Cosmology* (Cambridge: Cambridge University Press)

[87] Hoyle, F., Burbidge, G. and Narlikar, J. V. 1995, The basic theory underlying the quasi-steady state cosmology, *Proc. Roy. Soc.*, **A448**, 191

[88] Arp, H. C. 1987, *Quasars, Redshifts and Controversies* (Berkeley, CA: Interstellar Media)

[89] Arp, H. C. 1998, *Seeing Red* (Montreal: Apeiron)

[90] Sachs, R., Narlikar, J. V. and Hoyle, F. 1996, The quasi-steady state cosmology: analytical solutions of field equations and their relationship to observations, *Astron. Astrophys*, **313**, 703

[91] Narlikar, J. V., Burbidge, G. and Vishwakarma, R. G. 2007, Cosmology and cosmogony in a cyclic universe, *J. Astrophys. Astron.*, **28**, 67

[92] Burbidge, G. and Hoyle, F. 1998, The origin of helium and other light elements, *Ap. J. Lett.*, **509**, L1

[93] Narlikar, J. V., Burbidge, G. and Vishwakarma, R. G. 2002, Interpretations of the accelerating universe, *PASP*, **114**, 1092

[94] Narlikar, J. V. 1984, The vanishing likelihood of spacetime singularity in quantum conformal cosmology, *Foundations Phys.*, **14**, 443

[95] Hawking, S. W. 1974, Black hole explosions, *Nature*, **248**, 30

[96] Rovelli, C. 2004, *Quantum Gravity* (Cambridge: Cambridge University Press)

[97] Schwarz, J. H. 2007, *String Theory: Progress and Problems*, hep-th/0702219

Index

Printed in the United States
By Bookmasters